# Study on Building Energy Efficiency Related to Simulation Models

# Study on Building Energy Efficiency Related to Simulation Models

Guest Editors

**Pengyuan Shen**
**Yun Kyu Yi**
**Huilong Wang**

MDPI

Basel • Beijing • Wuhan • Barcelona • Belgrade • Novi Sad • Cluj • Manchester

*Guest Editors*
Pengyuan Shen
Institute of Future Human Habitats
Shenzhen International Graduate School
Tsinghua University
Shenzhen
China

Yun Kyu Yi
School of Architecture
University of Illinois Urbana-Champaign
Champaign
USA

Huilong Wang
Department of Construction Management and Real Estate
Shenzhen University
Shenzhen
China

*Editorial Office*
MDPI AG
Grosspeteranlage 5
4052 Basel, Switzerland

This is a reprint of the Special Issue, published open access by the journal *Buildings* (ISSN 2075-5309), freely accessible at: https://www.mdpi.com/journal/buildings/special_issues/P8JFSBW089.

For citation purposes, cite each article independently as indicated on the article page online and as indicated below:

Lastname, A.A.; Lastname, B.B. Article Title. *Journal Name* **Year**, *Volume Number*, Page Range.

**ISBN 978-3-7258-5425-7 (Hbk)**
**ISBN 978-3-7258-5426-4 (PDF)**
**https://doi.org/10.3390/books978-3-7258-5426-4**

# Contents

**Preface** . . . . . . . . . . . . . . . . . . . . . . . . . . . . . . . . . . . . . . . . . . . . . . . . **vii**

**Shihao Li, Meilin Wang, Pengyuan Shen, Xue Cui, Linqian Bu, Ruji Wei, et al.**
Energy Saving and Thermal Comfort Performance of Passive Retrofitting Measures for Traditional Rammed Earth House in Lingnan, China
Reprinted from: *Buildings* **2022**, *12*, 1716, https://doi.org/10.3390/buildings12101716 . . . . . . **1**

**Yongbao Chen, Yunyang Ye, Jingnan Liu, Lixin Zhang, Weilin Li and Soheil Mohtaram**
Machine Learning Approach to Predict Building Thermal Load Considering Feature Variable Dimensions: An Office Building Case Study
Reprinted from: *Buildings* **2023**, *13*, 312, https://doi.org/10.3390/buildings13020312 . . . . . . . **20**

**Shen Xu, Gaomei Li, Hailong Zhang, Mengju Xie, Thushini Mendis and Hu Du**
Effect of Block Morphology on Building Energy Consumption of Office Blocks: A Case of Wuhan, China
Reprinted from: *Buildings* **2023**, *13*, 768, https://doi.org/10.3390/buildings13030768 . . . . . . . **44**

**Xinyu Yang, Ying Ji, Jiefan Gu and Menghan Niu**
An Electricity Consumption Disaggregation Method for HVAC Terminal Units in Sub-Metered Buildings Based on CART Algorithm
Reprinted from: *Buildings* **2023**, *13*, 967, https://doi.org/10.3390/buildings13040967 . . . . . . . **68**

**Xueping Li, Luo Qin and Jingjing Li**
Research on Multiple Energy-Saving Strategies for Existing Coach Stations: A Case of the Xi'an Area, China
Reprinted from: *Buildings* **2023**, *13*, 1215, https://doi.org/10.3390/buildings13051215 . . . . . . **89**

**Darko Palaić, Ivan Štajduhar, Sandi Ljubic and Igor Wolf**
Development, Calibration, and Validation of a Simulation Model for Indoor Temperature Prediction and HVAC System Fault Detection
Reprinted from: *Buildings* **2023**, *13*, 1388, https://doi.org/10.3390/buildings13061388 . . . . . . **110**

**Shanshan Wang, Yun Kyu Yi and Nianxiong Liu**
The ANN Architecture Analysis: A Case Study on Daylight, Visual, and Outdoor Thermal Metrics of Residential Buildings in China
Reprinted from: *Buildings* **2023**, *13*, 2795, https://doi.org/10.3390/buildings13112795 . . . . . . **131**

**Carlos Pérez-Carramiñana, Aurelio de la Morena-Marqués, Ángel Benigno González-Avilés, Nuria Castilla and Antonio Galiano-Garrigós**
Influence of Balcony Thermal Bridges on Energy Efficiency of Dwellings in a Warm Semi-Arid Dry Mediterranean Climate
Reprinted from: *Buildings* **2024**, *14*, 703, https://doi.org/10.3390/buildings14030703 . . . . . . . **164**

**Sai Liu, Farid E. Mohamed Ghazali, Jingjing Yang, Zongkang Guo, Kejun Zeng and Yixing Chen**
Retrofit Analysis of Exterior Windows for Large Office Buildings in Different Climate Zones of China
Reprinted from: *Buildings* **2024**, *14*, 3904, https://doi.org/10.3390/buildings14123904 . . . . . . **190**

**Ezio Micelli, Giulia Giliberto and Eleonora Righetto**
Is the Energy Transition of Housing Financially Viable? Unlocking the Potential of Deep Retrofits with New Business Models
Reprinted from: *Buildings* **2025**, *15*, 1175, https://doi.org/10.3390/buildings15071175 . . . . . . **211**

**Jaeho Lee and Jaewan Suh**
Building Energy Assessment of Thermal and Electrical Properties for Compact Cities: Case Study of a Multi-Purpose Building in South Korea
Reprinted from: *Buildings* **2025**, *15*, 3023, https://doi.org/10.3390/buildings15173023 . . . . . . **237**

# Preface

Building energy efficiency represents one of the most significant opportunities for reducing global greenhouse gas emissions and achieving sustainable development goals. As buildings account for approximately 40% of global energy consumption, the development of sophisticated simulation models and computational tools has become crucial for optimizing energy performance across diverse building types and climate conditions.

This Reprint presents a comprehensive collection of research addressing the complex challenges in building energy simulation and efficiency optimization. The scope encompasses innovative methodologies ranging from machine learning approaches for thermal load prediction to advanced algorithms for HVAC system analysis. The motivation for compiling these studies stems from the urgent need to bridge the gap between theoretical modeling capabilities and practical implementation in the building industry.

The featured studies demonstrate the evolution of building energy simulation from traditional methods to cutting-edge computational approaches. Machine learning techniques, artificial neural networks, and sophisticated disaggregation algorithms are explored as powerful tools for enhancing prediction accuracy and system optimization. The research addresses various building typologies, from traditional rammed earth structures to modern office complexes, examining energy performance across different climate zones and retrofit scenarios.

This Reprint is addressed to researchers, practitioners, and policymakers working in building energy efficiency, sustainable construction, and computational modeling. We hope it can serves as a valuable resource for academics developing new simulation methodologies, engineers implementing energy optimization strategies, and industry professionals seeking evidence-based approaches to building performance enhancement. The compilation aims to accelerate the adoption of advanced simulation tools in practice while highlighting future research directions in this rapidly evolving field.

**Pengyuan Shen, Yun Kyu Yi, and Huilong Wang**

*Guest Editors*

buildings

MDPI

Article

# Energy Saving and Thermal Comfort Performance of Passive Retrofitting Measures for Traditional Rammed Earth House in Lingnan, China

**Shihao Li [1], Meilin Wang [1], Pengyuan Shen [1,*], Xue Cui [2], Linqian Bu [1], Ruji Wei [1], Longzhu Zhang [1] and Chengjia Wu [1]**

[1] School of Architecture, Harbin Institute of Technology, Shenzhen 518055, China
[2] Department of Building and Real Estate, Faculty of Construction and Environment, The Hong Kong Polytechnic University, Hong Kong, China
* Correspondence: pengyuan_pub@163.com; Tel.: +86-13764602660

**Abstract:** The traditional rammed earth houses sharing similar patterns in the Lingnan region, south China, and distributed in rectangular arrays, are gradually losing their vitality and becoming uninhabited under modern living conditions. This research examined a typical pattern called the "Four-point gold" house and analyzed the suitability of different retrofitting technologies by field measurements and building simulation. To optimize energy consumption, indoor thermal comfort, and the corresponding economic performance of the retrofitting measures for the prototypical house, five measures, including wall insulation, reflective roof coating, carpet, sunshade, and natural ventilation, are proposed after considering the status quo of the building envelope. It is found that the best performance in energy-saving, dynamic investment payback period, and annual indoor thermal comfort are 2192.27 kWh/a, 9.17 years, and 1766 h, respectively. Different parameters are included to be clustered by K means clustering technique, and the comprehensively optimized scheme consists of a regime of 30 mm XPS 30 mm, ZS-221 white coating, carpet, 0.5 m sunshade width, and turning off windows (doors). The proposed retrofitting strategy can be promoted to a wide range of traditional rammed earth houses in the Lingnan region in China and holds a conspicuous energy-saving potential for the suburban and rural residential sectors in the region.

**Keywords:** rammed earth house; indoor thermal comfort; building retrofit; clustering technique

## 1. Introduction

Rammed earth refers to a compacted mixture of mud blocks that are strong, dense, and have few gaps. It is used as a construction material and has very strong plasticity [1]. Since the 1970s, different groups around the world have begun to conduct scientific research on the properties of rammed earth [2]. Compared with conventional building materials, rammed earth is low-cost, readily available, recyclable, and considered environmentally friendly and sustainable [3], especially when the material is locally available [4,5]. Rammed earth building has good thermal performance under specific climatic and technical conditions [6]. This is due to its high thermal inertia [7] and good thermal insulation performance [8]. Rammed earth buildings do not need too much external energy input throughout the year [9], and indoor thermal comfort can be achieved during most of the time of the year [3,7]. In addition, it has been shown that due to its good bioclimatic properties, an earth envelope can lead to a reduction in energy used for heating, cooling, and humidity control [10]. In general, the quality of the indoor thermal environment in summer can be acceptable [11,12]. The comfortable time can be extended when other design measures, such as insulation, double glazing, shading, and ventilation, are taken into account [11,13]. Small and large window shadings lead to lower energy loads in the torrid-arid area [14]. Moreover, the results of the research show that a combination of large compacted earth walls, nighttime cross ventilation, and overhangs can effectively reduce the energy demand for space cooling [15]. However, thermal comfort in winter cannot be fully achieved without heating [12,13]. Thus, it is possible to archive acceptable comfort conditions by passive means [16–19], which contributes to reducing the heating/cooling energy demand of the traditional rammed earth buildings [20–22].

What is more, historical and traditional rammed earth houses in Lingnan account for a large part of the existing building stock, and they are quite distributed in the region inhabited by the clans in the Chaoshan area.. There are mainly two styles of "climbing lion" and "Four-point gold", which are generally distributed in a rectangular array [23]. As shown in Figure 1, the "Climbing lion" house looks like a lion poised to climb. The gate indicates its mouth, the two small rooms are the front claws, the hall is the belly, and the two other bedrooms on the sides of the hall are the rear claws. "Four-point gold" has two extra rooms near the gate compared with the "Climbing lion" house. The shape is like the Chinese character '金' (gold) and can be seen in Figure 2. Therefore, the architectural styles in traditional villages are the same and arranged in the same direction, and "cold lanes" are formed between buildings [24,25]. Research has shown that the residents in Tulou (earth building) had better satisfaction with the indoor thermal environment and overall indoor environmental quality than residents in normal rural buildings [26]. Building integrated photovoltaic was employed to optimize the solar energy application on the rooftop of traditional Lingnan buildings [27]. However, there are few studies on the indoor thermal environment and energy-saving renovation of rammed earth houses in Lingnan, and there are no rules, norms, or standards for reference [28].

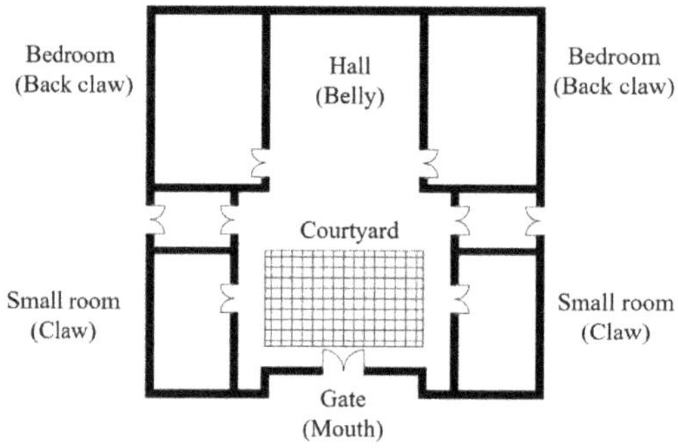


**Figure 1.** Archaic "climbing lion" rammed earth house.

The industrialization of the construction industry (materials, systems, etc.) led to the standardization of buildings. The performance of traditional buildings can gradually deteriorate over time. As a consequence, they became more dependent on heating, ventilation, and air conditioning (HVAC) systems to cater to thermal comfort requirements, which resulted in energy use that was not a necessity in previous vernacular buildings. This situation led to the disregard for the vernacular wisdom in passive design, changing not only the way buildings and indoor spaces were designed (commonly arranged by function and thermal needs) but also the occupants' living habits [29]. Coupled with the outflow of the rural population in China due to the urbanization process, many traditional rammed earth houses are now uninhabited. Traditional buildings are gradually abandoned by the descendants of their original residents in the process of modernization.

Therefore, the investigation of Lingnan rammed earth houses is not only conducive to the protection and inheritance of historical buildings but also of great significance to the sustainable development of society. In this study, the indoor and outdoor thermal environment parameters of the traditional rammed earth house "Four-point gold" in Lingnan were tested. The building model is established using DesignBuilder and calibrated by onsite measured data. Then the thermal performance of the rammed earth house is analyzed, and the passive energy-saving measures are put forward. Finally, the corresponding optimal scheme for each optimization objective and the comprehensively optimal scheme are proposed by the orthogonal experiment method. This research helps to better understand the characteristics of rammed earth houses in the Lingnan region and provides technical support for energy saving in the residential sector. It is anticipated that the renovation measures can be applied to not only the prototypical building in this research but also to the rammed earth buildings in the region so that these traditional houses in Lingnan will have the chance to rejuvenate in terms of modern living conditions and continue to maintain vitality in the future.

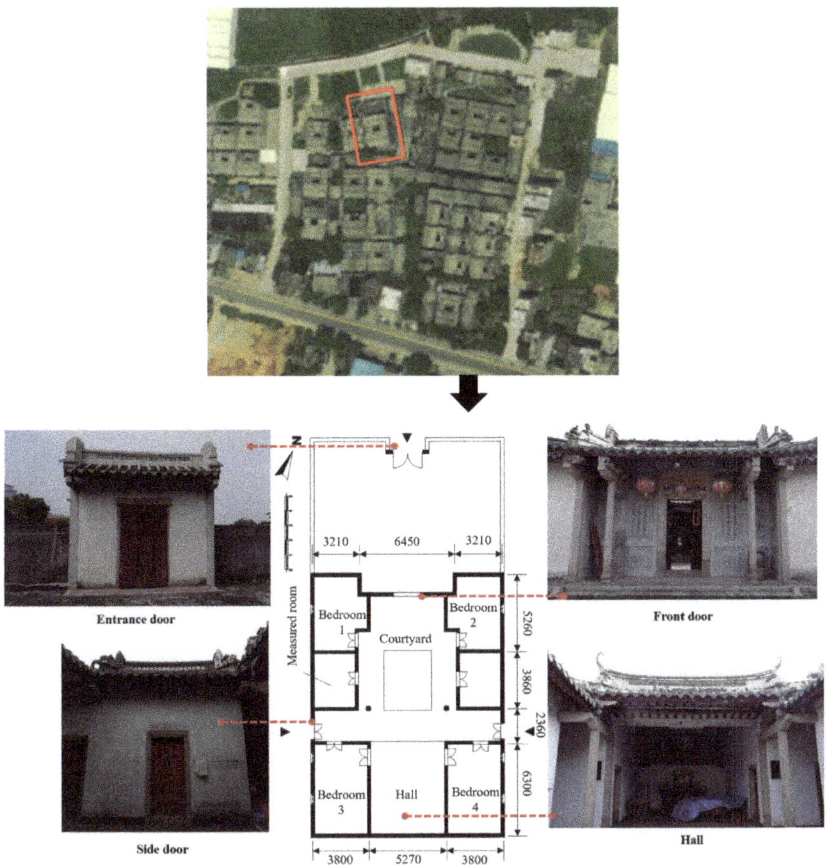


**Figure 2.** Location, status quo and the floor plan of the "Four-point gold" house.

## 2. Methods

### 2.1. Field Survey

This study selected a house in Jieyang, Guangdong Province, located in the north of Xinzhizhai. There are 14 "climbing lion" and 13 "Four-point gold" houses in this village. The "Four-point gold" houses are the target prototype in this research. Its floor plan and locations are shown in Figure 2. The house retains the traditional Lingnan architectural styles. The archaic rammed earth technique of Chaoshan is adopted, which is made of red mud, coarse sand, and other materials. The top of the wall directly supports the wooden beams, and the rooftop is covered with tiles. The overhanging eaves in front of the hall are undertaken by stone pillars in front of the courtyard. There are walls on the periphery of the building. Behind the entrance door, there lays the front hall, and the rooms on the left and right side are the front rooms (bedrooms). Further ahead is a courtyard, and the rooms on both sides are the kitchen and utility room. On the other side of the courtyard, a multifunctional space with a large room on each side (bedroom) is designed. The large room is divided into two floors by a wooden partition, and the attics are used for storage. Influenced by the traditional concept of "withholding wealth", "Four-point gold" generally does not open windows to the outside but only openings to the inner courtyard. The house only has external windows on the east and west sides, which is limited by the strength of the rammed earth walls. The window-to-wall ratios of the east and west external walls of the bedrooms are 0.017 and 0.011, according to onsite measurements.

*2.2. Onsite Test*

Field measurements and investigations were conducted from 2–9 March in winter and from 12–30 July in summer in 2019. The indoor temperature, relative humidity, and black bulb temperature were tested by HOBO temperature and humidity and the global temperature meter. The detailed parameters of the instruments are shown in Table 1. To ensure the accuracy of the tested data, the room was not occupied during the measurement period. Doors and windows are closed to test the condition without natural ventilation. The thermal environment of the utility room and hall was measured and monitored, as shown in Figure 3.

**Table 1.** The parameters of the measurement equipment used.

| Equipment | Specification | Measurement Range | Accuracy |
|---|---|---|---|
| HOBO temperature and humidity recorder | U12-012 | temperature:−20~70 °C relative humidity: 5~95% | ±2.5% |
| Global temperature recorder | testo480 | −20~+80 °C | ±0.3 °C |

(**a**) (**b**)

**Figure 3.** Instruments and their layouts: (**a**) Hall; (**b**)Utility room.

*2.3. Building Simulation Model*

The indoor environment and energy-saving effect of various retrofitting measures were simulated by DesignBuilder, which uses EnergyPlus as the simulation engine and energy use analysis. According to the onsite investigation, the geometric information of the house was collected and fed into DesignBuilder. As shown in Figure 4, the building has two stories with wide eaves, which can effectively shade and ensure ventilation and daylight in the courtyard. We manually tuned the model to fit the simulation results to the onsite measurements as model calibration. The calibrated thermophysical properties of building walls, roofs, and floors are shown in Table 2. The thermal resistance of the rammed earth wall is 0.57 $m^2 \cdot K/W$, which is higher than that of the tile roof and concrete floor. This research only considers the basic structure of the building due to its simple interior furnishings. There is no larger appliance, so the model does not account for heat loads other than lamps. An occupancy intensity of two people per bedroom is assumed, and the bedrooms are not used during 9:00–22:00. Due to the aging of the house and poor airtightness of doors and windows, the air infiltration of the house is calibrated to be 3.5 $h^{-1}$. Currently, the house is not equipped with an HVAC system. We propose a simple packaged terminal heat pump system for the building to meet the requirement of thermal comfort in the simulation. The parameters of the HVAC system are listed in Table 3. The cooling temperature is 25 °C, and the heating temperature is 18 °C. Location coordinates and meteorological parameters are also input parameters.

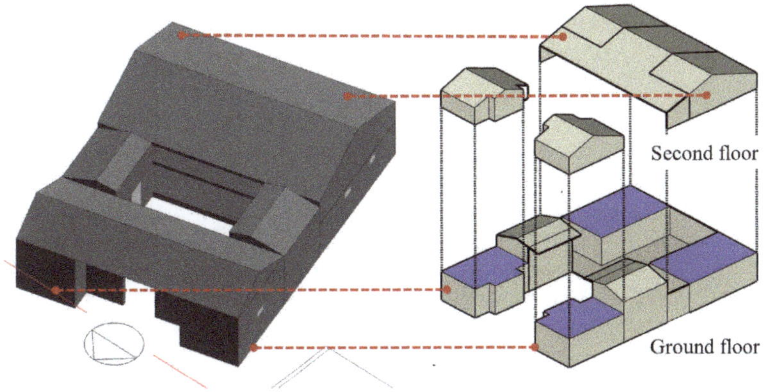


**Figure 4.** A 3D model of the house.

**Table 2.** Thermophysical properties of the building envelope.

| Structure | Heat Transfer Coefficient/W/($m^2 \cdot K$) | Thermal Resistance/($m^2 \cdot K/W$) | Thickness /m |
|---|---|---|---|
| Rammed earth wall | 1.754 | 0.570 | 0.24 |
| Roof | 2.790 | 0.358 | 0.3 |
| Floor | 2.098 | 0.477 | 0.3 |

**Table 3.** Parameters of the packaged terminal heat pump system.

| Parameters | Values |
|---|---|
| Design indoor temperature in winter | 18 °C |
| Design indoor temperature in summer | 25 °C |
| Heat Pump Heating Coil Gross Rated COP | 2.75 |
| Cooling Coil Gross Rated COP | 3 |
| Operational time | 22:00–9:00 |
| Meteorological parameters | Typical meteorological year data of Shantou |
| Air infiltration | 3.5 $h^{-1}$ |
| Indoor thermal disturbance | People and lighting |

Retrofitting measures could be put forward through the analysis of original building's heat transfer and indoor thermal environment. The corresponding thermophysical parameters of the measures are new inputs to simulate energy consumption and thermal comfort. The best retrofitting scheme can be found according to the characteristics of low HVAC system energy consumption, long comfortable period, and short dynamic investment payback period. Orthogonal experimental method is adopted to select suitable scheme with evenly dispersed study cases. Each average of sensitivity index ($\overline{K_{jm}}$) is the average experimental index of the corresponding cases with same retrofitting measure (factor j) and level (m), which can evaluate the performance of the measures. Maximum difference ($R_j$) of retrofitting measures is the difference between maximum and minimum $\overline{K_{jm}}$. Higher $R_j$ indicates more sensitivity of the specific retrofitting measure.

### 2.4. *Dynamic Investment Payback Period*

For building energy-saving retrofitting, economic evaluation is important to the feasibility of the renovation project. The payback period, incremental cost, and net present value are the usually used indicators of economic analysis. Incremental cost refers to the direct and indirect costs caused by transformation, including materials, construction measures, machinery, and labor costs. The dynamic payback period is used for economic analysis, as shown in Equations (1) and (2). The discount rate is 8% when that is calculated [30].

$$NPV = \sum_{t=1}^{n} \frac{C_t}{(1+r)^t} - C_0 \tag{1}$$

$$\Delta P_t = (N-1) + \frac{|A_{N-1}|}{C_N} \tag{2}$$

where $NPV$ refers to the net present value of year $t$, $n$ refers to the calculation cycle, $r$ refers to the discount rate, $C_t$ refers to cash flow in year $t$, $C_0$ refers to initial incremental investment cost, $\Delta P_t$ refers to the dynamic investment payback period, $N$ refers to the year of cumulative net cash flow first becoming positive, $A_{N-1}$ refers to the cumulative net cash flows with the last negative item, $C_N$ refers to the cumulative net cash flow of year $N$.

### 2.5. *K-Means Clustering Based Decision Making*

In this study, a combined indicator of building energy consumption, payback period, and indoor thermal comfort was used to evaluate the combined performance of different energy efficiency retrofit measures. However, since energy consumption, payback period, and indoor thermal comfort have different unit systems, it is not possible to further evaluate the merits of the optimized measures. Therefore, the K-means clustering method [31] was used to make the different optimization target values dimensionless and facilitate the comparative analysis of the integrated indexes. By clustering, the experimental results of the three optimization objectives were classified into 11 categories. The degree of merit of the categories is expressed by the magnitude of their values, i.e., measures with higher value have better improvement effects. The composite index is the sum of the values of the improvement targets corresponding to the clustering categories, and the final optimization scheme can be determined by assessing the magnitude of the composite index.

## 3. Building Simulation Results

### 3.1. *Model Validation*

#### 3.1.1. Winter

As shown in Figure 5, from 3–9 March, the overall outdoor temperature showed a downward trend. The maximum outdoor temperature was 25.13 °C, and the lowest temperature was 13.38 °C on 8 March. The average and minimum indoor temperatures are 18.7 °C and 16.3 °C, fluctuating with the outdoor temperature, and this is able to create a comfortable indoor thermal environment most of the time. The variation trend of the indoor global temperature is consistent with the indoor temperature, but that is always slightly higher than the indoor air temperature. The air temperature in the hall is almost the same as the ambient temperature, and the global temperature in the hall is also slightly higher than the hall temperature. Due to the strong solar radiation intensity on the 13th and 14th, and the semi-open structure of the hall, the global temperature was significantly higher than the air temperature, while that was still slightly higher than the air temperature in the measured room. As shown in Figure 5b, the indoor relative humidity ranges from 71.9% to 87.7%. Even in winter, the ambient relative humidity is high, with an average of 93.94%. The calibrated model can accurately simulate indoor temperature in winter. The difference between the simulated indoor temperature and the measured temperature is very small, and the average absolute error is only 0.47 °C with an average relative error of 2.51%. With the change of time, the simulated indoor relative humidity has roughly the same trend as the measured value. On the 3rd, 4th, and 9th, the model showed relatively higher accuracy. During other time periods, the absolute error of this model is relatively large, but the mean absolute error is still well controlled at 5.76%, and the mean relative error is 8.22%.

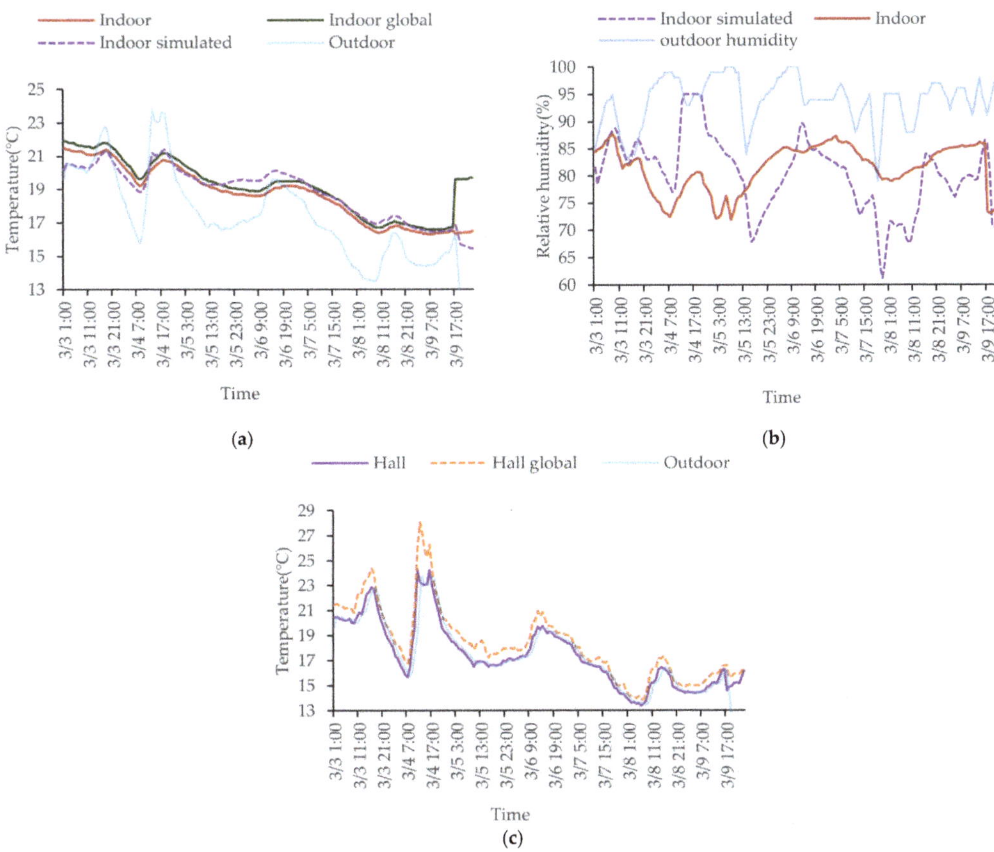


**Figure 5.** Comparison of measured and simulated values in winter: (**a**) Indoor temperature; (**b**) Indoor relative humidity; (**c**) hall temperature.

### 3.1.2. Summer

The measured and simulated temperatures from 17–30 July are shown in Figure 6a,c. It can be seen that the highest outdoor air temperature can reach 37.4 °C, the average value is 29.25 °C, and the daily temperature variation is large. During the day, due to the thermal resistance and thermal inertia of the envelope, the peaks of the hall temperature and indoor temperature are lower and delayed compared with the outdoor air temperature, but the average temperatures of them are 30.39 °C and 30.36 °C, respectively. However, on 21, 23, and 30 July, the peak air temperature of the hall was slightly higher than the outdoor air temperature. This was due to the thermal inertia of the house itself not reducing the air temperature as quickly as the rain did to the outdoor air. The wind speed was about 0.7 m/s at that time, so the heat accumulated inside the hall was dissipated slowly, resulting in higher air temperature. The air temperature at night is lower than the indoor air temperature, thanks to night flushing. The indoor global temperature is slightly higher than the air temperature, and the peak values of the hall are also significantly higher than that of the indoor. As shown in Figure 6b, the city of Puning has high humidity in summer. Even on sunny days, the outdoor relative humidity can reach about 90%, and the daily humidity fluctuates greatly. Although the hall is a semi-open space, the humidity change trend is quite different from that of the outdoors. The outdoor humidity peak usually occurs at 20:00–22:00, while the maximum humidity in the hall occurs at dawn. Since the doors and windows are closed, the indoor humidity changes are relatively less affected by the outdoor humidity, and the relative humidity fluctuates between 70% and 90%. After the rain from the 21st to the 30th, the indoor relative humidity was mostly above 80%. In such high humidity, people are more likely to feel stuffy in summer.

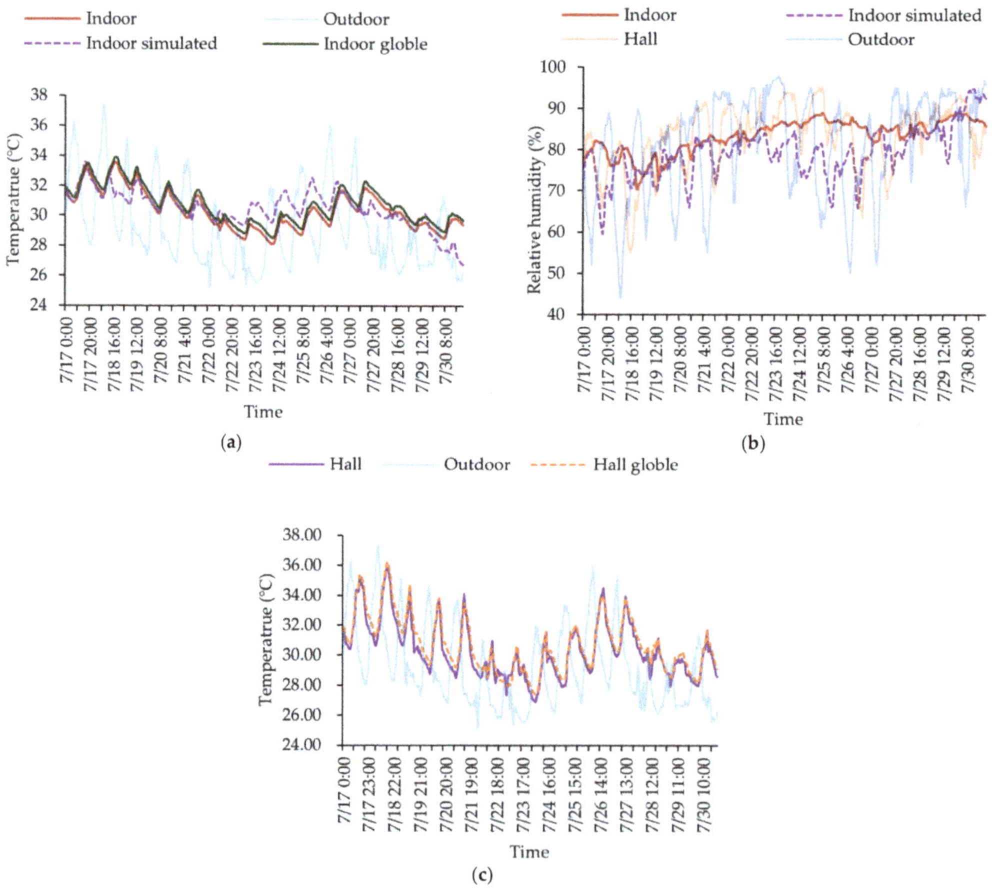


**Figure 6.** Comparison of measured and simulated values in summer: (**a**) Indoor temperature; (**b**) Relative humidity; (**c**) Hall temperature.

The weather station data of the city of Puning in 2019 was used as the weather input for building simulation. The variation trend of indoor air temperature predicted by simulation is consistent with the measured value, and it shows good prediction performance on sunny days. The measured relative humidity is close to the simulated value, as shown in Figure 6b. Overall, the manually calibrated model can accurately simulate indoor temperature and relative humidity most of the time, with an average absolute error of 0.78 °C and 5.17%, respectively. Therefore, this model is deemed suitable for the simulation of an indoor thermal environment and further calculation of energy use if an HVAC system is applied to the house.

### 3.2. Heat Gain and Loss

The heat gain and loss of the envelope are simulated by feeding the typical meteorological year (TMY) data to the calibrated EnergyPlus model, assuming an HVAC system is installed and turned on. As can be seen from Figure 7, in summer, most of the heat gain of the house comes from the roof, and the maximum hourly heat gain can be reached at 4.93 kW. In winter, the heat loss of the roof is also larger than the floor and walls, with the maximum hourly heat loss in November of 3.34 kW. Affected by the ambient temperature and solar radiation, the wall is constantly being heated by solar heat gain from June to October, with a maximum hourly heat gain of 2.66 kW. The daily temperature is low in February and March, and the indoor air temperature is higher than the ambient temperature resulting in the wall being the heat-loss surface most of the time. Puning is located in hot summer and warm winter zone, and the summer is hot and long. The exterior rammed earth wall is a heated surface most of the time during the year, and the hourly mean heat gain is 0.41 kW. In addition, the heat transfer through the floor is also large, which is conducive to the passive cooling of the house during summer. From May to October, the floor always transfers heat to the ground with an average hourly heat loss of 2.25 kW. Thus, in summer, the floor is the heat loss surface, and the roof is the main heat gain surface. In winter, the roof is also the main heat gain surface, which potentially raises the indoor air temperature and creates a warm indoor environment for the occupants.

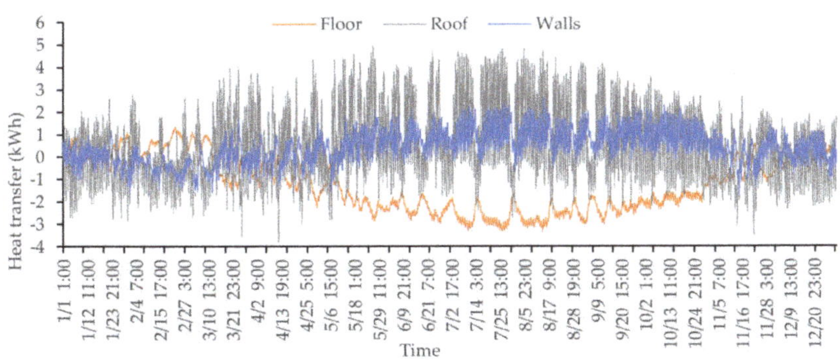


**Figure 7.** Hourly heat transfer of house envelope.

### 3.3. Indoor Thermal Comfort

#### 3.3.1. Natural Ventilation

The hourly indoor environmental condition under natural ventilation conditions are simulated by DesignBuilder, and the simulated results are compared with the Chinese thermal comfort standard GB/T 50785-2012 "Civil Building Thermal and Humid Environment Evaluation Standard" [32] and the American ASHRAE 55-2017 standard [33]. The uncomfortable duration of the whole indoor thermal environment is shown in Figure 8. Under the provisions of the Chinese thermal comfort standard, the uncomfortable durations under the I and II levels are 3310 h and 829 h, respectively, accounting for 37.8% and 9.5% of the whole year. Under the ASHRAE 55-2017 standard, the duration of not meeting the acceptable ranges for 90% and 80% are 4418 h and 2984 h, respectively, which is longer than the uncomfortable duration calculated by the Chinese standard. Considering the differences in climate, culture, and physiology, this study adopts the Chinese thermal comfort standard to evaluate indoor thermal comfort in Lingnan. Table 4 summarizes the uncomfortable durations for the four bedrooms under different criteria. Bedroom 1 has the best comfort, while bedroom 2 and bedroom 3 are less comfortable. On the whole, according to Chinese thermal comfort standard level II, the uncomfortable duration of each room is within 615~870 h. That is, the thermal comfort requirements are not met for about a month, and all of them appear in winter.

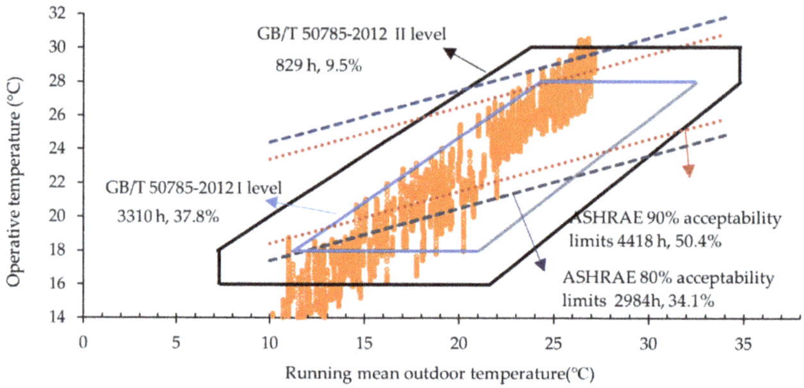


**Figure 8.** Overall indoor thermal comfort with natural ventilation.

**Table 4.** Uncomfortable durations with natural ventilation (unit: h).

| Bedroom | GB/T 50785-2012 | | ASHRAE | |
|---|---|---|---|---|
| | I Level | II Level | 90% Acceptability | 80% Acceptability |
| 1 | 2069 | 663 | 3847 | 2971 |
| 2 | 2567 | 870 | 4000 | 3126 |
| 3 | 2588 | 816 | 3881 | 3029 |
| 4 | 1852 | 615 | 3716 | 2777 |

The number of bedrooms corresponds to Figure 2.

### 3.3.2. HVAC

To further improve indoor thermal comfort, the bedroom thermal environment is simulated while the HVAC system is turned on. Indoor thermal comfort was evaluated by predicted mean vote (PMV). The clothing insulation of winter and summer is 1 clo and 0.5 clo, respectively, and the human metabolic index is 0.9. Figure 9 shows the changes in comfort indicators in each bedroom over time. It can be seen that about 95% of the time, the PMV is between −0.5 and 0.5, and the corresponding predicted dissatisfaction percentage (PPD) is within 10%. Indoor PMV that does not meet the comfort range appeared in April, May, and November. PMV is between −0.85 and 0.85 about 99% of the time, and PPD is within 20%, which meets the requirements of human comfort.

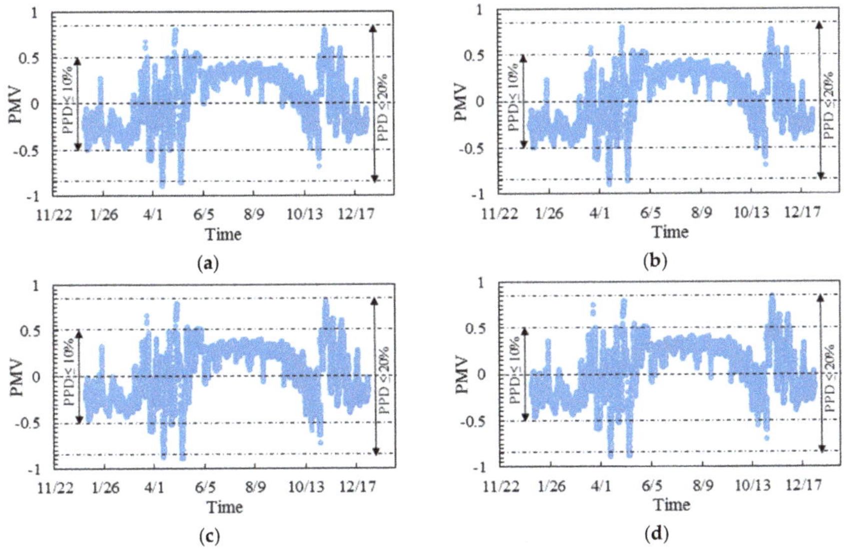


**Figure 9.** Bedroom thermal comfort with the HVAC system of (**a**) Bedroom 1; (**b**) Bedroom 2; (**c**) Bedroom 3; (**d**) Bedroom 4.

## 4. Retrofitting Measures

### 4.1. Energy Saving and Thermal Performance Retrofitting

From the heat gain and loss in Section 3.2, it can be known that in summer, heat is introduced into the room mainly through the roof and exterior walls, and the maximum hourly heat gain of the roof is close to 5 kW. Therefore, in summer, the optimization measures can be focused on improving the heat absorption by the roof. In winter, exterior walls are heat-loss surfaces most of the time because the indoor air temperature is higher than the ambient temperature. Therefore, in winter, the optimization measures can be focused on the exterior walls. Retrofitting measures of the envelope by reducing the roof heat gain and enhancing the wall insulation are included. In order to ensure that the original historical features of the "Four-point gold" are not damaged, an orthogonal experiment method is used to carry out the retrofitting. Natural ventilation, enhanced roof reflection and insulation, wall insulation, window shading, and carpet laying were selected as design variables. The orthogonal experiment design table of the retrofitting is shown in Table 5. The detail of the design rules and data processing method of the orthogonal experiments is shown in Appendix A.

**Table 5.** Factor-level table of the orthogonal experiment design of the retrofitting.

| | Factors | | | | |
|---|---|---|---|---|---|
| | **Natural Ventilation** | **Heat-Reflective and Thermal Insulation Coating** | **Wall Insulation** | **Sunshade Width** | **Carpet** |
| Level 1 | Turn on | ZS-221 white | EPS(50 mm) | 0.1 m | Carpet |
| Level 2 | Turn off | Water-based | XPS(30 mm) | 0.2 m | PVC carpet |
| Level 3 | | Cool glue | EPS(100 mm | 0.3 m | Textile carpet |
| Level 4 | | Kaisheng | XPS(50 mm) | 0.4 m | |
| Level 5 | | RLHY-A05 | EPS(20 mm) | 0.5 m | |

The physical properties of materials and construction costs of each measure are shown in Tables 6–8, respectively. Among them, carpets are only laid when the outdoor temperature is below 15 °C, which is from 2 December to 4 March in the typical meteorological year. In this research, carpets, polyvinyl chloride (PVC) carpets, and textile carpets are considered. The expandable polystyrene (EPS) and extruded polystyrene (XPS) boards of different thicknesses are used for wall insulation with a heat conductivity coefficient of 0.8 W/(m·k) and 0.034 W/(m·k), respectively. To minimize the heat gain of the roof, five coatings with high thermal reflectivity, solar reflectivity, and hemispheric emissivity were considered. In addition, aluminum sunshades of different widths are planned to be installed above the east and west windows.

**Table 6.** Physical parameters and prices of wall insulation and carpet materials.

| Component | Materials | Heat Conductivity Coefficient (W/(m·k)) | Specific Heat (J/(kg·K)) | Density (kg/m$^3$) | Unit Price (¥/m$^2$) | Area (m$^2$) | Total Price (¥) |
|---|---|---|---|---|---|---|---|
| Floor | Carpet | 0.28 | 1000 | 290 | 16.33 | 179.6 | 2932.87 |
| | PVC carpet | 0.04 | 750 | 100 | 34.47 | 179.6 | 6190.81 |
| | Textile carpet | 0.06 | 1300 | 200 | 29.1 | 179.6 | 5226.36 |
| Wall | EPS(50 mm) | 0.8 | 1300 | 16 | 39 | 316 | 12,324 |
| | XPS(30 mm) | 0.034 | 1400 | 35 | 18 | 316 | 5688 |
| | EPS(100 mm) | 0.8 | 1300 | 16 | 78 | 316 | 24648 |
| | XPS(50 mm) | 0.034 | 1400 | 35 | 30 | 316 | 9480 |
| | EPS(20 mm) | 0.8 | 1300 | 16 | 15.6 | 316 | 4929.6 |

**Table 7.** Physical parameters and prices of heat-reflective and thermal insulation coating of roof.

| Coating | Thermal Reflectivity | Tolar Reflectivity | Hemispheric Emissivity | Unit Price (¥/m$^2$) | Area (m$^2$) | Total Price (¥) |
|---|---|---|---|---|---|---|
| ZS-221 | 0.92 | 0.88 | 0.89 | 7.5 | 232.8 | 1746 |
| Water-based | 0.9 | 0.91 | 0.93 | 8.75 | 232.8 | 2037 |
| Cool glue | 0.85 | 0.89 | 0.89 | 4.25 | 232.8 | 989.4 |
| Kaisheng | 0.85 | 0.85 | 0.88 | 7.5 | 232.8 | 1746 |
| RLHY-A05 | 0.92 | 0.95 | 0.9 | 7.5 | 232.8 | 1746 |

**Table 8.** Widths and prices of aluminum panel.

| Width(m) | Unit Price (¥/m$^2$) | Area (m$^2$) | Total Price (¥) |
|---|---|---|---|
| 0.1 | 72 | 0.406 | 29.232 |
| 0.2 | 72 | 0.812 | 58.464 |
| 0.3 | 72 | 1.218 | 87.696 |
| 0.4 | 72 | 1.624 | 116.928 |
| 0.5 | 72 | 2.03 | 146.16 |

When the indoor temperature is 2 °C higher than the outdoor temperature, natural ventilation is used for cooling. The HVAC system is activated only when natural ventilation is insufficient to maintain comfort. The operating hours coincide with the rest period of the occupants, from 22:00 to 9:00 the next morning. In addition, the compressor was placed at the window on the second floor in conjunction with the bedroom layout of a traditional rammed house and its storage space on the second floor, which allows for more comfortable living conditions without destroying the historic character. Last but not least, the whole house is retrofitted, but the HVAC system was only used in the four bedrooms.

### 4.2. Sensitivity Analysis of Retrofitting Measures

#### 4.2.1. Energy Saving

The annual HVAC energy consumption before the retrofitting of the rammed earth house was calculated to be 4695.62 kWh, which was used as the benchmark for calculating energy savings. After taking retrofitting measures, the total energy consumption was significantly reduced. The sensitivity analysis of each energy-saving retrofitting measure is shown in Table 9. The maximum average energy saving in the orthogonal experiment is 2192.27 kWh/a. The maximum difference for energy saving of wall insulation, heat-reflective and thermal insulation coating, sunshade width, natural ventilation, and carpet is 1294.96 kWh/a, 197.49 kWh/a, 159.96 kWh/a, 91.51 kWh/and 79.53 kWh/a, respectively. Therefore, the impact of each retrofitting measure on energy savings can be ranked as follows: wall insulation, heat-reflective and thermal insulation coating, sunshade width, natural ventilation, and carpet. Among wall insulations, the energy-saving effect of level 4 (XPS 50 mm) is higher than other wall insulation materials. For heat-reflective and thermal insulation coatings and sunshade width, the energy-saving effect of level 5 coating material (RLHY-A05) and level 5 sunshade width (0.5 m) is better than other measures in each group. For natural ventilation and carpets, turning off the window (door) and PVC carpet have better performance in energy saving compared with other measures. To sum up, the combined design of 50 mm XPS panel, RLHY-A05 coating, 0.5 m sunshade width, turning off the window (door), and PVC carpet can be considered as the optimal scheme for energy saving.

**Table 9.** Sensitivity analysis of the energy saving.

| | | Factors | | | | |
|---|---|---|---|---|---|---|
| | | **Natural Ventilation** | **Heat-Reflective and Thermal Insulation Coating** | **Wall Insulation** | **Sunshade Width** | **Carpet** |
| $\overline{K_{jm}}$ | Level 1 | 1358.55 | 1395.93 | 1138.04 | 1278.21 | 1413.01 |
| | Level 2 | 1450.05 | 1444.99 | 1888.74 | 1422.72 | 1430.45 |
| | Level 3 | | 1288.47 | 873.58 | 1409.33 | 1350.92 |
| | Level 4 | | 1360.40 | 2168.54 | 1427.32 | |
| | Level 5 | | 1485.96 | 906.85 | 1438.17 | |
| $R_j$ | | 91.51 | 197.49 | 1294.96 | 159.96 | 79.53 |
| Rank of impact level | | Wall insulation > Heat-reflective and thermal insulation coating > Sunshade width > Natural ventilation > Carpet | | | | |
| Best scheme | | 2 | 5 | 4 | 5 | 2 |

#### 4.2.2. Economic Performance

The study has shown that the total cost of retrofitting measures is approximately 1.2–1.5 times the cost of energy-saving materials [34]. Therefore, the total cost of each retrofitting measure can be calculated based on the unit price and area of each material shown in Tables 6–8. In this research, the dynamic investment payback period was selected as the criteria to evaluate the economic performance of each retrofitting measure. The sensitivity analysis of the dynamic investment payback period for each retrofitting measure is shown in Table 10. Natural ventilation can be achieved by manually opening windows and doors without investment costs. Therefore, there is no need to discuss the economic performance of natural ventilation. The minimum average dynamic investment payback period in the orthogonal experiment is 9.17 years. When EPS materials are used, the dynamic investment payback period cannot be calculated. The maximum differences for a dynamic investment payback period of sunshade, carpet, and coating are 2.99, 2.86, and 2.83, respectively. Therefore, the impact of each retrofitting measure on the dynamic investment payback period can be ranked in order as follows: wall insulation, sunshade width, carpet, and heat-reflective and thermal insulation coating. For wall insulation, the economic performance of level 2 (XPS 30 mm) is better than other wall insulation materials. Among all heat-reflective and thermal insulation coatings, the water-based coating has the best economic performance. For carpets, the economic performance of the carpet is the best. For sunshade width, the 0.5 m sunshade width has the best performance in economic performance. In general, the combined design of 30 mm XPS panel, water-based coating, carpet, and 0.5 m sunshade width can achieve optimal economic performance.

**Table 10.** Sensitivity analysis of the dynamic investment payback period.

| | | Factors | | | | |
|---|---|---|---|---|---|---|
| | | **Natural Ventilation** | **Heat-Reflective and Thermal Insulation Coating** | **Wall Insulation** | **Sunshade Width** | **Carpet** |
| $\overline{K_{jm}}$ | Level 1 | N/A | 6.47 | N/A | 8.07 | 4.29 |
| | Level 2 | N/A | 5.33 | 15.97 | 6.58 | 7.12 |
| | Level 3 | | 8.16 | N/A | 7.04 | 7.15 |
| | Level 4 | | 6.85 | 16.86 | 6.07 | |
| | Level 5 | | 6.02 | N/A | 5.07 | |
| $R_j$ | | N/A | 2.83 | N/A | 2.99 | 2.86 |
| Rank of impact level | | Wall insulation > Sunshade width > Carpet > Heat-reflective and thermal insulation coating | | | | |
| Best scheme | | N/A | 2 | 2 | 5 | 1 |

#### 4.2.3. Indoor Thermal Environment

The annual indoor unacceptable thermal comfort hours of the rammed earth house are selected as the criteria to evaluate the indoor thermal environment improvement. The sensitivity analysis of the annual indoor unacceptable thermal comfort hours for each retrofitting measure is shown in Table 11. The maximum average annual indoor unacceptable thermal comfort hours in the orthogonal experiment is 1766 h. The maximum difference for the annual indoor unacceptable thermal comfort hours of wall insulation, heat-reflective and thermal insulation coating, sunshade width, carpet, and natural ventilation is 276.60 h, 253.00 h, 116.80 h, 85.50 h, and 5.37 h, respectively. Therefore, the impact of each retrofitting measure on the annual indoor unacceptable thermal comfort hours can be ranked in positive order as follows: wall insulation, heat-reflective, and thermal insulation coating, sunshade width, carpet, and natural ventilation. Among wall insulations, the annual indoor unacceptable thermal comfort hours of level 3 (EPS 100 mm) are higher than other wall insulation materials. For all heat-reflective and thermal insulation coatings, ZS-221 white coating results in the largest increase in the annual indoor thermal comfort period. For carpets, the annual indoor unacceptable thermal comfort hours of the textile carpet are the longest. For the sunshade width, the 0.3 m sunshade width has the best performance for the annual indoor unacceptable thermal comfort hours. For natural ventilation, opening windows (doors) can have the best performance in shrinking the annual indoor unacceptable thermal comfort hours. In general, the combined design of 100 mm EPS panel, ZS-221 white coating, textile carpet, 0.3 m sunshade width, and opening windows (doors) can achieve the optimal scheme for the indoor thermal environment.

**Table 11.** Sensitivity analysis of the annual indoor unacceptable thermal comfort hours.

| | | Factors | | | | |
|---|---|---|---|---|---|---|
| | | **Natural Ventilation** | **Heat-Reflective and Thermal Insulation Coating** | **Wall Insulation** | **Sunshade Width** | **Carpet** |
| $\overline{K_{jm}}$ | Level 1 | 1243.47 | 1322.00 | 1258.80 | 1217.40 | 1199.40 |
| | Level 2 | 1238.10 | 1204.20 | 1173.20 | 1208.00 | 1218.70 |
| | Level 3 | | 1307.20 | 1382.00 | 1324.80 | 1284.90 |
| | Level 4 | | 1304.20 | 1105.40 | 1225.00 | |
| | Level 5 | | 1069.00 | 1287.20 | 1231.40 | |
| $R_j$ | | 5.37 | 253.00 | 276.60 | 116.80 | 85.50 |
| Rank of impact level | | Wall insulation > Heat-reflective and thermal insulation coating > Sunshade width > Carpet > Natural ventilation | | | | |
| Best scheme | | 1 | 1 | 3 | 3 | 3 |

### 4.3. Optimized Scheme

The above retrofitting schemes focus on a single optimization objective without considering the combined impact of various measures on building energy consumption, economic performance, and the indoor thermal environment. However, various retrofitting measures will be applied simultaneously for a single building. Therefore, it is necessary to propose a comprehensively optimized scheme which can be used in the actual retrofitting. The K-means clustering algorithm is used to find the comprehensively optimized scheme. According to the results of the dynamic investment payback period, different indicators of twenty-five cases in Appendix A are divided into eleven grades, and the number 11 represents the best grade. The comprehensive indicator is the sum of grade values for energy saving, economic performance, and indoor thermal environment. The sensitivity analysis of the comprehensively optimized scheme is shown in Table 12. The maximum difference for the performance of the comprehensively optimized scheme of wall insulation, heat-reflective and thermal insulation coating, sunshade width, carpet, and natural ventilation is 10.00, 3.60, 1.80, 1.50, and 1.13, respectively. Therefore, the impact of each retrofitting measure on the performance of the comprehensively optimized scheme can be ranked in positive order as follows: wall insulation, heat-reflective and thermal insulation coating, sunshade width, carpet, and natural ventilation. Among them, 30 mm XPS, ZS-221 white coating, carpet, 0.5 m sunshade width, and turning off windows (doors) can be regarded as the combined design to achieve the comprehensively optimized scheme.

**Table 12.** Sensitivity analysis of the comprehensively optimized scheme.

| | | Factors | | | | |
|---|---|---|---|---|---|---|
| | | **Natural Ventilation** | **Heat-Reflective and Thermal Insulation Coating** | **Wall Insulation** | **Sunshade Width** | **Carpet** |
| $\overline{K_{jm}}$ | Level 1 | 15.67 | 17.80 | 14.00 | 15.20 | 17.00 |
| | Level 2 | 16.80 | 16.40 | 21.80 | 15.20 | 15.50 |
| | Level 3 | | 15.60 | 11.80 | 16.60 | 16.30 |
| | Level 4 | | 16.60 | 21.20 | 16.60 | |
| | Level 5 | | 14.20 | 11.80 | 17.00 | |
| $R_j$ | | 1.13 | 3.60 | 10.00 | 1.80 | 1.50 |
| Rank of impact level | | Wall insulation > Heat-reflective and thermal insulation coating > Sunshade width > Carpet > Natural ventilation | | | | |
| Best scheme | | 2 | 1 | 2 | 5 | 1 |

In summary, improving wall insulation is the most effective way for energy saving and indoor thermal comfort mitigation, but EPS is too expensive for the dynamic investment payback period. Heat-reflective and thermal insulation coating and sunshades with low initial investments also have good performance in reducing the heat gains from solar radiation, resulting in lower energy consumption and longer comfort period. Carpet is only used in winter, so the annual benefits can be relatively insignificant. Moreover, the enhancement of natural ventilation has not yielded desired effect because the transition season in this climate is short. Based on the comprehensive index, 30 mm XPS wall insulation, ZS-221 heat-reflective and thermal insulation coating, sunshade width of 0.5 m, carpet, and natural ventilation are a recommended retrofitting scheme.

## 5. Conclusions

This study analyzed the thermal performance and environment of traditional rammed earth houses in Lingnan, south China. Energy-saving retrofitting measures are proposed to improve the indoor thermal environment by taking the building conditions and local customs into account. It not only retains the historical features and cultural value of traditional rammed earth houses but also improves thermal comfort and achieves energy saving.

The heat transfer of rammed earth houses behaves differently by seasons. In summer, most of the heat gain comes from the roof and exterior walls. During winter, the heat loss is mainly due to the exterior walls. Hence, the retrofitting measures mainly focus on reducing the heat gain of the roof and enhancing the insulation of the wall.

Five energy-saving retrofitting measures, comprised of natural ventilation, enhanced roof reflection and insulation, wall insulation, window shading, and carpet laying, are proposed. The highest performance of retrofitting scheme on energy saving, dynamic investment payback period, and the annual indoor thermal comfort increase table are 2192.27 kWh/a, 9.17 years, and 1766 h, respectively. The research puts forward a comprehensively optimized scheme, which consists of 30 mm XPS 30 mm, ZS-221 white coating, carpet, 0.5 m sunshade width, and turning off windows (doors). The optimized scheme can be practically used in renovation, which balances energy saving, economic performance, and indoor thermal environment.

## 6. Limitations

This work takes the "four-point gold" house as a sample to analyze its thermal performance and search for a suitable retrofitting scheme. The conclusions and the application of retrofitting measures are limited to this specific type of traditional rammed earth residential buildings in the Lingnan region, and it is also one of the most popular types. We will study other types of traditional rammed earth buildings in the future and achieve more universal and comprehensive results for the traditional residential houses in the region.

**Author Contributions:** Conceptualization, S.L. and M.W.; methodology, P.S.; software, S.L., L.Z. and C.W.; validation, X.C., C.W. and L.B.; formal analysis, X.C., L.B. and R.W.; investigation, P.S.; resources, P.S.; data curation, M.W. and L.Z.; writing—original draft preparation, S.L. and M.W.; writing—review and editing, X.C., P.S. and M.W.; visualization, R.W.; supervision, P.S.; project administration, P.S.; funding acquisition, P.S. All authors have read and agreed to the published version of the manuscript.

**Funding:** This research was funded by the National Natural Science Foundation of China (NSFC) [Grant number. 52008132]; Harbin Institute of Technology (Shenzhen) Innovation Research Course Construction Project; and Shenzhen Science and Technology Program [Grant number. RCBS202007141 14921062].

**Institutional Review Board Statement:** Not applicable.

**Informed Consent Statement:** Not applicable.

**Data Availability Statement:** Not applicable.

**Conflicts of Interest:** The authors declare no conflict of interest.

## Appendix A

**Table A1.** Orthogonal experiment table.

| Case | Factors | | | | | | | | |
|---|---|---|---|---|---|---|---|---|---|
| | Natural Ventilation | Roof Reflection | Thermal Insula-tion | Sunshade Width | Carpet | Energy Saving (kWh/a) | Dynamic Investment Payback Period Year | Annual Indoor Unacceptable Thermal Comfort Hours h | Comprehensive Indicator |
| 1 | 1 | 1 | 1 | 1 | 1 | 1083.34 | N/A | 1361 | 16 |
| 2 | 1 | 2 | 2 | 2 | 2 | 2032.26 | 14.83 | 1110 | 20 |
| 3 | 1 | 3 | 3 | 3 | 3 | 847.24 | N/A | 1766 | 14 |
| 4 | 1 | 4 | 4 | 4 | 4(2) | 2157.65 | 21.19 | 1244 | 21 |
| 5 | 1 | 5 | 5 | 5 | 5(3) | 991.42 | N/A | 1176 | 10 |
| 6 | 2 | 1 | 2 | 3 | 4(2) | 2035.69 | 14.27 | 1307 | 25 |
| 7 | 2 | 2 | 3 | 4 | 5(3) | 909.33 | N/A | 1305 | 12 |
| 8 | 2 | 3 | 4 | 5 | 1 | 2192.27 | 12.29 | 1071 | 23 |
| 9 | 2 | 4 | 5 | 1 | 2 | 833.85 | N/A | 1307 | 11 |
| 10 | 2 | 5 | 1 | 2 | 3 | 1227.55 | N/A | 1110 | 12 |
| 11 | 3(1) | 1 | 3 | 5 | 2 | 838.8 | N/A | 1351 | 13 |
| 12 | 3(1) | 2 | 4 | 1 | 3 | 2150.63 | 11.82 | 1074 | 24 |
| 13 | 3(1) | 3 | 5 | 2 | 4(2) | 900.47 | N/A | 1290 | 12 |
| 14 | 3(1) | 4 | 1 | 3 | 5(3) | 1034.91 | N/A | 1341 | 15 |
| 15 | 3(1) | 5 | 2 | 4 | 1 | 2049.45 | 9.17 | 963 | 23 |
| 16 | 4(2) | 1 | 4 | 2 | 5(3) | 2175.06 | 18.07 | 1226 | 22 |
| 17 | 4(2) | 2 | 5 | 3 | 1 | 961.73 | N/A | 1298 | 13 |
| 18 | 4(2) | 3 | 1 | 4 | 2 | 1173.4 | N/A | 1248 | 14 |
| 19 | 4(2) | 4 | 2 | 5 | 3 | 1997.36 | 13.08 | 1325 | 26 |
| 20 | 4(2) | 5 | 3 | 1 | 4(2) | 994.28 | N/A | 1184 | 10 |
| 21 | 5(1) | 1 | 5 | 4 | 3 | 846.76 | N/A | 1365 | 13 |
| 22 | 5(1) | 2 | 1 | 5 | 4(2) | 1170.98 | N/A | 1234 | 13 |
| 23 | 5(1) | 3 | 2 | 1 | 5(3) | 1328.95 | 28.51 | 1161 | 15 |
| 24 | 5(1) | 4 | 3 | 2 | 1 | 778.24 | N/A | 1304 | 10 |
| 25 | 5(1) | 5 | 4 | 3 | 2 | 2167.08 | 20.95 | 912 | 16 |

## References


1. Minke, G. *Building with Earth: Design and Technology of a Sustainable Architecture*; Birkhäuser: Basel, Switzerland, 2012.
2. Jaquin, P.A.; Augarde, C.E.; Gerrard, C.M. Chronological description of the spatial development of rammed earth techniques. *Int. J. Archit. Herit.* **2008**, *2*, 377–400. [CrossRef]
3. Arrigoni, A.; Beckett, C.; Ciancio, D.; Dotelli, G. Life cycle analysis of environmental impact vs. durability of stabilised rammed earth. *Constr. Build. Mater.* **2017**, *142*, 128–136. [CrossRef]
4. Reddy, B.V.V.; Kumar, P.P. Embodied energy in cement stabilised rammed earth walls. *Energy Build.* **2010**, *42*, 380–385. [CrossRef]
5. Venkatarama Reddy, B.V.; Leuzinger, G.; Sreeram, V.S. Low embodied energy cement stabilised rammed earth building—A case study. *Energy Build.* **2014**, *68*, 541–546. [CrossRef]
6. Strazzeri, V.; Karrech, A. Energy and thermal performance of a typical rammed earth residential building in Western Australia. *Energy Build.* **2022**, *260*, 111901. [CrossRef]
7. Fernandes, J.; Mateus, R.; Gervásio, H.; Silva, S.M.; Bragança, L. Passive strategies used in Southern Portugal vernacular rammed earth buildings and their influence in thermal performance. *Renew. Energy* **2019**, *142*, 345–363. [CrossRef]
8. Soudani, L.; Woloszyn, M.; Fabbri, A.; Morel, J.-C.; Grillet, A.-C. Energy evaluation of rammed earth walls using long term in-situ measurements. *Sol. Energy* **2017**, *141*, 70–80. [CrossRef]
9. Soebarto, V. Analysis of indoor performance of houses using rammed earth walls. In Proceedings of the Eleventh International IBPSA Conference, Glasgow, Scotland, 27–30 July 2009; pp. 1530–1537.
10. Dong, X.; Soebarto, V.; Griffith, M. Strategies for reducing heating and cooling loads of uninsulated rammed earth wall houses. *Energy Build.* **2014**, *77*, 323–331. [CrossRef]
11. Taylor, P.; Fuller, R.J.; Luther, M.B. Energy use and thermal comfort in a rammed earth office building. *Energy Build.* **2008**, *40*, 793–800. [CrossRef]
12. Beckett, C.; Cardell-Oliver, R.; Ciancio, D.; Huebner, C. Measured and simulated thermal behaviour in rammed earth houses in a hot-arid climate. *Part B Comfort. J. Build. Eng.* **2017**, *13*, 146–158.
13. Xiang, D.; Soebarto, V.; Griffith, M. Achieving thermal comfort in naturally ventilated rammed earth houses. *Build. Environ.* **2014**, *82*, 588–598.
14. Dong, X.; Soebarto, V.; Griffith, M. Design optimization of insulated cavity rammed earth walls for houses in Australia. *Energy Build.* **2015**, *86*, 852–863. [CrossRef]
15. Giuffrida, G.; Detommaso, M.; Nocera, F.; Caponetto, R. Design optimisation strategies for solid rammed earth walls in Mediterranean climates. *Energies* **2021**, *14*, 325. [CrossRef]
16. Fernandes, J.; Pimenta, C.; Mateus, R.; Monteiro Silva, S.; Bragança, L. Contribution of Portuguese Vernacular Building Strategies to Indoor Thermal Comfort and Occupants' Perception. *Buildings* **2015**, *5*, 1242–1264. [CrossRef]
17. Fernandes, J.; Mateus, R.; Braganca, L.; Silva, J.J.C.D. Portuguese vernacular architecture: The contribution of vernacular materials and design approaches for sustainable construction. *Archit. Sci. Rev.* **2015**, *58*, 324–336. [CrossRef]
18. Martin, S.; Mazarron, F.R.; Canas, I. Study of thermal environment inside rural houses of Navapalos (Spain): The advantages of reuse buildings of high thermal inertia. *Constr. Build. Mater.* **2010**, *24*, 666–676. [CrossRef]
19. Priya, R.S.; Sundarraja, M.C.; Radhakrishnan, S.; Vijayalakshmi, L. Solar passive techniques in the vernacular buildings of coastal regions in Nagapattinam, TamilNadu-India—a qualitative and quantitative analysis. *Energy Build.* **2012**, *49*, 50–61. [CrossRef]
20. Saljoughinejad, S.; Sharifabad, S.R. Classification of climatic strategies, used in Iranian vernacular residences based on spatial constituent elements. *Build. Environ.* **2015**, *92*, 475–493. [CrossRef]
21. Fernandes, J.; Mateus, R.; Bragana, L. The Potential of Vernacular Materials to the Sustainable Building Design. In Proceedings of the International Conference on Vernacular Heritage and Earthen Architecture (CIAV)/VerSus/7th ATP, Vila Nova de Cerveira, Portugal, 16–20 October 2013; pp. 623–629.
22. Singh, M.K.; Mahapatra, S.; Atreya, S.K. Solar passive features in vernacular architecture of North-East India. *Sol. Energy* **2011**, *85*, 2011–2022. [CrossRef]
23. Gu, Y.; Xu, H. Spatial Representation of Rural Clan-Organized Society Structure: The Case of Chaoshan Culture Area. *Urban Plan. Forum* **2017**, *3*, 103–109.
24. Archaic Pashi House. Available online: http://dz.cppfoto.com/activity/showG.aspx?works=1807770&page-=1 (accessed on 4 January 2022).
25. Zeng, Z.; Li, L.; Pang, Y. Analysis on climate adaptability of traditional villages in Lingnan, China–World Cultural Heritage Site of Majianglong Villages as example. *Procedia Eng.* **2017**, *205*, 2011–2018. [CrossRef]
26. Li, Q.; You, R.; Chen, C.; Yang, X. A field investigation and comparative study of indoor environmental quality in heritage Chinese rural buildings with thick rammed earth wall. *Energy Build.* **2013**, *62*, 286–293. [CrossRef]
27. Zhang, W.; Chen, Z. Studies on Optimization and Integration of Photovoltaics in traditional Lingnan buildings: A case study in Guangzhou. *Procedia Eng.* **2017**, *205*, 3003–3010. [CrossRef]
28. Mazzarella, L. Energy retrofit of historic and existing buildings. The legislative and regulatory point of view. *Energy Build.* **2015**, *95*, 23–31.
29. Healy, S. Air-conditioning and the 'homogenization' of people and built environments. *Build. Res. Inf.* **2008**, *36*, 312–322. [CrossRef]

30. China Development and Reform Commission. *Construction Project Economic Evaluation Approaches and Parameters*; China Plan Press: Beijing, China, 2006.
31. Jain, A.K. Data Clustering: 50 Years Beyond K-means. *Pattern Recognit. Lett.* **2010**, *31*, 651–666. [CrossRef]
32. Ministry of Housing and Urban-Rural Development. *Civil Building Thermal and Humid Environment Evaluation Standard*; China Construction Industry Press: Beijing, China, 2012; pp. 5–10.
33. *ANSI/ASHRAE Standard 55-2017*; Thermal Environmental Conditions for Human Occupancy. ASHRAE: Atlanta, GA, USA, 2017; pp. 4–18.
34. Li, Z.; Wang, Z.Z.; Sun, Y.; Shao, Z.B. Economic Evaluation of Energy Saving Technology for Building Envelope: Taking a residential building in Henan as an example. *Build. Energy Effic.* **2017**, *7*, 123–126. [CrossRef]

buildings

MDPI

*Article*

# Machine Learning Approach to Predict Building Thermal Load Considering Feature Variable Dimensions: An Office Building Case Study

**Yongbao Chen [1], Yunyang Ye [2,*], Jingnan Liu [1], Lixin Zhang [1], Weilin Li [3] and Soheil Mohtaram [1]**

1 School of Energy and Power Engineering, University of Shanghai for Science and Technology, Shanghai 200093, China
2 Department of Civil, Environmental and Architectural Engineering, University of Colorado Boulder, Boulder, CO 80309, USA
3 School of Civil Engineering, Zhengzhou University, Zhengzhou 450001, China
* Correspondence: yunyang.ye@colorado.edu

**Abstract:** AbstractAn accurate and fast building load prediction model is critically important for guiding building energy system design, optimizing operational parameters, and balancing a power grid between energy supply and demand. A physics-based simulation tool is traditionally used to provide the building load demand; however, it is constrained by its complex model development process and requirement for engineering judgments. Machine learning algorithms (i.e., data-driven models) based on big data can bridge this gap. In this study, we used the massive energy data generated by a physics-based tool (EnergyPlus) to develop three data-driven models (i.e., LightGBM, random forest (RF), and long-short term memory (LSTM)) and compared their prediction performances. The physics-based models were developed using office prototype building models as baselines, and ranges were provided for selected key input parameters. Three different input feature dimensions (i.e., six-, nine-, and fifteen-input feature selections) were investigated, aiming to meet different demands for practical applications. We found that LightGBM significantly outperforms the RF and LSTM algorithms, not only with respect to prediction accuracy but also in regard to computation cost. The best prediction results show that the coefficient of variation of the root mean squared error (CVRMSE), squared correction coefficient ($R^2$), and computation time are 5.25%, 0.9959, and 7.0 s for LightGBM, respectively, evidently better than the values for the algorithms based on RF (18.54%, 0.9482, and 44.6 s) and LSTM (22.06%, 0.9267, and 758.8 s). The findings demonstrate that a data-driven model is able to avoid the process of establishing a complicated physics-based model for predicting a building's thermal load, with similar accuracy to that of a physics-based simulation tool.

**Keywords:** load prediction; feature engineering; machine learning; LightGBM; grid-integrated buildings

## 1. Introduction

### 1.1. Literature Study

Building thermal load prediction plays an important role in energy management and efficiency. It has wide applications, such as in determining the capacity of heating, ventilation, and air conditioning (HVAC) systems in the phase of building design [1]; providing operational optimization control in building energy systems for existing or retrofitted buildings [2]; and determining the demand response baseline for grid-integrated buildings [3–5]. In the field of building load prediction, the thermal loads for heating and cooling are the most difficult parts to accurately predict, owing to the complex, nonlinear relationships among the influencing factors (input feature variables), such as the weather conditions, building physics, and different operational behaviors. Therefore, previous studies have focused on the thermal load of buildings, especially on the loads of HVAC systems. Prediction approaches are widely categorized into three types: physics-based models (white-box), data-driven models (black-box), and reduced-order models (gray-box) [6,7], while the data-driven models have been increasingly discussed in recent years.

White-box models, such as professional software, can predict building loads with high accuracy. Several mature software tools, including EnergyPlus, TRNSYS, and Modelica, are widely used in the engineering industry [8]. However, developing a physics-based model is time-consuming, and prior knowledge is required. Gray-box models simplify the complicated functions to a resistance and capacity model. The key shortcoming of gray-box models is that they do not consider internal heat gains and occupant behaviors [9]. A black-box model is a data-driven model based on historical data. Nowadays, with the rapid development of building-based big data and machine learning algorithms, developing a black-box model is a promising approach for load prediction. A rich set of studies have focused on machine learning algorithms, such as support vector regression (SVR) [4,10], random forest (RF) [11,12], extreme gradient boosting (XGBoost) [13–15], long-short term memory (LSTM) [16–18], artificial neural networks (ANNs) [19], and convolutional neural networks (CNN) [20]. In previous studies, the two most popular fields concerned sensitivity analyses of different dimensions of input features and hyperparameter tuning. Based on a comprehensive literature review, Table 1 lists the papers investigating different data-driven models. We found that LightGBM (similar to XGBoost but faster), RF, and LSTM have been reported as the most popular algorithms for building load predictions.

**Table 1.** Detailed descriptions and comparisons of data-driven models to predict load demands in papers.

| Reference | Input Features | Data-Driven Model | Prediction Results |
|---|---|---|---|
| [13] | Day of week; Hour of day; Holiday; Outdoor dry bulb temperature; Outdoor relative humidity | XGBoost; RF; SVR; LSTM | Metrics CVRMSE<br>XGBoost: 21.1%<br>RF: 23.7%<br>SVR: 25.0%<br>LSTM: 20.2% |
| [21] | Total gross floor area; year of build; building height; shape form factor; vertical to horizontal ratio; length of the building; and width of the building; building morphology | LightGBM; XGBoost; RF; SVR | Metrics $R^2$<br>LightGBM: 0.8608<br>XGBoost: 0.8137<br>RF: 0.7959<br>SVR: 0.7363 |
| [22] | Historical load; weather data; calendar rules | LightGBM; XGBoost; SVM; RF | Stacking method<br>XGBoost and LightGBM have obtained the higher accuracy |
| [23] | Outdoor dry bulb temperature; Schedules | XGBoost; RF; SVR; ANN | Metrics CVRMSE<br>XGBoost: 62%<br>RF: 64%<br>SVR: 64%<br>ANN: 73% |
| [24] | Outdoor dry bulb temperature; Outdoor relative humidity; Wind speed; Solar radiation; Hour of day | XGBoost; RF; SVR; ANN | Metrics CVRMSE<br>XGBoost: 4.5%<br>RF: 4.6%<br>SVR: 5.5%<br>ANN: 5.1% |
| [25] | Relative compactness; Surface area; Wall area; Roof area; Number of floors; Orientation; Glazing area; Outdoor dry bulb temperature; Outdoor relative humidity; Solar radiation | ANN; SVM; RF; XGBoost | Metrics $R^2$<br>XGBoost: 0.998<br>RF: 0.973<br>SVR: 0.972<br>ANN: 0.968 |
| [26] | Aspect ratio; Relative compactness; Glazing area; Roof area; Surface area; Wall area; Orientation; Number of floors; Glazing area | ANN; SVR; RF | Metrics MAE (kW)<br>ANN: 1.15<br>SVR: 0.90<br>RF: 1.45 |

XGBoost was first released in 2014 and has become a powerful algorithm; most Kaggle competitions have reported it as the final winner [27,28]. XGBoost is based on a gradient boosting algorithm for assembling weak learners into a strong learner. XGBoost can be easily implemented using the Python, R, Julia, and Scala platforms [29]. Wang et al. [13] predicted long-term building thermal loads using XGBoost. Five input feature variables were considered: the day of the week, hour of the day, holiday status, temperature, and relative humidity. They found that, in shallow machine learning, XGBoost is the best algorithm. Yan et al. [23] obtained similar results. In the cooling season, they found that 11 input features could represent the main factors influencing cooling energy consumption. The prediction performances of XGBoost (CVRMSE: 62%), RF (CVRMSE: 64%), SVR (CVRMSE: 64%), and ANNs (CVRMSE: 73%) were not good relative to the results in [13]. Wang et al. [24] investigated models including XGBoost, RF, ANN, and SVR for predicting consumption from the thermal load of a residential building in Tianjin, China. The CVRMSE values of the prediction results from these models were as follows: RF (5.0%), XGBoost (5.8%), SVR (6.2%), and ANN (7.0%). Although the prediction accuracy was good, the computation cost is huge when the input feature dimensions and data size are large. In recent years, Light Gradient Boosting Machine (LightGBM) has been proposed as a novel and promising gradient boosting framework; it is similar to XGBoost. XGBoost and LightGBM are various tree-boosting methods. Shi et al. [22] concluded that LightGBM obtained a higher prediction accuracy compared to SVR in electric load forecasting. Zhang et al. [21] proposed a model of LightGBM integrated with the Shapley Additive exPlanation algorithm to predict energy usage and greenhouse gas emissions. The results show that the proposed LightGBM can achieve a higher prediction accuracy compared to XGBoost, RF, and SVR. Through the literature review, we can find that the tree-based algorithm including LightGBM and XGBoost is a promising method to obtain a better prediction result [30].

RF is a supervised learning algorithm based on decision trees. Compared with other algorithms, fewer parameters need to be tuned when using the RF model [31]. Ahmad et al. [11] compared three different algorithms for energy predictions. They selected the ambient temperature and relative humidity ratio as the input feature variable, and they found that the RF model (MAPE: 2.64%) outperforms LMSR (MAPE: 3.10%) and NARM (MAPE: 4.21%). Except the advantages of less overfitting and higher accuracy, RF presents the importance of the input features, which can be used in the model training and testing processes. A feature importance analysis chooses the main features and skips the weak features; this is critical to accelerating the computational process and ensuring the prediction accuracy. LSTM is a type of recurrent neural network (RNN) algorithm. It was first introduced by Hochreiter and Schnidhuber in 1997 [32]. Differing from a traditional neural network, LSTM passes the last step's information to the next time step (i.e., backpropagation). With these merits, LSTM comprises an inborn network for processing sequential data. It has advantages in solving complex and long time lag tasks, whereas traditional RNN algorithms are not good at this. LSTM has performed better in short-term load predictions than the linear regression, SVM, RF, and XGBoost algorithms [13,33]. The detailed theory of LightGBM, RF, and LSTM can be seen in Section 2.2.

Different algorithms have their merits for different building energy datasets, according to the previous studies, and the generalization performance of a prediction model is mainly based on the quality of data. For a specific building case, the prediction accuracy of a data-driven model is good enough now [30], while the generalization performance for exogenous buildings is still poor in the field of building energy predictions. More papers have investigated data-driven models using a specific building case, and these models are usually biased and can lead to a poor generalization [34,35]. For this purpose, the data source should cover all the possible ranges of each main variable and represent the overall energy consumption patterns of buildings, because a data-driven model cannot deduce the result for unseen data. To do that, previous studies had some attempts to acquire big data from numerous buildings. For instance, 5000 residential buildings from the Ministry of Housing Communities and Local Government (MHCLG) repository have been used to develop a data-driven model [36]. Meteorological variables with specific ranges and a 5-min interval energy dataset from EnerNOC were used to develop a data-driven interval forecasting model for building energy predictions [37]. Despite more dimensions of information having been considered, the diversity and depth of each main variable are still worth investigating for building a well-generalized model.

Although many studies have already investigated the prediction performances of different data-driven algorithms, there are still two research gaps. The first gap is whether the data-driven model is good enough to represent the physics-based tools. It is urgent to investigate the feasibility of using a data-driven model in the field of building load forecasting. To the best knowledge of the authors, the previous studies mostly focused on the existing building with acquired historical data, which means that the model is well developed on this specific building while it usually has a poor generalization for other buildings, especially for design phase buildings. Therefore, using physics-based tools to generate a massive dataset that covers all the common energy use scenarios could be a promising way to develop a good generalization data-driven model. Another research gap concerns providing a sufficient number of key input features to determine the building energy use in data-driven models. For building owners, the difficulty in obtaining the different dimensions of input features is not equal, and some main information might be unavailable. Thus, presenting the prediction performance based on the number of main feature variables is quite useful for building owners to estimate energy consumption with a certain accuracy. Researchers have made different conclusions in the past. For instance, Yan et al. [23] concluded that 11 input features were sufficient, whereas Wang et al. [13] stated that 5 features were adequate. How many input features should be used in the model to obtain a satisfying prediction result? This is still a challenge. To address these research gaps, first, we acquired data by running massive EnergyPlus building models that represented different buildings. Note that the data can also be acquired from any physics-based tools or on-site data. Second, different dimensions of the key feature variables were selected to develop three widely used models: LightGBM, RF, and LSTM in the context of forecasting the HVAC electrical load. Notably, the HVAC electrical load was predicted as an equivalent of a thermal load, as the electricity load is more commonly used in modern grid-integrated buildings.

### *1.2. Motivations and Contributions*

In practice, it is extremely difficult to measure all of the required inputs to a physics-based building energy model. These models require thousands of input parameters derived from prior knowledge, and their simulations are typically computationally intensive. Thus, a data-driven model could be a promising approach to solve this problem. However, developing a data-driven model to represent a physics-based model well is still a big challenge. First, the value range of each input variable should fully cover the practical situations; second, the simulation scenarios and dataset sizes should be big enough; finally, the computational costs should be acceptable for the practical engineering applications. According to previous studies, weather information is widely used in data-driven models; furthermore, building physical and thermophysical variables, such as the window-to-wall ratio and total heat transfer coefficient of the envelope, have been increasingly considered. In data-driven models, the number of input features can influence the prediction accuracy and computation speed. There is a tradeoff in constructing different sizes of input features under practical conditions. Therefore, it is intriguing to conduct an overall input feature dimension investigation when building managers can obtain different dimension building information. In this context, developing a data-driven model by using the massive energy data from simulation tools to represent a physics-based model is an appealing approach to building load predictions. The main contribution of this work is developing a data-driven model by using the massive energy data from simulation tools to represent a physics-based model to building load predictions. These developed models are of high practical value despite that building energy managers have enough or limited building information, especially for design phase buildings where the available information is lacking. The goal of this paper is to develop a suitable data-driven model to estimate the thermal load demand of buildings promptly and accurately, especially in the phase of building design. For existing building, we can calibrate and evaluate the proposed model using real data.

The remainder of this paper is organized as follows. Section 2 introduces the methodology, the theories of the selected algorithms and evaluation indexes. Section 3 presents the overall comparison results and discussion. Finally, the main conclusions and future applications are presented in Section 4.

## 2. Methodology

Data-driven models do not require establishing thermal equilibrium equations; usually, fewer inputs are required compared to physics-based simulation tools. A data-driven model uses data to deduce the hidden relationships between output (e.g., cooling load) and input feature variables (e.g., weather and building physics information) using a statistical approach. This approach is well adapted to buildings in the design phase, where detailed input parameters may be lacking. The research outline is shown in Figure 1. The methodology was designed with four main steps. The first step was EnergyPlus model development, which was the main step when using the co-simulation method of Python and EnergyPlus to obtain the required dataset. Note that the development of building energy models can be replaced by any other physics-based simulation tools. The second step processed the data and analyzed the key input features. The widely used input variables included these three types: (1) time-related information, such as the day type, occupancy, and equipment schedules; (2) weather conditions, such as the temperature, humidity, and solar radiation; and (3) building physical parameters, such as the window-to-wall ratio and R-value of the wall. The output targets are generally thermal loads or electricity consumption [38,39]. The third step selected the data-driven algorithm by reviewing previous studies and tuned the selected models. Data-driven models have gained great interest in the buildings field because of their simplicity and flexibility [40]. In this section, three promising data-driven methods are presented, including LightGBM, RF, and LSTM. Additionally, some important hyperparameters are tuned, and they can be seen in Tables A1–A3. The last step evaluated the developed models and listed the future applications. It is worth noting that simulation data was used, not real data, because we needed to change each input feature variable within a wider range to obtain enough data points (thousands of building types and millions of data points) to train a well-generalization model, which is nearly impossible to attain from real buildings.

### 2.1. Seed Model and Energy Data Source Description

#### 2.1.1. Seed Model Description

This section describes the office building models used in EnergyPlus (Version 9.0.1) and the ranges of the input features used in these cases to obtain energy data. Three categories of building energy models were developed: (1) small office building, (2) medium office building, and (3) large office building. DOE Commercial prototype building models were used as the starting point [41]. The geometry of these three building types is shown in Table 2; all of the models have a rectangular footprint. Table 2 provides key information on these three models. The small office building has 1 floor, the medium office building has 3 floors, and the large office building has 12 above-grade floors and 1 basement. Furthermore, the envelope types and HVAC system types are different. Table 2 also shows the types of exterior walls, roofs, heating and cooling systems, and HVAC system operation schedules.

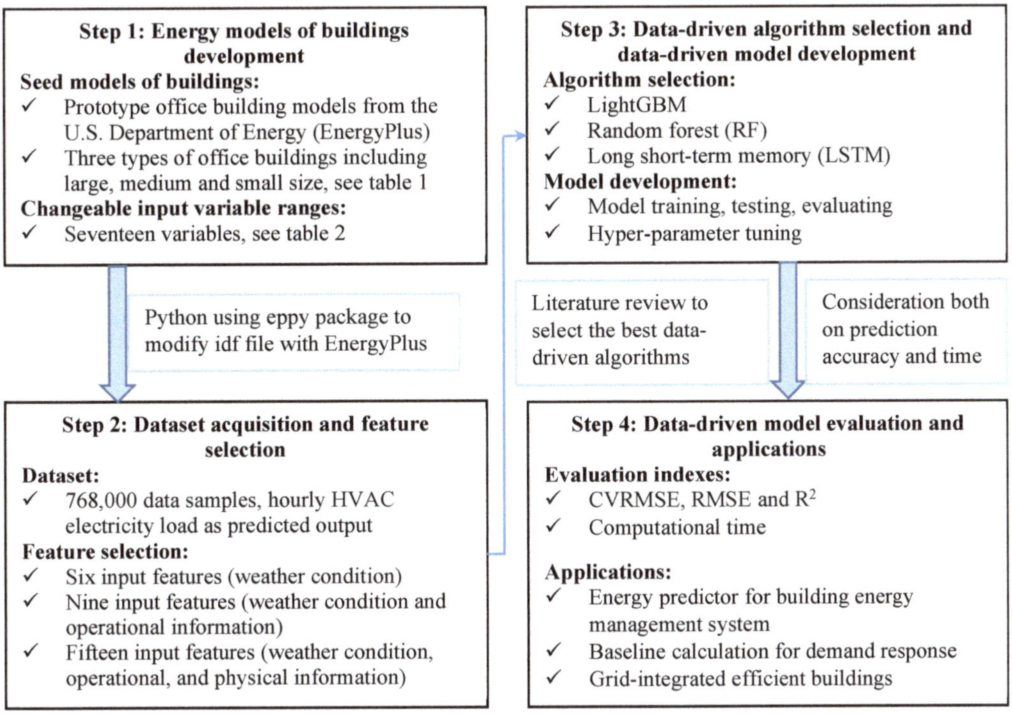


**Figure 1.** Research methodology and its outline.

**Table 2.** Seed models description in EnergyPlus [41].

| Item | Description<br>Small Office | <br>Medium Office | <br>Large Office |
|---|---|---|---|
| Geometry | 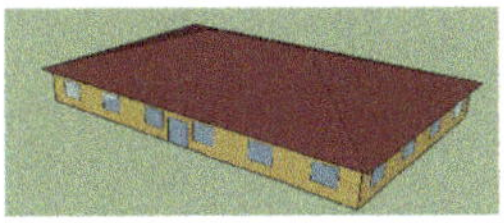 | 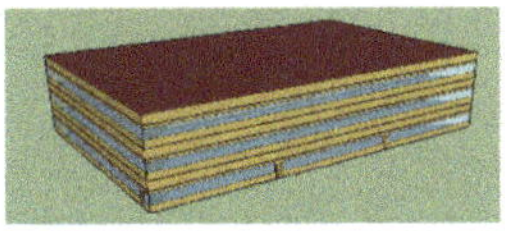 | 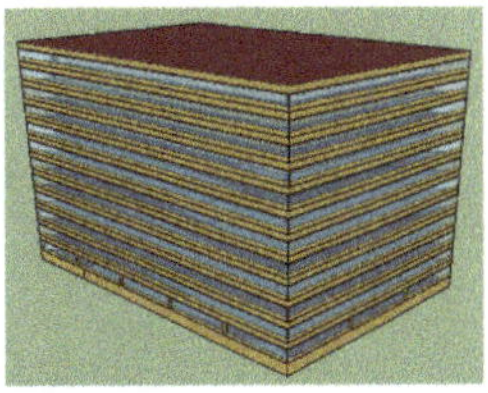 |
| Total Floor Area | 511 $m^2$ | 4980 $m^2$ | 46,321 $m^2$ |
| Exterior Wall Type | Wood frame walls | Steel frame walls | Mass walls |
| Roof Type | Attic roof with wood joint | Built-up roof | Built-up roof |
| Heating Type | Air source heat pump with gas furnace as backup | Gas furnace | One gas-fired boiler |
| Cooling Type | Air source heat pump | Packaged air conditioning | Water direct expansion cooling coil<br>Two water-cooled centrifugal chillers |
| HVAC Operation Schedule | Weekdays: 6:00 am–7:00 pm | Weekdays: 6:00 am–10:00 pm<br>Saturdays: 6:00 am–6:00 pm | Weekdays: 6:00 am–10:00 pm<br>Saturdays: 6:00 am–6:00 pm |

#### 2.1.2. Energy Data Source Description

Seventeen key input variables were changed with the appropriate steps in each EnergyPlus model. Notably, for each EnergyPlus model, the first four variables (No. 1–4 in Table 3 are EnergyPlus standard TMY3 weather data) are time series data that change for each simulation time step, whereas the other variables are constant in each step once determined. Based on the default values provided by the office prototype building models, we determined the ranges of these input variables. We used ±20% of the default values as the lower and upper limits for most of the input variables [42], except for the cooling and heating temperature set points. For these two input variables, we used ±1.11 °C (±2 °F) of the default values as the lower and upper limits, so as to ensure that the indoor temperature was within a preferable range. Table 3 lists the input variable ranges for these three types of office buildings in detail. In order to simplify the data-driven model, we selected the most important input feature variables in Table 3, and some not important variables such as airtightness were neglected. A total of 768,000 valid data samples were generated from the above-mentioned three EnergyPlus seed models, and the dataset and full code to develop the data-driven model can be downloaded freely in [43]. When we generated the dataset, the Python package named eppy was used to co-simulate with EnergyPlus. All the variables listed in Table 3 were changed respectively in different steps, and the total simulation time to obtain all the data was about 700 h (running on a Dell Precision 7920 Tower, 20 kernel CPU).

**Table 3.** Input variable ranges for the three types of office buildings.

| No. | Input Feature Variables | Unit | Range | | |
|---|---|---|---|---|---|
| | | | Small Office | Medium Office | Large Office |
| 1 | Dry Bulb | °C | [−32.8, 37.0] | [−32.8, 37.0] | [−32.8, 37.0] |
| 2 | Relative Humidity | % | [4, 100] | [4, 100] | [4, 100] |
| 3 | Global Horizontal Radiation | $Wh/m^2$ | [0, 964] | [0, 964] | [0, 964] |
| 4 | Wind Speed | m/s | [0, 14.9] | [0, 14.9] | [0, 14.9] |
| 5 | Total Floor Area | $m^2$ | [409, 613] | [3986, 5979] | [37,056, 55,584] |
| 6 | Aspect Ratio | - | [1.2, 1.8] | [1.2, 1.8] | [1.2, 1.8] |
| 7 | Window-to-Wall Ratio | - | [0.16, 0.24] | [0.26, 0.40] | [0.32, 0.48] |
| 8 | Floor Height | m | [2.44, 3.66] | [2.19, 3.29] | [2.20, 3.29] |
| 9 | Exterior Wall Insulation R-value | $(m^2{\cdot}K)/W$ | [2.46, 3.68] | [2.25, 3.38] | [1.31, 1.97] |
| 10 | Roof Insulation R-value | $(m^2{\cdot}K)/W$ | [6.48, 9.72] | [4.25, 6.37] | [4.25, 6.37] |
| 11 | Specific Heat for Internal Thermal Mass | $J/(kg{\cdot}K)$ | [968, 1452] | [968, 1452] | [968, 1452] |
| 12 | Cooling Temperature Set Point | °C | [22.78, 25.00] | [22.89, 25.11] | [22.89, 25.11] |
| 13 | Heating Temperature Set Point | °C | [20.00, 22.22] | [19.89, 22.11] | [19.89, 22.11] |
| 14 | Fresh air volume | $m^3/s\text{-}m^2$ | [0.000345, 0.000518] | [0.000345, 0.000518] | [0.000345, 0.000518] |
| 15 | People Density | $m^2$/person | [13.27, 19.91] | [14.86, 22.29] | [14.86, 22.29] |
| 16 | Lighting Power Density | $W/m^2$ | [6.80, 10.20] | [6.80, 10.20] | [6.80, 10.20] |
| 17 | Electric Equipment Power Density | $W/m^2$ | [5.42, 8.14] | [6.46, 9.68] | [6.46, 9.68] |

#### 2.1.3. Input Feature Selection

Feature selection is critical for data-driven models. External weather conditions, physical parameters, and operational schedules for equipment and occupant behavior are three common input feature types [44–46]. The feature selection of each building could be different from each other. We selected the input feature variables due to three reasons: (1) the importance of input variables in physics-based thermal equilibrium equations, (2) prior knowledge in building energy consumption estimation (including the literature study and engineering experience), and (3) the difficulty to obtain in practice. Through a literature study, we found that the outdoor dry bulb temperature, outdoor relative humidity, solar radiation, day of the week, and hour of the day were the five most frequently used features in data-driven models. Except for the external climate data, the physical information of the buildings (such as the number of floors, wall area, glazing area, and window-to-wall ratio) was used to improve the prediction accuracy. Therefore, we selected 17 widely used variables that are easier to obtain and have a great impact on building energy consumption.

The complexity of the input feature size influences the computational cost and prediction accuracy. To study this impact, three input feature scenarios were investigated, aiming to meet different demands for practical applications. Table 4 shows the details of these scenarios. The key input variables represent the main weights in the machine learning models. The use of limited input information to achieve satisfactory prediction results and an acceptable time cost is a concern for building energy managers.

**Table 4.** Input feature scenarios used in our three algorithms.

| Scenarios | Scenario 1: Six Input Features | Scenario 2: Nine Input Features | Scenario 3: Fifteen Input Features |
|---|---|---|---|
| Description | Weather condition | Weather condition and operational information | Weather condition, operational information, and physical parameters |
| Input features | Hour of the day<br>Historical load data<br>Dry bulb temperature (°C)<br>Relative humidity (%)<br>Global horizontal radiation (W·h/m$^2$)<br>Wind speed (m/s) | Scenario 1<br>Cooling temperature Set point (°C)<br>Heating temperature Set point (°C)<br>Fresh air volume [m$^3$/(s·m$^2$)] | Scenario 2<br>R-value of wall [m$^2$·K/W] Internal mass (average specific heat of the walls) [J/(kg·K)]<br>Window to wall ratio<br>Floor height (m)<br>Shape coefficient (1/m)<br>Aspect ratio |

*2.2. Selected Data-Driven Algorithm Description*

2.2.1. LightGBM

LightGBM is a gradient-boosting framework comprising a tree-based learning algorithm, i.e., a gradient-boosting decision tree (GBDT). The GBDT is a widely used algorithm in machine learning, owing to its efficiency and accuracy; XGBoost is a typical framework employing this algorithm. However, when the input feature dimensions and/or data size are large, as in modern buildings, the predictable data scale and computation speed remain unsatisfactory. To tackle these deficiencies, LightGBM was proposed based on two novel techniques: gradient-based one-sided sampling (GOSS) and exclusive feature bunding (EFB). These two techniques speed up the training process by up to 20 times, with almost the same accuracy as the traditional GBDT algorithm [47]. LightGBM was first released on 17 October 2016 as a part of Microsoft Corporation's "Distributed Machine Learning Toolkit" project [48]. It was designed to be distributed and efficient, with the advantages of a faster training speed, higher efficiency, lower memory usage, parallel support, and the ability to handle large-scale data. LightGBM is a promising algorithm for big data [47,49]. GBDT is a mature algorithm, and the detailed theory thereof is discussed in other references [13,28]; thus, in this study, we only introduce the theories for GOSS and EFB in detail.

GOSS is a technique for balancing data information reduction and prediction performance. GOSS reduces the computation costs by distinguishing between different gradients of instances, retaining larger gradient instances while randomly sampling smaller gradients and thereby reducing the computation memory costs. The gradient magnitude of the instance represents the training error; thus, an instance with a small gradient can be eliminated, as it is already well trained. To avoid large changes in the training data distribution from the elimination of some instances, GOSS also randomly samples small gradient instances to secure the integrity of the original data. This way, although GOSS reduces the number of instances, the generalization error is close to that calculated using the full data instances. To prove that, the variance gain $V_j(d)$ of feature $j$ at splitting point $d$ is defined as shown in Equation (1).

$$V_{j|O}(d) = \frac{1}{A}\left(\frac{\left(\Sigma\{x_i \epsilon O : x_{ij} \le d\} g_i\right)^2}{n^j_{l|O}(d)} + \frac{\left(\Sigma\{x_i \epsilon O : x_{ij} > d\} g_i\right)^2}{n^j_{r|O}(d)}\right) \quad (1)$$

In the above, $A = \sum I[x_i \epsilon O]$, $n^j_{l|O}(d) = \sum I\left[x_i \epsilon O : x_{ij} \leq d\right]$, and $n^j_{r|O}(d) = \sum I\left[x_i \epsilon O : x_{ij} > d\right]$. $x$ is the training set with $i$ instances, $g_i$ is the negative gradient of the loss function, and O is the training dataset on a fixed node of the decision tree.

The training instances are initially ranked by their absolute gradients in descending order; next, the top $a$% of the larger gradients are selected as subset $A$; then, $b$% of the remaining gradients are randomly selected as subset $B$. Thus, the estimated variance gain $\widetilde{V}_j(d)$ over the subset $A \cup B$ can be defined as shown in Equation (2).

$$\widetilde{V}_j(d) = \frac{1}{n}\left(\frac{\left(\sum\{x_i \epsilon A_l\} g_i + \frac{1-a}{b} \sum\{x_i \epsilon B_l\} g_i\right)^2}{n^j_l(d)} + \frac{\left(\sum\{x_i \epsilon A_r\} g_i + \frac{1-a}{b} \sum\{x_i \epsilon B_r\} g_i\right)^2}{n^j_r(d)}\right) \quad (2)$$

Here, $A_l = \{x_i \in A : x_{ij} \leq d\}$, $A_r = \{x_i \in A : x_{ij} > d\}$, $B_l = \{x_i \in B : x_{ij} \leq d\}$, and $B_r = \{x_i \in B : x_{ij} > d\}$. There is a theory for proving that GOSS would not lose much training accuracy as compared with the full dataset [47], i.e., $\varepsilon^{GOSS}_{gen}(d) = \left|\widetilde{V}_j(d) - V_*(d)\right| \leq \left|\widetilde{V}_j(d) - V_j(d)\right| + \left|V_j(d) - V_*(d)\right| \triangleq \varepsilon_{GOSS}(d) + \varepsilon_{gen}(d)$.

EFB is another technique for reducing feature dimensions to improve the computational efficiency and is based on feature bundling. Usually, the bundled features are mutually exclusive, e.g., one feature is zero and the other is non-zero; therefore, these two features can be bundled together without losing information. In the case where two features are not mutually exclusive, a "conflict ratio" can be used to measure the degree of non-exclusion. When this ratio is small, the two features can be bound without excessively affecting the final accuracy. There are three steps in the EFB method. In Step 1, the features are sorted according to the total number of non-zero values; in Step 2, the conflict ratio between different features is calculated; and in Step 3, the conflict ratio is minimized by iterating through each feature and then binding the features. In this way, the time complexity is reduced from $O\left(N_{data} * N_{feature}\right)$ to $O(N_{data} * N_{bundle})$, where $N_{bundle} \ll N_{feature}$.

### 2.2.2. Random Forest (RF)

The diagrammatic prediction process of RF is shown in Figure 2. Each decision tree is randomly formed with different features and training samples, and the trees can be trained in parallel. Thus, the prediction accuracy is higher than that of a single decision tree. In the RF model, the number of trees and depth of a tree are the two key parameters that must be tuned; therefore, fewer parameters must be set than in other algorithms [12,31,50]. The RF algorithm includes four main processes: bootstrap resampling, bagging and out-of-bag error (OOBE) estimation, random feature selection, and full-depth decision tree growth [51], as shown in Figure 2. First, $N$ samples from the training dataset $S_n$ are randomly selected as bootstrap samples chosen with replacements, i.e., the same sample $(X_i, Y_i)$ may appear repeatedly. Second, the bagging technique selects samples from the bootstrap samples $N$; the remaining samples comprise the out-of-bag dataset. Third, there is a random selection of a predefined number $p$ of total features $k$, and RF attempts to search for the best cutting among these $p$ features. Finally, the best cutting is set by minimizing the cost function until the full-depth decision tree grows. The OOBE technique, or generalization error, is highly effective for estimating the generalization ability of the constructed model. In view of these technologies, the main advantage of the RF algorithm is its immunity to noise [51,52].

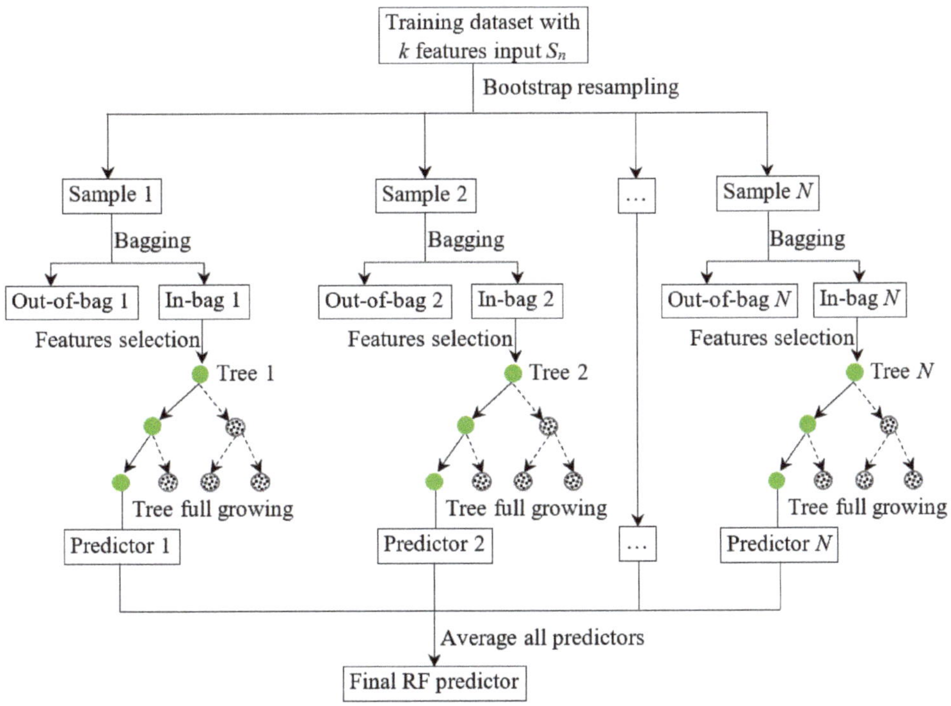


**Figure 2.** Diagrammatic process of the random forest (RF) algorithm.

2.2.3. Long Short-Term Memory (LSTM)

LSTM is a type of RNN algorithm. Figure 3 shows the principle of LSTM. It is an inborn network capable of accurately modeling complex multivariate sequences (such as building energy demands), although this increases its computation costs [53]. It has advantages in solving complex and long time lag tasks that traditional RNN does not. In one study [13], LSTM performed better for the load prediction than the SVM and XGBoost algorithms.

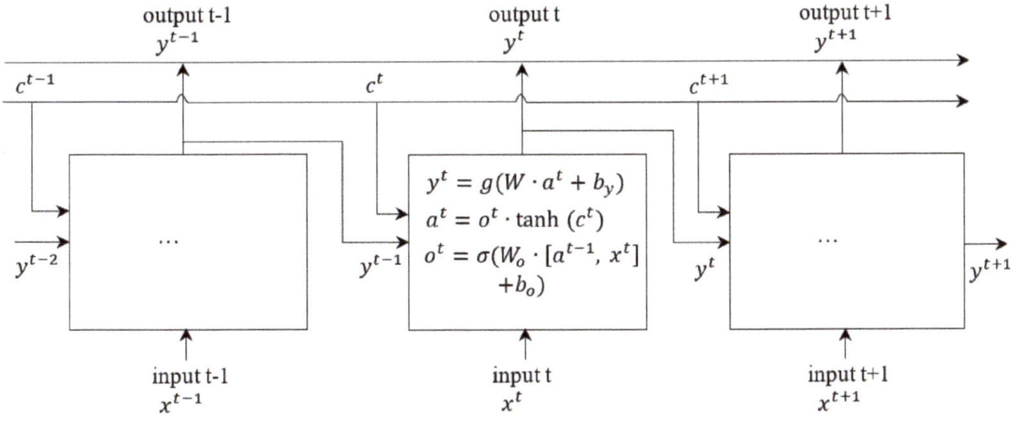


**Figure 3.** Principles of long short-term memory (LSTM).

Equations (3) and (4) define the architecture of the basic RNN algorithm. In this algorithm, only the memory of the last time step $t-1$ can be passed to time step $t$. However, the longer memory of the past time steps $t-n$ can be passed by introducing three special gates and two memory cells: input gate $i^t$, forget gate $f^t$, and output gate $o^t$ are defined as shown in Equations (5)–(7), respectively; the candidate memory cell $\tilde{c}^t$ and memory cell $c^t$ are defined in Equations (8) and (9), respectively.

$$a^t_{RNN} = g\left(W_a \cdot \left[a^{t-1},\ x^t\right] + b_a\right) \tag{3}$$

$$y^t_{RNN} = g'\left(W_a{}' \cdot a^t_{RNN} + b_y\right) \tag{4}$$

$$i^t = \sigma\left(W_i \cdot \left[a^{t-1},\ x^t\right] + b_i\right) \tag{5}$$

$$f^t = \sigma\left(W_f \cdot \left[a^{t-1},\ x^t\right] + b_f\right) \tag{6}$$

$$o^t = \sigma\left(W_o \cdot \left[a^{t-1},\ x^t\right] + b_o\right) \tag{7}$$

$$\tilde{c}^t = \tanh\left(W_c \cdot \left[a^{t-1},\ x^t\right] + b_c\right) \tag{8}$$

$$c^t = f^t \cdot \tilde{c}^{t-1} + i^t \cdot \tilde{c}^t \tag{9}$$

$$a^t_{LSTM} = o^t \cdot \tanh\left(c^t\right) \tag{10}$$

$$y^t_{LSTM} = g'\left(W_a{}' \cdot a^t_{LSTM} + b_y\right) \tag{11}$$

In the above, $g(x)$ and $g'(x)$ are two activation functions; $a^t$ is the activation function at time step $t$; $x^t$ and $y^t$ are the input and output at time step $t$, respectively; $b$ is the bias; $W$ is the weight factor; $\sigma(x)$ is the sigmoid function, which is defined in Equation (13); and tanh($x$) is the hyperbolic tangent function, as defined in Equation (12).

$$\tanh(x) = \frac{e^{2x} - 1}{e^{2x} + 1} \tag{12}$$

$$\sigma(x) = \frac{e^x}{e^x + 1} \tag{13}$$

### 2.3. Data-Driven Model Development Process

The development process (i.e., Step 3 in Figure 1) of our proposed data is shown in Figure 4. First, the massive dataset is divided into two sets, including training and test datasets after data pre-processing. Second, feature engineering is implemented for the model inputs, which includes the weather conditions (e.g., temperature, humidity, and solar radiation); building physical parameters (e.g., R-value of the wall, floor height, internal mass, and Shape coefficient); and operational information (e.g., temperature setting and fresh air volume). The model output is the HVAC electrical load. Then, different models can be trained, and the hyperparameters are required to tune for better results. Last is using the test dataset to test and evaluate the developed model.

### 2.4. Prediction Performance Indices

To evaluate the prediction performances of different algorithms, three indices are generally used: the CVRMSE, root mean squared error (RMSE), and squared correlation coefficient ($R^2$). The CVRMSE is a scale-independent indicator that is normalized by averaging the RMSE. The CVRMSE has been used in studies [13,54] and is recommended by the American Society of Heating, Refrigerating and Air-conditioning Engineers' (ASHRAE) Guidelines 14 [55]; the RMSE is a scale-dependent indicator and thus maintains the same scale as the original data; and $R^2$ is the coefficient of determination, which ranges from 0 to 1 and can reflect the goodness of fit. These three indices are defined in Equations (14)–(16), respectively.

$$CVRMSE = \frac{\sqrt{\sum_{i=1}^{n}(\check{y}_i - y_i)^2/n}}{\sum_{i=1}^{n} y_i/n} \tag{14}$$

$$RMSE = \sqrt{\sum_{i=1}^{n}(\check{y}_i - y_i)^2/n} \tag{15}$$

$$R^2 = 1 - \frac{\sum_{i=1}^{n}(y_i - \check{y}_i)^2}{\sum_{i=1}^{n}(y_i - \overline{y})^2} \tag{16}$$

where $\check{y}_i$, $y_i$, and $\overline{y}$ represent the predicted value of sample $i$, the actual value of sample $i$, and the mean value of all sample datasets, respectively; $n$ denotes the number of samples.

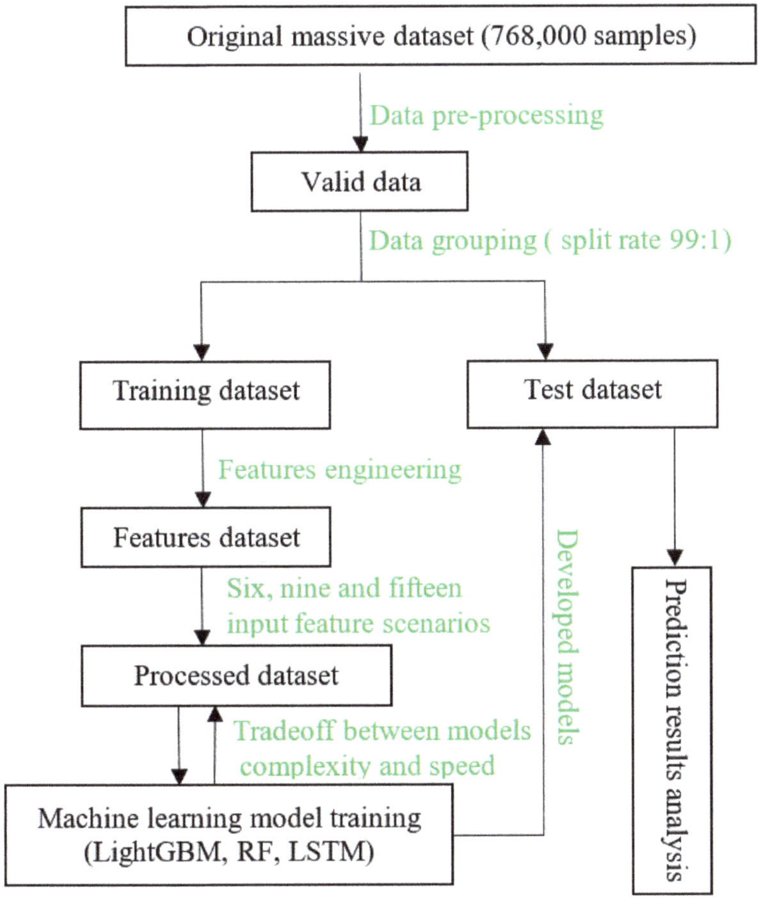


**Figure 4.** Flow chart of the data-driven model development process.

## 3. Results and Discussion

The prediction performances were calculated and compared under the above-mentioned scenarios, and the prediction granularity is 1 h. For each scenario, it is worth noting that we first split the 768,000 data samples into training and test sets at a ratio of 99:1 because of the big size of the dataset [56], although the commonly used ratio is 80:20 or 70:30 for normal-sized datasets [36] Second, different feature scenarios were selected as the inputs, and the hourly predicted HVAC electrical load was the output. Finally, we applied the five-fold cross-validation approach implemented in the scikit-learn Python package for the hyperparameter tuning in our models. The hyperparameter searching range and optimum results are shown in Tables A1–A3. All of the models were trained and tested on identical datasets. We compared three different machine learning algorithms (LightGBM, RF, and LSTM) under these three scenarios. To address cases with (such as existing buildings) and with no (such as buildings in the design phase) historical energy load data available, the input features were applied with and without the historical HVAC electricity load. In this paper, we obtained all the required HVAC electricity loads at the stage where we simulated all the seed models mentioned in Section 2.1. It is worth noting that we categorized the cases in the next three scenarios by distinguishing the use of historical load data or not. In all scenarios, three prediction performance indices were used to evaluate the prediction performance on the testing dataset: RMSE, CVRMSE, and $R^2$.

### 3.1. Scenario 1: Six Input Features

The simplified scenario used only six key input features to develop the model. As shown in Table 5, LightGBM has the highest prediction accuracy (CVRMSE: 7.14%) and fastest computational speed (5.4 s); the results from LSTM are the worst, not only in regard to accuracy (CVRMSE: 26.15%) but also to computational cost (716.5 s). If the CVRMSE is below 30% when the prediction step is hourly, the model is considerably acceptable and sufficiently close to physical reality for engineering purposes [57]. By this criterion, the LSTM result without the historical load is unacceptable. Additionally, the computation cost of the LSTM is more than a hundred times larger than that of LightGBM; this is consistent with our theoretical analysis in Section 2 and corresponds to the conclusions from the literature, i.e., that LSTM is recommended for small datasets and short-term predictions. Figure 5 presents the hourly predicted and actual HVAC electrical load profiles on the testing dataset; ten days are randomly selected from the testing data for visualization purposes. It is evident that the prediction performance of the LSTM algorithm is quite poor. All the models provide better prediction results when the historical load data is considered. The LSTM algorithm that does not use historical load data performs the worst, and the prediction deviation is large during the peak and valley load times.

**Table 5.** Comparison results of Scenario 1 with and without historical load data.

| Scenario 1 | Long Short-Term Memory (LSTM) (with) | LSTM (without) | LightGBM (with) | LightGBM (without) | Random Forest (RF) (with) | RF (without) |
|---|---|---|---|---|---|---|
| Root mean square error (RMSE) | 1.07 | 2.26 | 0.29 | 0.71 | 0.76 | 1.02 |
| Coefficient of variance of RMSE (CVRMSE) | 26.15% | 55.32% | 7.14% | 17.40% | 18.59% | 24.85% |
| $R^2$ | 0.896968 | 0.538813 | 0.992327 | 0.954380 | 0.947891 | 0.906964 |
| Computation Time(s) | 716.5 | 719.4 | 5.4 | 6.1 | 19.1 | 10.7 |

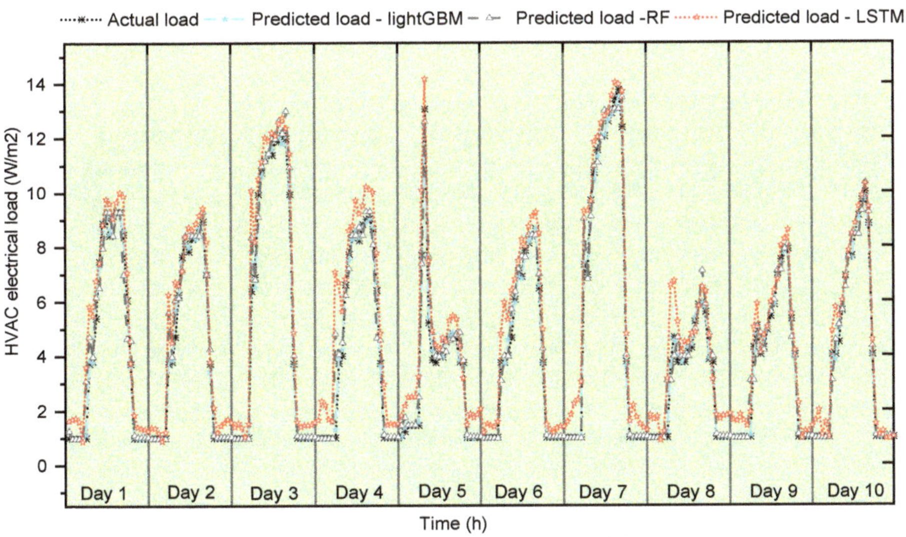


(**a**) With historical HVAC electricity load data

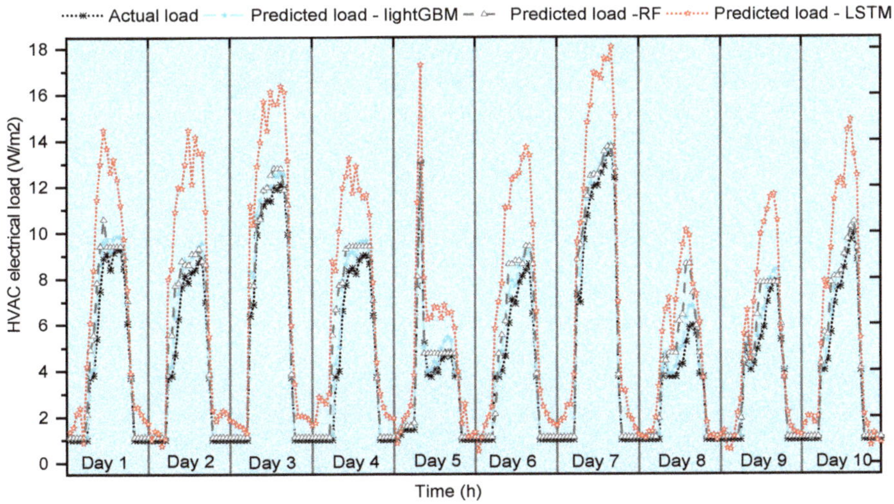


(**b**) Without historical HVAC electricity load data

**Figure 5.** Hourly prediction performances using six input features.

### *3.2. Scenario 2: Nine Input Features*

As mentioned in Section 1, HVAC energy consumption is influenced by multiple factors, such as weather conditions, building physics, and operational parameters. Only six input features were used in Scenario 1, and the prediction results might not be convincing. Therefore, we added operational parameters (room temperature setting and fresh air volume) to boost the knowledge learning level in the training models. From comparing Tables 5 and 6, it can be seen that the prediction accuracy is slightly improved in all three models, although the computation cost also increases. By adding these operational parameters, the mean improvement percentages in the CVRMSE are approximately 10.9%, 17.6%, and 0.5% for LSTM, LightGBM, and RF, respectively. This improvement is evident in the LightGBM and LSTM models. As shown in Figure 6, the prediction deviation remains large during the peak and valley load times for the LSTM.

**Table 6.** Comparison results of Scenario 2 with and without historical load data.

| Scenario 2 | LSTM (with) | LSTM (without) | LightGBM (with) | LightGBM (without) | RF (with) | RF (without) |
|---|---|---|---|---|---|---|
| RMSE | 1.04 | 1.83 | 0.25 | 0.57 | 0.76 | 1.01 |
| CVRMSE | 25.42% | 44.79% | 6.04% | 13.94% | 18.58% | 24.61% |
| $R^2$ | 0.902582 | 0.697689 | 0.994506 | 0.970706 | 0.947979 | 0.908734 |
| Computation Time(s) | 765.0 | 780.0 | 6.5 | 5.7 | 29.2 | 22.6 |

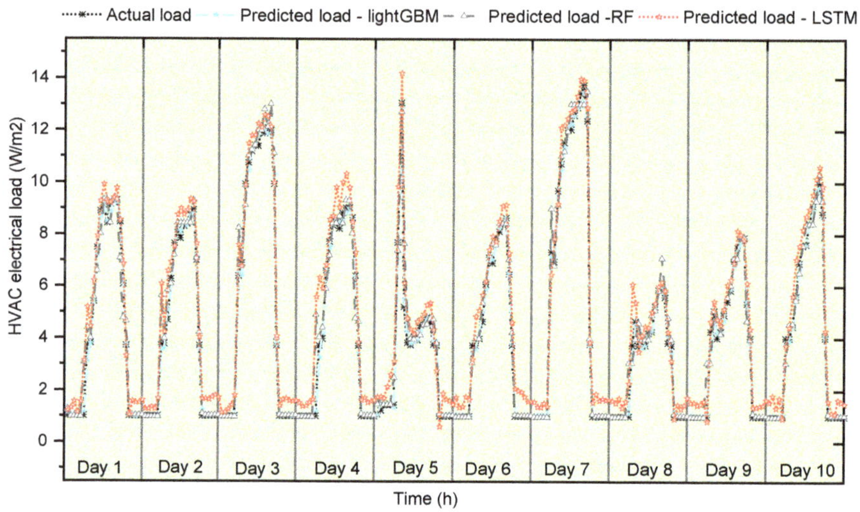


(**a**) With historical HVAC electricity load data

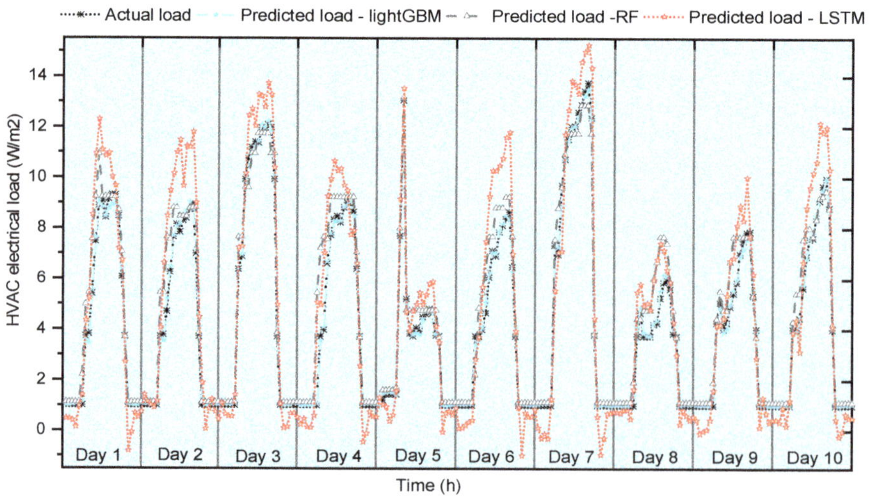


(**b**) Without historical HVAC electricity load data

**Figure 6.** Hourly prediction performances using nine input features.

### 3.3. Scenario 3: Fifteen Input Features

In addition to the features in Scenario 2, the physical information of the building can be used to improve the prediction accuracy. We used 15 key input features to train the models in Scenario 3; three physical aspects (building shape, R-values of walls, and building thermal mass) were considered. Table 7 shows the hourly results from the different models with and without the historical load data. The best result for the CVRMSE was 5.25% in LightGBM, a promising result for the field of thermal load prediction. Furthermore, the computation time was only 7 s. The best results from the LSTM and RF approaches were 22.06% and 18.54%, respectively, i.e., close to the results from the previous study discussed in the Introduction. By adding six more building physical parameters, the mean improvement percentage of the CVRMSE was approximately 12.2%, 30.3%, and 1.6% for the LSTM, LightGBM, and RF approaches, respectively. The CVRMSE values of all three models were lower than 30%, indicating that they were all acceptable and sufficiently close to physical reality for engineering purposes. Figures 5–7 show that the prediction is gradually improved by employing additional input features.

**Table 7.** Comparison results of Scenario 3 with and without historical load data.

| Scenario 3 | LSTM (with) | LSTM (without) | LightGBM (with) | LightGBM (without) | RF (with) | RF (without) |
|---|---|---|---|---|---|---|
| RMSE | 0.90 | 1.63 | 0.21 | 0.3 | 0.76 | 0.98 |
| CVRMSE | 22.06% | 39.75% | 5.25% | 7.31% | 18.54% | 23.88% |
| $R^2$ | 0.926666 | 0.761820 | 0.995854 | 0.991937 | 0.948189 | 0.914036 |
| Computation Time(s) | 756.8 | 751.4 | 7.0 | 7.4 | 44.6 | 36.5 |

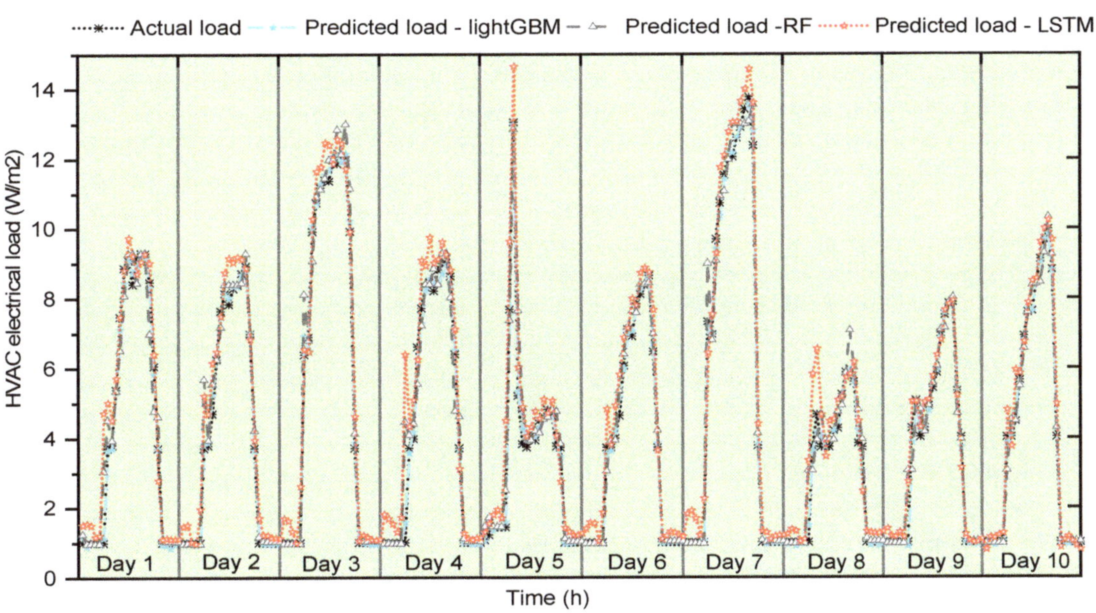


(**a**) With historical HVAC electricity load data

**Figure 7.** *Cont.*

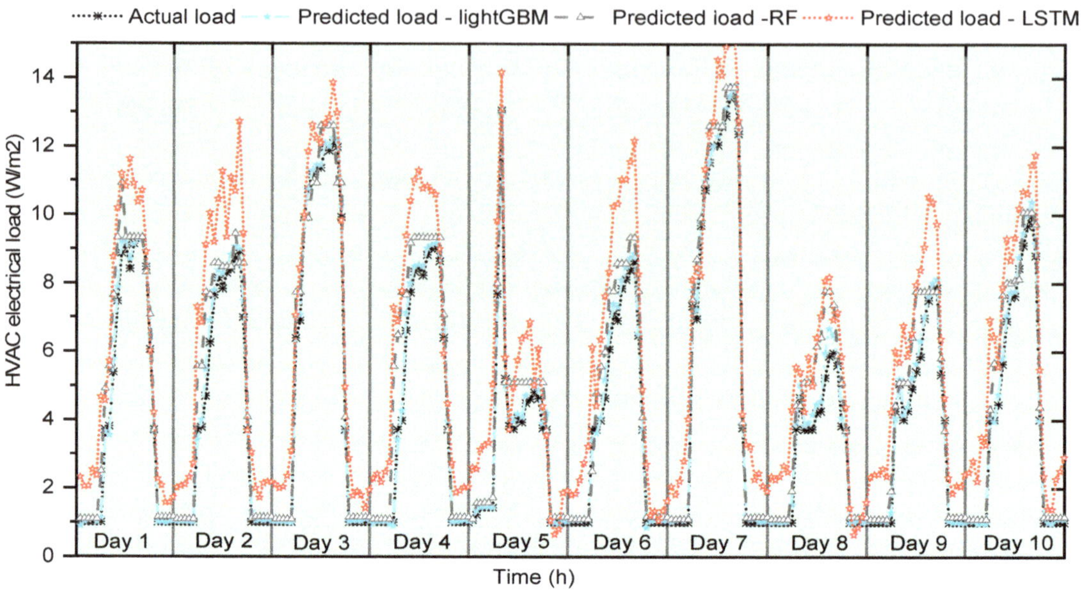


(**b**) Without historical HVAC electricity load data

**Figure 7.** Hourly prediction performances using fifteen input features.

### *3.4. Discussion*

Generally, the prediction accuracy can be improved when more information of the building is used in the data-driven models. As shown in Figure 8, the CVRMSE of scenario 3 with the historical HVAC load data is the best in these three models, although the improvement is higher in LightGBM and LSTM. In the model of RF, different scenarios have a close prediction accuracy that means additional building information is not necessary to improve its accuracy. We can also find that the historical HVAC load data is important to improve the accuracy of the models. All in all, a CVRMSE of 7.1% can be achieved when only the weather information is used, and the highest accuracy of 5.3% reached when the weather and operational and physical information of the building structure are considered. Except for these fifteen input features, adding more information such as the occupant's behavior is worthwhile as a further study in the future.

As shown in Figure 8, LightGBM is the best algorithm in building thermal energy prediction. Generally, the more data samples trained in the model training process, the better the prediction accuracy, as more hidden knowledge between the inputs and outputs can be learned. We investigated this effect by increasing the size of the training dataset sample in LightGBM, and Figure 9 shows the results. When we used 76,800 data samples, the prediction was the worst; the prediction accuracy generally improved as the sample size increased from 76,800 to 768,000. Additional data samples improved the prediction accuracy but also increased the computation cost.

A quantified investigation of the feature importance is also interesting for researchers. Therefore, we investigated the feature importance in the LightGBM model. Figure 10 ranks the importance of these fifteen input features. In our building's case, the results showed that the day of hours, outdoor dry bulb temperature, historical load data, global horizontal radiation, and relative humidity are the five most important features for building thermal load prediction. However, other features that have lower importance values can still be used to further improve the prediction accuracy.

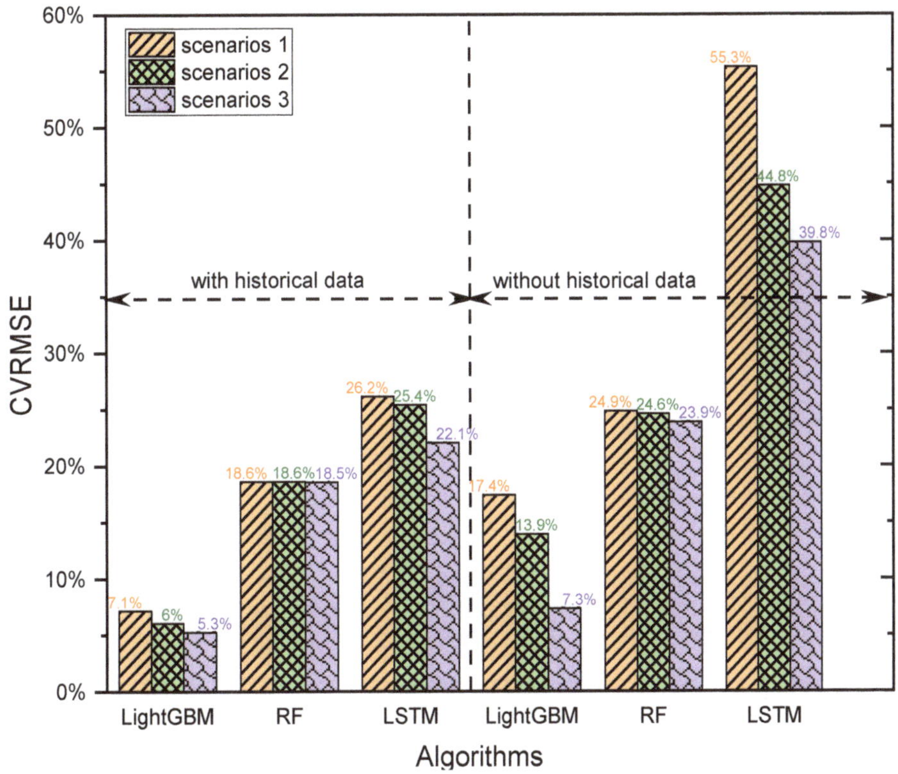


**Figure 8.** Prediction performances in different scenarios.

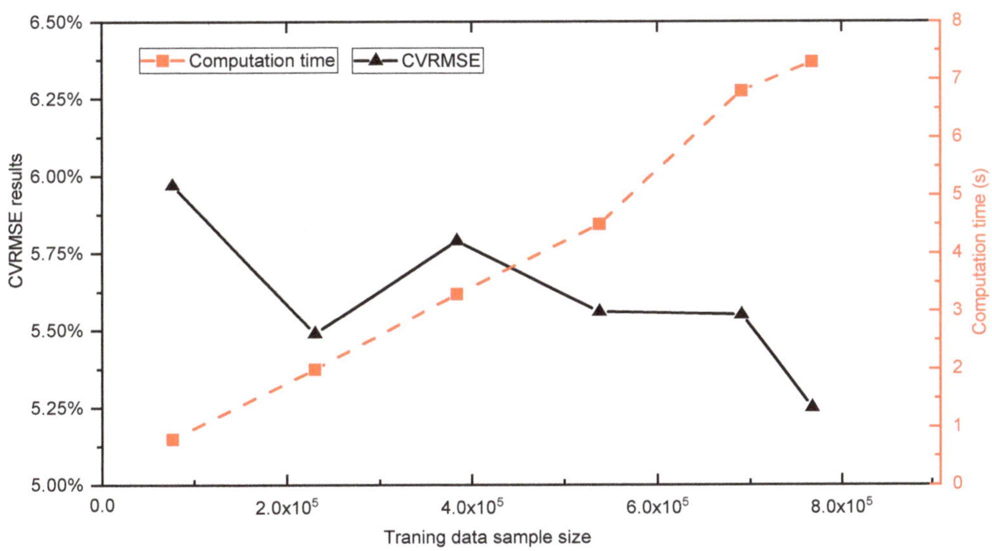


**Figure 9.** Prediction performance using different sizes of data samples in LightGBM.

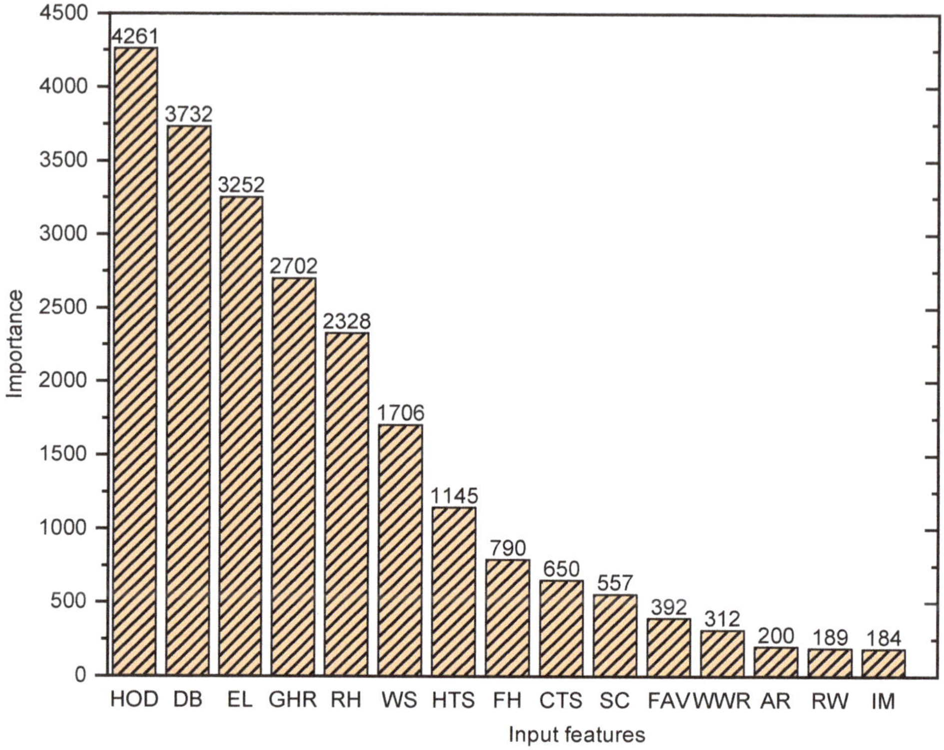


**Figure 10.** Feature importance ranking in LightGBM. (HOD: Hour of the day, DB: Dry bulb temperature, EL: Electrical load, GHR: Global horizontal radiation, RH: Relative humidity, WS: Wind speed, HTS: Heating temperature set point, FH: Floor height, CTS: Cooling temperature set point, SC: Shape coefficient, FAV: Fresh air volume, WWR: Window-to-wall ratio, AR: Aspect ratio, RW: R-value of wall, and IM: Internal mass.)

As using more features can improve the predictive accuracy, we investigated the effect of the number of features on the accuracy and computation speed in these three algorithms. The computational expense comparison of this study was performed on Windows 10 with a 2.6 GHz processor (Intel Corporation Core i7-10700) and 16 GB RAM memory. The Spyder scientific Python-integrated development environment was used to implement the prediction tasks. As shown in Figure 11, LightGBM requires the least computation time for model training and testing. In contrast, LSTM spends much more time. Generally, the computation expense increases with the input feature size. The model development time reaches the maximum when all 15 input features are considered. The LSTM model takes much longer time than the other two models, which is not suitable for a real-time energy management control system. The computational time of LightGBM is very short (several seconds), which makes LightGBM a suitable model for a real-time control system. Note that the acquisition of the original massive dataset is time-consuming, as we spent about 700 h to obtain the dataset using a power machine (Dell Precision 7920 Tower, 20 kernel CPU). The acquisition of the dataset process is time-consuming; however, this developed model achieves higher prediction accuracy, and it can be easily generalized for various energy use scenarios and building types once it is well trained. In this way, the model developers do not need to develop a specific model for different buildings and energy use scenarios.

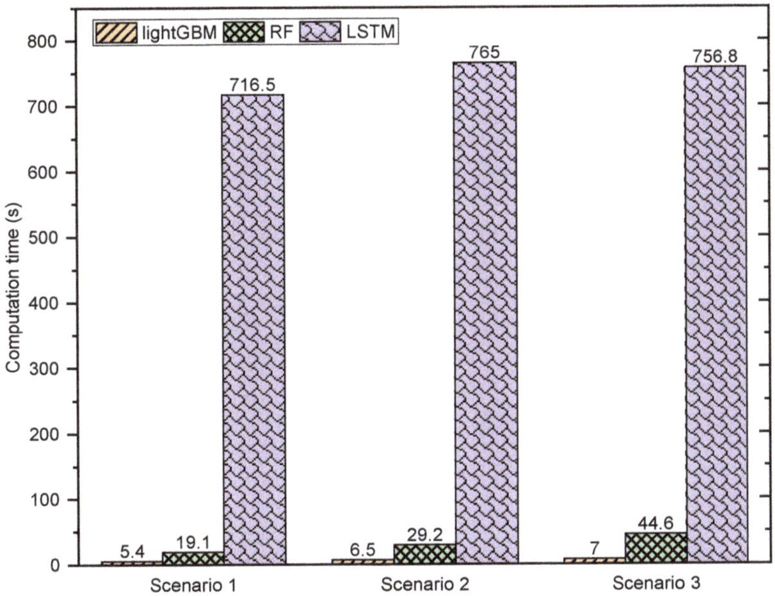


**Figure 11.** Computation time of the models in different scenarios.

## 4. Conclusions

A well-developed data-driven model to represent the physics-based tools is a challenge in both academic and practical fields. Traditionally, well-designed physics-based models, i.e., white-box models, have been widely applied. However, a white-box model requires a massive amount of detailed input parameters, which can be troublesome and difficult for engineers, especially for a building in the design phase. A fast and accurate building thermal load prediction method is critically important for optimal HVAC control, energy demand-side management, smart building management, and other tasks. In this study, therefore, we ran a big amount of EnergyPlus simulations to obtain massive energy data that covers the common energy use scenarios to develop a good generalization data-driven model. Using this data source, three machine learning models were developed and compared in three different input feature scenarios. Upon completion of the investigation, the following conclusions were reached.

(1) LightGBM is the most accurate and fastest prediction model. In the best scenario, the CVRMSE and $R^2$ of LightGBM are 5.25% and 0.99, respectively. Compared with the results of the other two algorithms and those in the existing literature, LightGBM is the most promising and best algorithm for building thermal load prediction.
(2) By training with the large amount of energy data generated by physics-based tools or on-site data, a data-driven model is able to represent a physics-based tool with comparable accuracy.
(3) The dimensions of the input features influence the prediction performance. Compared with a scenario using only weather information, the CVRMSE can be further improved when physical and operational information are considered. Although better accuracy is achieved with bigger dimensions of input features, it impacts the computational speed. Therefore, there will always be a tradeoff between the prediction accuracy demand and prediction speed tolerance.

The findings and the proposed models in this study are useful for real applications, such as smart building energy management, baseline calculation of demand response programs, and grid-integrated efficient building improvements. LightGBM is strongly recommended when dealing with large amounts of data, as it is faster and more robust. Building managers and engineers do not need to build a sophisticated physical model (such as the EnergyPlus model) to calculate the energy demand. Several basic inputs are able to reduce the tiresome tasks ordinarily required beforehand and can be used to obtain the energy demand with acceptable accuracy. In the early design phases of buildings with a lack of building information, only basic building and weather information are required to implement the prediction task. For existing buildings, not only considering the 17 variables mentioned above but more detailed building physics information, the operational schedule, and more historical load data can be used to rebuild the data-driven model and further improve the prediction performance.

**Author Contributions:** Conceptualization, Y.C. and Y.Y.; methodology, Y.C.; software, Y.Y.; writing—original draft preparation, Y.C. and Y.Y.; writing—review and editing, J.L., L.Z. and S.M.; supervision, Y.Y.; funding acquisition, Y.C. and W.L. All authors have read and agreed to the published version of the manuscript.

**Funding:** The research was funded by the National Natural Science Foundation of China (No. 52208116), the China Postdoctoral Science Foundation (No. 2020M681347) and the Key R&D and Promotion Project of the Department of Science and Technology of Henan Province, China (No. 222102320113).

**Data Availability Statement:** The dataset and full code to develop the data-driven model can be downloaded freely on GitHub: https://github.com/Bob05757/Key-inputs-setting-and-Energy-Data.

**Conflicts of Interest:** The authors declare no conflict of interest.

## Nomenclature

| | |
|---|---|
| ANN | Artificial neural networks |
| AR | Aspect ratio |
| CTS | Cooling temperature set point |
| CV | Coefficient of variation |
| CVRMSE | Coefficient of variation of root mean square error |
| DB | Dry bulb temperature |
| EL | Electrical load |
| ELM | Extreme learning machine |
| FAV | Fresh air volume |
| FH | Floor height |
| GHR | Global horizontal radiation |
| HOD | Hour of the day |
| HTS | Heating temperature set point |
| IM | Internal mass |
| MAPE | Mean absolute percentage error |
| MARS | Multivariate adaptive regression splines |
| MLP | Multilayer layer perceptron |
| NARM | Nonlinear autoregressive model |
| RF | Random forest |
| RH | Relative humidity |
| RNN | Recurrent neural networks |
| RW | R-value of wall |
| SC | Shape coefficient |
| SVM | Support vector machine |
| SVR | Support vector regression |
| WS | Wind speed |
| WWR | Window to wall ratio |
| XGBoost | Extreme gradient boosting |

## Appendix A

Hyperparameter tuning information with five-fold cross-validation in LSTM, LightGBM, and RF.

**Table A1.** Hyperparameter grid search and tuning results in LSTM.

| Hyperparameter | Description | Grid Searching Range | Selected |
|---|---|---|---|
| □activation | Activation functions | Sigmoid; Tanh; Relu | Relu |
| □optimizer | Optimization algorithms | Adam; RMSprop; Adagrad; SGD | RMSprop |
| □loss | Loss function | Mean Square Error; Mean Absolute Error; Mean Squared Logarithmic Error | Mean Square Error |
| □units | Number of memory cells | range (20, 200, 20) | 60 |
| □epochs | Number of epochs | range (20, 200, 20) | 100 |
| □batch_size | Number of batch size | [20, 32, 60, 100, 500] | 60 |

**Table A2.** Hyperparameter grid search and tuning results in LightGBM.

| Hyper-Parameters | Description | Grid Searching Range | Selected |
|---|---|---|---|
| learning_rate | Step size shrinkage used in the update to prevent overfitting | range (0.02, 0.12, 0.02) | 0.1 |
| n_estimators | Number of estimators | range (50, 400, 50) | 350 |
| max_depth | The depth of tree model | range (3, 10, 1) | 9 |
| num_leaves | the main parameter to control the complexity of the tree model | range (5, 500, 5) | 65 |
| max_bin | the maximum number of bins stored in feature | range (5, 256, 10) | 95 |

**Table A3.** Hyperparameter grid search and tuning results in RF.

| Hyperparameters | Description | Grid Searching Range | Selected |
|---|---|---|---|
| n_estimators | The number of trees in the forest | range (10, 110, 10) | 20 |
| max_depth | The maximum depth of the tree | range (2, 20, 2) | 8 |
| min_sample_split | The minimum number of samples required to split an internal node | range (1, 11, 1) | 6 |
| max_features | The number of features to consider when looking for the best split | ['auto', 'sqrt', 'log2'] | 'auto' |

## References


1. Fan, C.; Liao, Y.; Zhou, G.; Zhou, X.; Ding, Y. Improving cooling load prediction reliability for HVAC system using Monte-Carlo simulation to deal with uncertainties in input variables. *Energy Build.* **2020**, *226*, 110372. [CrossRef]
2. Li, W.; Gong, G.; Fan, H.; Peng, P.; Chun, L.; Fang, X. A clustering-based approach for "cross-scale" load prediction on building level in HVAC systems. *Appl. Energy* **2021**, *282*, 116223. [CrossRef]
3. Jang, Y.; Byon, E.; Jahani, E.; Cetin, K. On the long-term density prediction of peak electricity load with demand side management in buildings. *Energy Build.* **2020**, *228*, 110450. [CrossRef]
4. Chen, Y.; Xu, P.; Chu, Y.; Li, W.; Wu, Y.; Ni, L.; Bao, Y.; Wang, K. Short-term electrical load forecasting using the Support Vector Regression (SVR) model to calculate the demand response baseline for office buildings. *Appl. Energy* **2017**, *195*, 659–670. [CrossRef]
5. Chen, Y.; Chen, Z.; Xu, P.; Li, W.; Sha, H.; Yang, Z.; Li, G.; Hu, C. Quantification of electricity flexibility in demand response: Office building case study. *Energy* **2019**, *188*, 116054. [CrossRef]
6. Foucquier, A.; Robert, S.; Suard, F.; Stephan, L.; Jay, A. State of the art in building modelling and energy performances prediction: A review. *Renew. Sustain. Energy Rev.* **2013**, *23*, 272–288. [CrossRef]
7. Luo, X.J.; Lukumon, O.O.; Anuoluwapo, O.A.; Olugbenga, O.A.; Hakeem, A.O.; Ashraf, A. Feature extraction and genetic algorithm enhanced adaptive deep neural network for energy consumption prediction in buildings. *Renew. Sustain. Energy Rev.* **2020**, *131*, 109980. [CrossRef]

8. Nageler, P.; Schweiger, G.; Pichler, M.; Brandl, D.; Mach, T.; Heimrath, R.; Schranzhofer, H.; Hochenauer, C. Validation of dynamic building energy simulation tools based on a real test-box with thermally activated building systems (TABS). *Energy Build.* **2018**, *168*, 42–55. [CrossRef]
9. Wang, Z.; Hong, T.; Piette, M.A. Data fusion in predicting internal heat gains for office buildings through a deep learning approach. *Appl. Energy* **2019**, *240*, 386–398. [CrossRef]
10. Wu, J.; Wang, Y.-G.; Tian, Y.-C.; Burrage, K.; Cao, T. Support vector regression with asymmetric loss for optimal electric load forecasting. *Energy* **2021**, *223*, 119969. [CrossRef]
11. Ahmad, T.; Chen, H. Nonlinear autoregressive and random forest approaches to forecasting electricity load for utility energy management systems. *Sustain. Cities Soc.* **2019**, *45*, 460–473. [CrossRef]
12. Lahouar, A.; Ben Hadj Slama, J. Day-ahead load forecast using random forest and expert input selection. *Energy Convers. Manag.* **2015**, *103*, 1040–1051. [CrossRef]
13. Wang, Z.; Hong, T.; Piette, M.A. Building thermal load prediction through shallow machine learning and deep learning. *Appl. Energy* **2020**, *263*, 114683. [CrossRef]
14. Cao, L.; Li, Y.; Zhang, J.; Jiang, Y.; Han, Y.; Wei, J. Electrical load prediction of healthcare buildings through single and ensemble learning. *Energy Rep.* **2020**, *6*, 2751–2767. [CrossRef]
15. Moon, J.; Park, S.; Rho, S.; Hwang, E. Robust building energy consumption forecasting using an online learning approach with R ranger. *J. Build. Eng.* **2022**, *47*, 103851. [CrossRef]
16. Wang, Y.; Gan, D.; Sun, M.; Zhang, N.; Lu, Z.; Kang, C. Probabilistic individual load forecasting using pinball loss guided LSTM. *Appl. Energy* **2019**, *235*, 10–20. [CrossRef]
17. Somu, N.; Raman, G.M.R.; Ramamritham, K. A hybrid model for building energy consumption forecasting using long short term memory networks. *Appl. Energy* **2020**, *261*, 114131. [CrossRef]
18. Xu, L.; Hu, M.; Fan, C. Probabilistic electrical load forecasting for buildings using Bayesian deep neural networks. *J. Build. Eng.* **2022**, *46*, 103853. [CrossRef]
19. Khwaja, A.S.; Anpalagan, A.; Naeem, M.; Venkatesh, B. Joint bagged-boosted artificial neural networks: Using ensemble machine learning to improve short-term electricity load forecasting. *Electr. Power Syst. Res.* **2020**, *179*, 106080. [CrossRef]
20. Zhou, Y.; Liang, Y.; Pan, Y.; Yuan, X.; Xie, Y.; Jia, W. A Deep-Learning-Based Meta-Modeling Workflow for Thermal Load Forecasting in Buildings: Method and a Case Study. *Buildings* **2022**, *12*, 177. [CrossRef]
21. Zhang, Y.; Teoh, B.K.; Wu, M.; Chen, J.; Zhang, L. Data-driven estimation of building energy consumption and GHG emissions using explainable artificial intelligence. *Energy* **2023**, *262*, 125468. [CrossRef]
22. Shi, J.; Li, C.; Yan, X. Artificial intelligence for load forecasting: A stacking learning approach based on ensemble diversity regularization. *Energy* **2023**, *262*, 125295. [CrossRef]
23. Lu, Y.; Meng, L. A simplified prediction model for energy use of air conditioner in residential buildings based on monitoring data from the cloud platform. *Sustain. Cities Soc.* **2020**, *60*, 102194.
24. Wang, R.; Lu, S.; Li, Q. Multi-criteria comprehensive study on predictive algorithm of hourly heating energy consumption for residential buildings. *Sustain. Cities Soc.* **2019**, *49*, 101623. [CrossRef]
25. Seyedzadeh, S.; Pour Rahimian, F.; Rastogi, P.; Glesk, I. Tuning machine learning models for prediction of building energy loads. *Sustain. Cities Soc.* **2019**, *47*, 101484. [CrossRef]
26. Kumar, S.; Pal, S.K.; Singh, R.P. A novel method based on extreme learning machine to predict heating and cooling load through design and structural attributes. *Energy Build.* **2018**, *176*, 275–286. [CrossRef]
27. Kaggle Competitions. Available online: https://www.kaggle.com/competitions (accessed on 12 January 2023).
28. Butch, Q. *Machine Learning with Spark-Covers XGBoost, LightGBM, Spark NLP.; Distributed Deep Learning with Keras, and More*; Springer Science, Business Media New York: New York, NY, USA, 2020.
29. Get Started with XGBoost. Available online: https://xgboost.readthedocs.io/en/latest/get_started.html (accessed on 12 January 2023).
30. Chen, Y.; Guo, M.; Chen, Z.; Chen, Z.; Ji, Y. Physical energy and data-driven models in building energy prediction: A review. *Energy Rep.* **2022**, *16*, 2656–2671. [CrossRef]
31. Dudek, G. Short-Term Load Forecasting Using Random Forests. *Adv. Intell. Syst. Comput.* **2015**, *323*, 821–828.
32. Hochreiter, S.; Schmidhuber, J. Long Short-Term Memory. *Neural Comput.* **1997**, *9*, 1735–1780. [CrossRef]
33. Zang, H.; Xu, R.; Cheng, L.; Ding, T.; Liu, L.; Wei, Z.; Sun, G. Residential load forecasting based on LSTM fusing self-attention mechanism with pooling. *Energy* **2021**, *229*, 120682. [CrossRef]
34. Zhang, L.; Wen, J. Active learning strategy for high fidelity short-term data-driven building energy forecasting. *Energy Build.* **2021**, *244*, 111026. [CrossRef]
35. Hu, Y.; Cheng, X.; Wang, S.; Chen, J.; Zhao, T.; Dai, E. Times series forecasting for urban building energy consumption based on graph convolutional network. *Appl. Energy* **2022**, *307*, 118231. [CrossRef]
36. Olu-Ajayi, R.; Alaka, H.; Sulaimon, I.; Sunmola, F.; Ajayi, S. Building energy consumption prediction for residential buildings using deep learning and other machine learning techniques. *J. Build. Eng.* **2022**, *45*, 103406. [CrossRef]
37. Li, Y.; Tong, Z.; Tong, S.; Westerdahl, D. A data-driven interval forecasting model for building energy prediction using attention-based LSTM and fuzzy information granulation. *Sustain. Cities Soc.* **2022**, *76*, 103481. [CrossRef]

38. Do, H.; Cetin, K.S. Evaluation of the causes and impact of outliers on residential building energy use prediction using inverse modeling. *Build. Environ.* **2018**, *138*, 194–206. [CrossRef]
39. Guo, Y.; Wang, J.; Chen, H.; Li, G.; Liu, J.; Xu, C.; Huang, R.; Huang, Y. Machine learning-based thermal response time ahead energy demand prediction for building heating systems. *Appl. Energy* **2018**, *221*, 16–27. [CrossRef]
40. Wang, Z.; Srinivasan, R.S. A review of artificial intelligence based building energy use prediction: Contrasting the capabilities of single and ensemble prediction models. *Renew. Sustain. Energy Rev.* **2017**, *75*, 796–808. [CrossRef]
41. Commercial Prototype Building Models. Available online: https://www.energycodes.gov/development/commercial/prototype_models (accessed on 12 January 2023).
42. Bryan, E.; Zheng, O.; Vladimir, A.F.; Igor, M. Uncertainty and sensitivity decomposition of building energy models. *J. Build. Perform. Simu.* **2012**, *5*, 171–184.
43. Key-Inputs-Setting-and-Energy-Data. Available online: https://github.com/Bob05757/Key-inputs-setting-and-Energy-Data (accessed on 12 January 2023).
44. Li, Q.; Meng, Q.; Cai, J.; Yoshino, H.; Mochida, A. Applying support vector machine to predict hourly cooling load in the building. *Appl. Energy* **2009**, *86*, 2249–2256. [CrossRef]
45. Leung, M.C.; Tse, N.C.F.; Lai, L.L.; Chow, T.T. The use of occupancy space electrical power demand in building cooling load prediction. *Energy Build.* **2012**, *55*, 151–163. [CrossRef]
46. Luo, X.J.; Oyedele, L.O.; Ajayi, A.O.; Monyei, C.G.; Akinade, O.O.; Akanbi, L.A. Development of an IoT-based big data platform for day-ahead prediction of building heating and cooling demands. *Adv. Eng. Inform.* **2019**, *41*, 100926. [CrossRef]
47. Guolin, K.; Qi, M.; Thomas, F.; Taifeng, W.; Wei, C.; Weidong, M.; Ye, Q.; Liu, T.-Y. LightGBM: A Highly Efficient Gradient Boosting Decision Tree. In Proceedings of the 31st Conference on Neural Information Processing Systems, Long Beach, CA, USA, 4–9 December 2017.
48. Distributed Machine Learning Toolkit-Big Data, Big Model, Flexibility, Efficiency. Available online: http://www.dmtk.io/ (accessed on 12 January 2023).
49. Jin, R.M.; Agrawal, G. Communication and Memory Efficient Parallel Decision Tree Construction. In Proceedings of the 2003 SIAM International Conference on Data Mining, San Francisco, CA, USA, 1–3 May 2003; pp. 119–129.
50. Moon, J.; Kim, Y.; Son, M.; Hwang, E. Hybrid Short-Term Load Forecasting Scheme Using Random Forest and Multilayer Perceptron. *Energies* **2018**, *11*, 3283. [CrossRef]
51. Ahmad, M.W.; Mourshed, M.; Rezgui, Y. Trees vs Neurons: Comparison between random forest and ANN for high-resolution prediction of building energy consumption. *Energy Build.* **2017**, *147*, 77–89. [CrossRef]
52. Ahmad, M.W.; Reynolds, J.; Rezgui, Y. Predictive modelling for solar thermal energy systems: A comparison of support vector regression, random forest, extra trees and regression trees. *J. Clean. Prod.* **2018**, *203*, 810–821. [CrossRef]
53. Shi, Y.; Song, X.; Song, G. Productivity prediction of a multilateral-well geothermal system based on a long short-term memory and multi-layer perceptron combinational neural network. *Appl. Energy* **2021**, *282*, 116046. [CrossRef]
54. Ding, Y.; Zhang, Q.; Yuan, T.; Yang, K. Model input selection for building heating load prediction: A case study for an office building in Tianjin. *Energy Build.* **2018**, *159*, 254–270. [CrossRef]
55. American Society Of Heating VAAC. Measurement of Energy and Demand Savings. *ASHRAE Guidel* **2014**, *4*, 1–150.
56. Geron, A. *Hands-On Machine Learning with Scikit-Learn, Keras, and TensorFlow: Concepts, Tools, and Techniques to Build Intelligent Systems;* O'Reilly Media: Sevastopol, CA, USA, 2017.
57. Fan, C.; Xiao, F.; Zhao, Y. A short-term building cooling load prediction method using deep learning algorithms. *Appl. Energy* **2017**, *195*, 222–233. [CrossRef]

buildings

MDPI

Article

# Effect of Block Morphology on Building Energy Consumption of Office Blocks: A Case of Wuhan, China

**Shen Xu [1,2], Gaomei Li [1,2], Hailong Zhang [3], Mengju Xie [1,*], Thushini Mendis [4] and Hu Du [5]**

1 School of Architecture & Urban Planning, Huazhong University of Science and Technology, Wuhan 430073, China
2 Hubei New Urbanization Engineering and Technology Research Center, Wuhan 430073, China
3 China Railway Development and Investment Group Co., Ltd., Kunming 650500, China
4 Department of Architecture and Built Environment, University of Nottingham, Ningbo 315100, China
5 School of Civil Engineering and Built Environment, Liverpool John Moores University, Cherie Booth Building, Byrom St., Liverpool L3 3AF, UK
* Correspondence: xmj_d@hust.edu.cn; Tel.: +86-18271409136

**Abstract:** Block morphology refers to critical parameters influencing building energy performance on the block scale. However, analysis of the combined effect of block morphological parameters on building energy consumption with real blocks is lacking. In this paper, the aim is to evaluate the combined effect of office block morphology on building energy consumption in the context of the Hot-summer and Cold-winter zone in China. First, a workflow for the energy assessment of office buildings with the coupled block morphology on the block scale was proposed with evaluation tools. Seventy office blocks in Wuhan were taken as examples and then classified based on building layout typology and building height. Afterwards, the morphological parameters and building energy use intensity (EUI) for different blocks were calculated. Then, the combined effect of block morphology on the buildings' energy consumption was evaluated and the model on predicting the building energy consumption of office blocks was proposed. Finally, based on the results, low-energy design strategies were projected for office blocks. The results illustrated that the effect of block morphology on building cooling, heating, and lighting is EUI 28.83%, 28.56%, and 23.23%, respectively. Building shape factor (BSF), floor area ratio (FAR), average building height of block (BH), and average building depth of block (BD) are effective block morphological parameters. The key morphological parameters which combined affect the building energy consumption of office blocks are BSF and FAR; BSF has 1.24 times the effect on building energy consumption than FAR. The workflow built in this paper can be applied to other cities around the world for promoting sustainable cities.

**Keywords:** office blocks; building energy consumption; block morphology; combined effect; multiple regression analysis; energy-saving strategies

## 1. Introduction

### 1.1. Background

The world is in the midst of its first global energy crisis, and to tackle it, the IEA has set out its analysis of energy supply per capita in three scenarios. The three scenarios are the stated policies scenario (STEPS), announced pledges scenario (APS), and net zero emissions (NZE) by 2050 scenario [1]. Regional trends in per capita energy demand under these three scenarios are shown in Figure 1. In the STEPS scenario, energy demand in developed economies declines at a rate of 0.5% per year. Renewable energy applications and energy electrification accelerate in the APS scenario, with a 17% reduction in fossil fuel demand in 2030 compared to the STEPS scenario. By 2030, energy demand in developing economies will grow at more than 1.4% per year under the STEPS scenario. China and India alone account for nearly 50% of energy growth [1]. Regional trends in per capita $CO_2$ emissions in 2020 and 2030 of the APS are shown in Figure 2. By 2030, the US' $CO_2$ emissions are 30%

lower in the APS than the STEPS scenario and per capita $CO_2$ emissions are reduced to 8 tonnes of $CO_2$ per capita, while China's $CO_2$ emissions will be cut by 6.25% by 2030 [2].

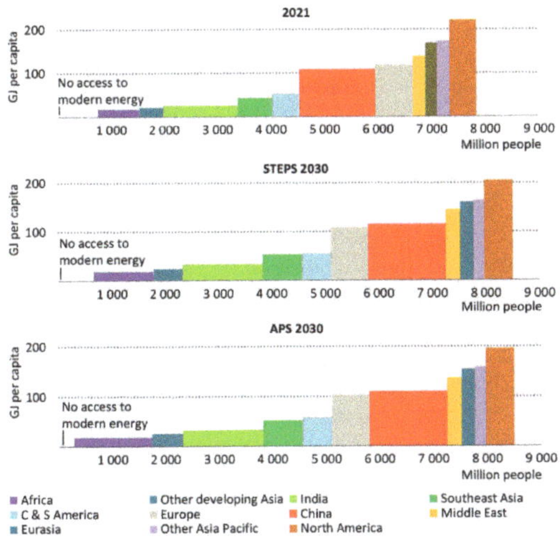


**Figure 1.** Regional trends in energy demand per capita under three scenarios [1].

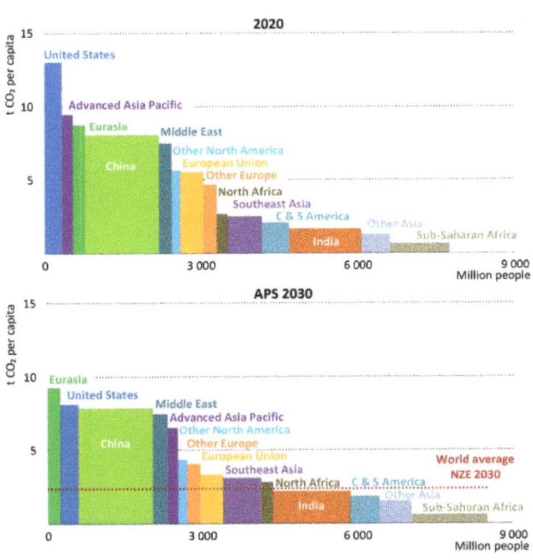


**Figure 2.** Regional trends in per capita CO2 emissions in 2020 and 2030 of the APS [2].

According to the International Energy Agency, the sectors of building and building construction together are accountable for nearly a third of the total global energy consumption, and almost 15% of direct $CO_2$ emissions in 2021 [2]. Therefore, there is a global consensus to save energy consumption in buildings. With the support of national policies [3,4], China is showing a trend towards the development of green buildings on a larger scale. Urban blocks in China consist of a multitude of functions, which include commercial, residential, industrial, office, etc. Office blocks account for greater than 50% of the floor area of public buildings, and the energy consumption per square meter of floor area of office buildings is 2.4 times higher than that of urban residential buildings [5]. Office blocks play a vital role in the development of urban green blocks.

### 1.2. Literature Review

#### 1.2.1. Urban Building Energy Modeling

With the continuous expansion of research on building energy consumption and the further development of computer technology, building energy consumption models at the building level are increasingly unable to meet the application requirements of the quantitative energy consumption assessment of building groups in complex urban environments, and it is difficult to make technical support available for the formulation of energy-saving targets and energy policies at the macro level of cities [6]. Therefore, it is urgent to study urban building energy consumption.

City information modelling (CIM) has become a research hotspot in recent years, as it can be used as a support decision tool for urban management, sustainable urban development, and smart city construction [7]. Urban building energy modeling (UBEM) is a domain of CIM implementation [8]. UBEM is a bottom-up, physics-based approach for simulating the thermal performance of new or existing neighborhoods and cities [9]. Reinhart and Cerezo Davila [10] defined urban building energy modeling (UBEM), which is an expanding area in modelling building energy, which covers a spatial scale from an urban block to a district for a whole city [11]. It is presently understood that urban building energy performance is dependent upon four dimensional factors [12], which are, respectively, (1) urban block morphology, (2) building design, (3) systems efficiency, and (4) occupant behavior (Figure 3). Block morphology refers to the critical parameters affecting building energy consumption on the block scale.

**Figure 3.** Factors that affect building energy consumption [12].

#### 1.2.2. The Effect of Block Morphology on Building Energy Consumption

Block morphology is an effective means of controlling the energy consumption of buildings [13,14], which can affect around 10% to 30% or more of a building's energy consumption [12,15]. Shareef and Altan [16] carried out a study on the impact of sinuous morphology on building energy consumption in Dubai, the results of which showed that the building energy for cooling can be reduced by 4.9% when buildings are arranged to be alternating in urban blocks. Bat et al. [17] researched the relationship between urban courtyard morphology and building energy consumption in different climates (cold, temperate, hot, and arid climates), the results showed that, in temperate climates, a courtyard with less width and a medium-depth shape allowed a reduction of building energy demand of about 58%. Bansal and Quan [18] studied the relationship between the urban form and energy consumption of residential buildings in different local climate zone contexts in Seoul, with the electricity and gas dataset; the results demonstrated that the local climate zone explained 16.2% and 13.2% of the variance in electricity and gas use, respectively. Deng et al. [19] carried out the effect of the residential block layout morphology on the building energy consumption for heating in cold II B zones in China; the research results showed that increasing the spacing between residential buildings can reduce heating energy consumption by 4%. Li et al. [20] studied the relationship between block morphology and household building energy consumption in Ningbo, China, with a real-block morphology and utility bill data of 534 household in 46 residential blocks; the results showed that the building energy consumption increases by 9.1%, when the floor area ratio increases per unit. In addition, other scholars have conducted empirical studies on the prediction of building energy consumption [21] and the identification of key morphological parameters affecting building energy consumption [22] using real-world data.

Some scholars have begun to study the correlation between block morphology and building energy consumption based on block-prototype morphology. Vartholomaios [23] carried out the effect of block morphology on the domestic energy consumption of heating and cooling in a Mediterranean city based on a parametric sensitivity analysis with the block prototype. The results showed that the difference in cooling and heating energy consumption due to morphological parameters in different urban density blocks ranges from 13.96% to 22.08%. Shi et al. [24] came up with a parametric method to investigate the relationship between block morphology and building energy consumption in a Singapore urban context using the block prototype. Other scholars have studied the correlation between block morphology and building energy consumption from different perspectives, using the block prototype in different climate zones [16,25–27]. Due to the complexity of real urban blocks, the use of the block prototype cannot fully restore the characteristics of real blocks and thus guide urban design and architectural practice. It is therefore necessary to carry out a study of the building energy consumption based on a real-block morphology.

Most of the current scholarly research on the impact of block morphology on building energy consumption is focused on the relationship between a single block morphological parameter and building energy consumption. Andersena and Sattrup [15] studied the effect of the urban canyon morphology on the energy consumption of office buildings using a qualitative approach in the context of Copenhagen, a temperate marine climate zone. The findings demonstrated that the geometry morphology of urban canyons had an impact on the total energy consumption of office buildings by up to 30%. Shareef [28] studied the impact of urban block morphology and the building's height diversity on the building energy consumption of residential blocks, using a qualitative approach in the desert climate zone of Dubai city. The results indicated that the main effect on the cooling energy consumption in urban blocks was from the orientation of the building, with a 6.4% reduction in the N–S orientation compared to NW–SE . Mangan et al. [29] evaluated the effect of morphological parameters such as building height to street width, orientation, and building type on the energy consumption of residential buildings, using a qualitative approach in temperate-humid zones. The findings indicated that increasing the building height of rectangular-pavilion, rectangular-slab, and square-pavilion residential buildings can reduce building energy consumption by 14%, 8%, and 18%, respectively. Zhang et al. [30] investigated the relationship between urban block morphology and building energy use in the urban context of Singapore, with thirty generic urban residential block cases in six typologies. The results showed that differences in urban block types can lead to up to 12 times the rate of reduction in building cooling loads and 25% reduction in the net energy use intensity of buildings, under identical planning conditions and design properties.

It has been found that building energy consumption is affected by multi-block morphological parameters [18,31–33], However, the mechanisms of the combined effect of block morphological parameters on building energy consumption have not been evaluated in detail, especially using a real-block morphology in the Hot-summer and Cold-winter zone in China. Such knowledge will play a vital role in optimal design approaches in the future for the low-energy design of office blocks. Immediate improvements in understanding the effect of block morphological parameters on building energy consumption are required. Particularly, preceding research which provides accurate evaluations of the building energy consumption of office blocks in China is very limited, and few applicable works of research have been carried out in relation to climatic conditions such as those in central China. Studies carried out under real climatic conditions are essential, because the existing studies have shown that the building energy consumption of office blocks in different climate zones show different characteristics of results [34].

### 1.3. Research Aim

In this context, this study aims to quantify the combined effect of block morphology on office building energy consumption, focusing on the following three issues:

- Are there differences in building energy consumption distribution characteristics among different office blocks? If so, to what extent?

- Do block morphological parameters have an effect on building energy consumption? If so, what morphological parameters? To what extent?
- What are the key morphological parameters that have a combined effect on building energy consumption in office blocks?

## 2. Methodology

The research workflow for this study is shown in Figure 4. It can be divided into four major steps: (1) access to 3D model data of office blocks; (2) building energy simulation (BES) workflow; (3) BES model validation; and (4) statistical analysis and proposal of design strategies.

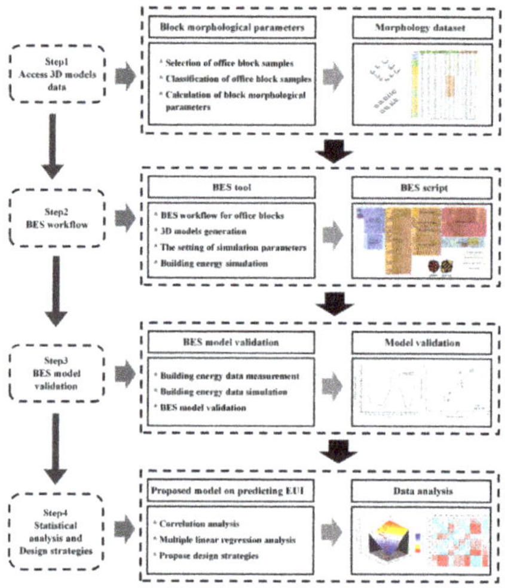


**Figure 4.** Workflow of holistic approach with four major steps.

### *2.1. Access to 3D Model Data of Office Blocks*

A combination of Open Street Map and satellite maps was used to obtain information about the floor plan and dimensions of office blocks through field research combined with street maps to obtain information about building heights and window-to-wall ratios, and then to build a 3D model. This method has been proven to be scientifically sound and feasible by academics [35]. To ensure the accuracy of the energy simulation results, the acquisition of 3D model data of office blocks includes three aspects: (1) selection of samples: The selected samples cover the whole city to make the selected samples representative. (2) calculation of morphological parameters: Morphological parameters include average building width of block (BW), average building depth of block (BD), average building height of block (BH), width-to-depth ratio of block (W/D), height-to-depth ratio of block (H/D), building shape factor (BSF), building coverage ratio (BCR), and floor area ratio (FAR). (3) acquire data used for the building energy simulation: The data include window-to-wall ratio of buildings, building envelope, occupancy rate, the operation rate of lighting and equipment, etc.

#### 2.1.1. Selection of Samples

In order to ensure that the selection of samples reflects the overall characteristics of office blocks in Wuhan, the block samples selected for this paper should satisfy several principles: (1) the land area of sample should be less than 1 $km^2$; (2) the building function of the block is mainly office; (3) the number of buildings in the block should be no less than 3, so as to give the block a distinct group character; and (4) the samples selected are built after the year 2000, in order to avoid excessive differences in thermal performance among office block buildings.

Based on the above principles, seventy office block samples in Wuhan were selected. Of these, 60 samples were used to develop building energy prediction model and 10 samples were used for model validation. The geographical distribution of the selected samples in Wuhan is shown in Figure 5.

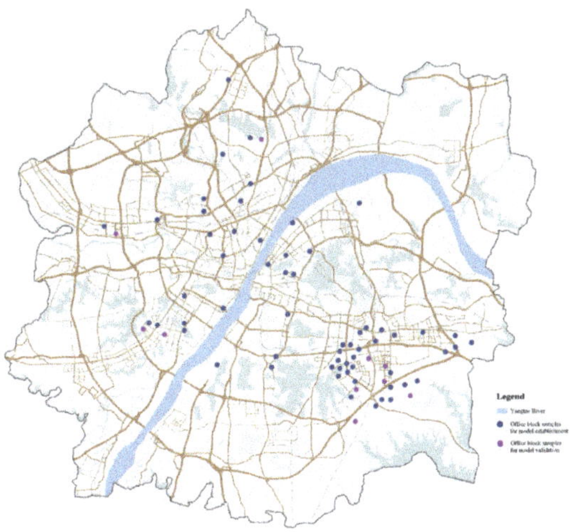


**Figure 5.** The distribution of office block samples.

### 2.1.2. Classification of Office Block Samples

The research aims to find the combined effect of block morphology on building energy consumption on the block scale. In addition, the purpose of classifying office blocks is to reconcile the research findings with the process of planning and design. Through field surveys and literature review [36], it can be found that office blocks in China show a certain consistency in terms of floor plan layout. Office blocks in China are usually composed of multiple buildings, mostly in the form of pavilion, slab, and courtyard layouts. The office buildings are mainly in the range of 1–6 storeys, 7–12 storeys, and 13–25 storeys. This is mainly due to the mandatory national code requirements for lighting and fire evacuation in office buildings [37,38]. Based on the above description, this study uses a combination of building layout and building height to classify office blocks.

Three layout typologies of blocks, namely, pavilion, slab, and courtyard [12,39–41], and three building height of blocks, namely, multi-storey, mid-rise, and high-rise [42], were selected to analyze and compare block morphological parameters and building energy consumption. Based on building layout type and building height type, this paper classified office blocks into the following seven categories: pavilion multi-storey, slab multi-storey, courtyard multi-storey, mid-rise pavilion, mid-rise slab, high-rise pavilion, and high-rise slab. The detailed classification results are shown in Table 1.

### 2.1.3. Calculation of Block Morphological Parameters

The block morphology is in reference to the spatial configuration of urban land use within a block area [43]. After the urban block design is put forward, several block morphological parameters can be defined to describe the block, and this also corresponds greatly to energy demand. The block morphology parameters selected in this paper should be considered in block design practice. The calculation equation of each morphological parameter is shown in Figure 6. This paper accurately studied the building energy consumption of office blocks with different morphological typologies through the analysis of block morphological parameters. The morphological parameters of different block samples are shown in Appendix A.

**Table 1.** The classification results of real office block samples.

| Building Height Type | Building Height | Building Layout Type | Block Typology | Block 3D Model | Block Sample |
|---|---|---|---|---|---|
| Multi-storey office blocks | H ≤ 24 m | Courtyard | Courtyard multi-storey |  |  |
| | | Pavilion | Pavilion multi-storey | 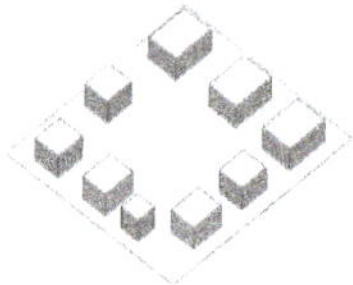 |  |
| | | Slab | Slab multi-storey | 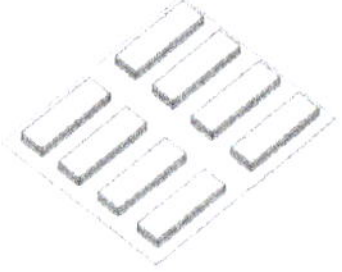 |  |
| Mid-rise office blocks | 24 < H ≤ 50 m | Pavilion | Mid-rise pavilion | 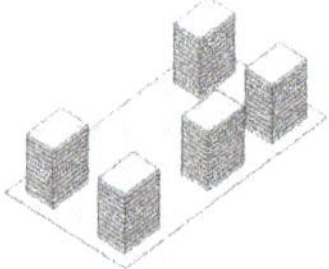 |  |
| | | Slab | Mid-rise slab | 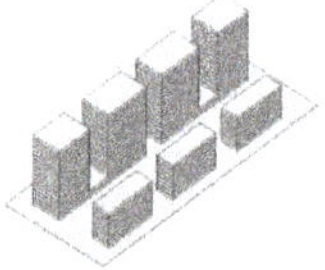 |  |
| High-rise office blocks | 50 < H ≤ 100 m | Pavilion | High-rise pavilion | 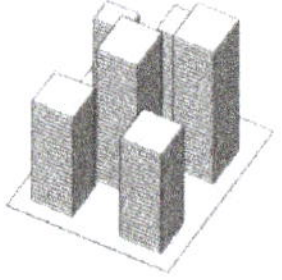 |  |
| | | Slab | High-rise slab | 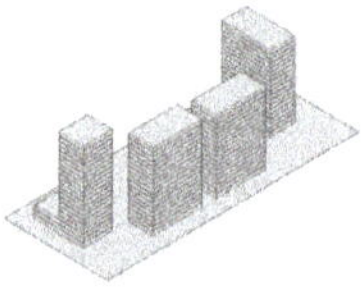 |  |

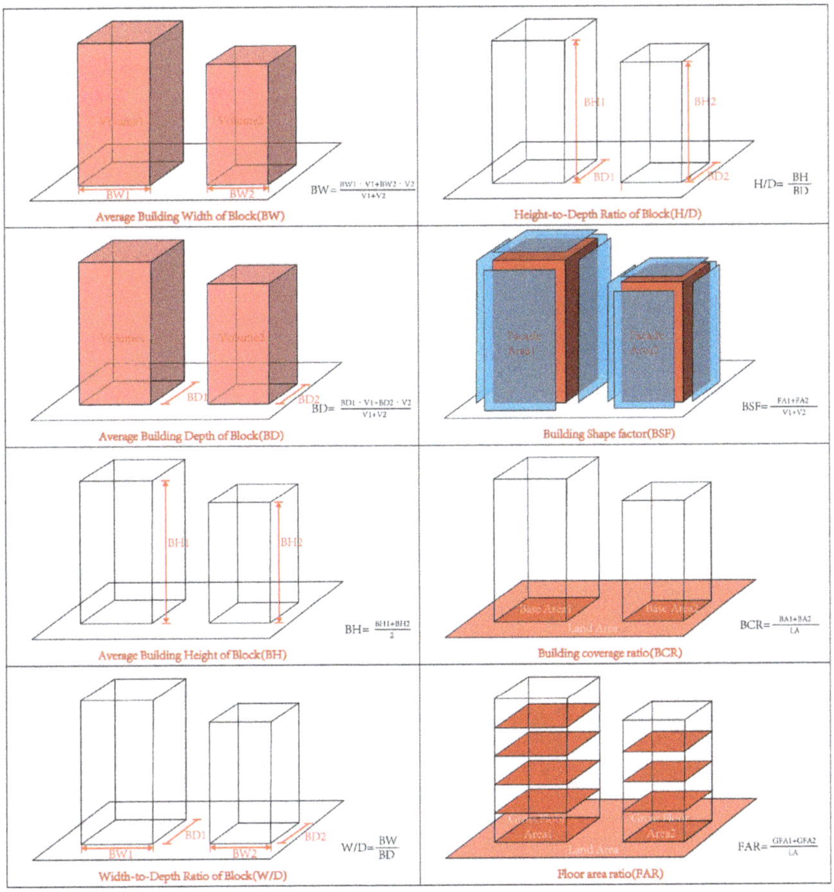


**Figure 6.** Calculation equation of block morphological parameters.

## *2.2. Building Energy Simulation (BES) Workflow*

### 2.2.1. BES Workflow for Office Blocks

The building energy simulation workflow(Figure 7) was built based on Rhino and Grasshopper platform, invoking the Energy Plus [44,45] simulation core, which was run in the Rhino Ladybug module [33,39,46]. The workflow was divided into four sections: (a) 3D model generation; (b) the setting of simulation parameters; (c) building energy simulation; and (d) result output.

### 2.2.2. 3D Model Generation

Firstly, the building plans were drawn in Rhino based on satellite images, GIS platforms, and geometric information obtained from field survey. Secondly, 3D building models were created based on the storey number of buildings, at a height of 4 m. Thirdly, the window-to-wall ratio script was used to automatically divide the windows of each building facade. Finally, the office buildings' thermal zone models of the block were established, based on Honeybee Tools. The process of the 3D model generation is shown in Figure 8.

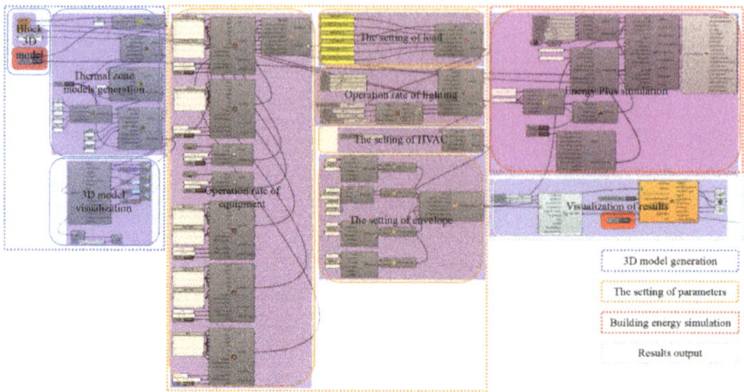


**Figure 7.** The workflow of building energy simulation.

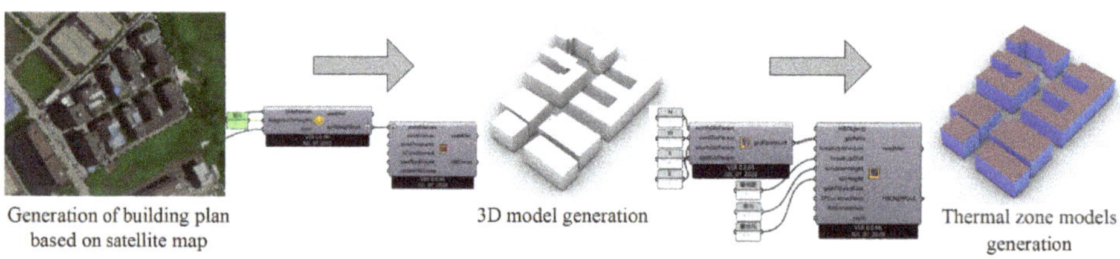


**Figure 8.** The process of 3D model generation.

### 2.2.3. The Setting of Simulation Parameters

The building energy simulation parameters required the setting of external meteorological conditions, building geometry parameters, and building physical parameters. The meteorological data were obtained from China Meteorological Data Network. The building geometry parameters were obtained through field survey combined with GIS. The data about the building envelope of the blocks were obtained through literature survey combined with energy-saving report of typical office block project. The occupancy rate, operation rate of lighting, and equipment of the office blocks were obtained from questionnaire survey. The simulation parameters were set as shown in Tables 2 and 3.

### 2.2.4. Building Energy Simulation

The building energy simulation used the Honeybee module Energy Plus as the calculation kernel to calculate the building energy consumption of the office blocks. Energy use intensity (EUI) [47,48] was used to denote the efficiency of building energy consumption in this study, which is in reference to the annual energy consumption per unit area of a building, typically expressed in $kWh/m^2/y$.

Due to the fact that equipment energy consumption varies greatly from office buildings, but it is not affected by the block morphology parameters [14]; therefore, equipment energy consumption was excluded from the results and analysis. The building energy consumption and the sum of cooling, heating, and lighting energy consumption, for each office block buildings, were simulated and the annual EUI was calculated by Equation (1):

$$\text{Total EUI} = \frac{E}{S_A} \tag{1}$$

Total EUI—the sum of cooling, heating, and lighting EUI
E—building energy consumption
$S_A$—total floor area of office block buildings

**Table 2.** Parameter setting in the BES model-1.

| Item | | Parameter Setting | | | | | |
|---|---|---|---|---|---|---|---|
| Occupancy Rate | 8 a.m.–7 p.m. (Mon.–Fri.) | 8 a.m.–9.a.m. 0.17 | 9 a.m.–12 a.m. 0.96 | 12 a.m.–1 p.m. 0.04 | 1 p.m.–2 p.m. 0.81 | 2 p.m.–6 p.m. 0.96 | 6 p.m.–7.p.m. 0.23 |
| | 9 a.m.–5 p.m. (Sat.–Sun.) | 8 a.m.–9 a.m. 0.10 | 9 a.m.–12 a.m. 0.18 | 12 a.m.–1 p.m. 0.04 | 1 p.m.–2 p.m. 0.04 | 2 p.m.–6 p.m. 0.18 | 6 p.m.–7 p.m. 0.10 |
| Operation Rate of Lighting | 8 a.m.–7 p.m. (Mon.–Fri.) | 8 a.m.–9 a.m 0.06 | 9 a.m.–12 a.m. 0.96 | 12 a.m.–1 p.m. 0.86 | 1 p.m.–2 p.m. 0.92 | 2 p.m.–6 p.m. 0.96 | 6 p.m.–7 p m. 0.75 |
| | 9 a.m.–5 p.m. (Sat.–Sun.) | 8 a.m.–9 a.m. 0.06 | 9 a.m.–12 a.m. 0.18 | 12 a.m.–1 p.m. 0.14 | 1 p.m.–2 p.m. 0.14 | 2 p.m.–6 p.m. 0.18 | 6 p.m.–7 p.m. 0.10 |
| Operation Rate of Equipment | 8 a.m.–7 p.m. (Mon.–Fri.) | 8 a.m.–9 a.m. 0.18 | 9 a.m.–12 a.m. 0.96 | 12 a.m.–1 p.m. 0.88 | 1 pm–2 pm 0.93 | 2 p.m.–6 p.m. 0.96 | 6 p.m.–7 p.m. 0.16 |
| | 9 a.m.–5 p.m. (Sat.–Sun.) | 8 a.m.–9 a.m 0.10 | 9 a.m.–12 a.m. 0.18 | 12 a.m.–1 p.m. 0.14 | 1 p.m.–2 p.m. 0.16 | 2 p.m.–6 p.m. 0.18 | 6 p.m.–7 p.m. 0.10 |
| Temperature set | Cooling Set Point (°C) | 8 a.m.–6 p.m. (Mon.–Fri.) | 26 | Heating Set Point (°C) | 8 a.m.–6 p.m. (Mon.–Fri.) | Heating Set Point (°C) | 18 |
| Density | | Occupancy density/($m^2$/person) | | | | 8 | |
| | | Lighting power density/($W/m^2$) | | | | 15 | |
| | | Equipment power density/($W/m^2$) | | | | 15 | |

**Table 3.** Parameter setting in the BES model-2.

| Item | | Parameter Setting | | | |
|---|---|---|---|---|---|
| Transparent Envelope | Window-to-Wall Ratio | N 0.5 | E 0.3 | S 0.5 | W 0.3 |
| | Solar Heat Gain Coefficient | N 0.48 | E 0.44 | S 0.44 | W 0.44 |
| Opaque Envelope | Heat Transfer Coefficient ($W/(m^2{\cdot}K)$) | Exterior wall 0.98 | Interior wall 0.79 | Roof 0.48 | Floor slabs 0.98 |
| Heat Transfer Coefficient ($W/(m^2{\cdot}K)$) | | 3.0 | Floor-to-Floor Height(m) | | 4 |

### 2.3. *BES Model Validation*

#### 2.3.1. Building Energy Data Measurement

The actual month-by-month building energy consumption of the office building in Wuhan was measured by smart meter for the whole year of 2019. The investigation revealed that the winter heating and summer cooling demand of the office building were met by air conditioning; i.e., the electricity consumption data were representative of the annual building energy consumption of the building. In addition, the month-by-month EUI data for this building is shown in Figure 9.

#### 2.3.2. BES Model Validation

The results simulated from the building energy model were compared with the measurements that were recorded during field experiments. Figure 10 exhibits variations of building energy consumption of office buildings month by month for a year (2019) and establishes clearly that the simulated results agree soundly with the measurements, with a deviation of less than 8.14%, which is within a reasonable margin of error [49,50]. Other comparisons in fitness were also carried out, and the results demonstrate good agreement, with an $R^2$ value of 0.87 (Figure 11).

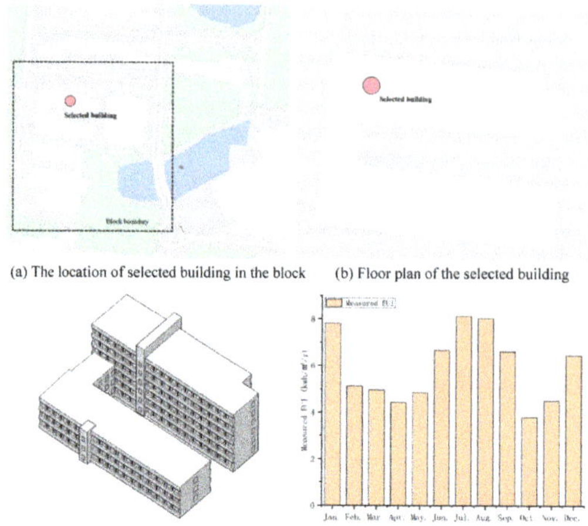


(a) The location of selected building in the block (b) Floor plan of the selected building

(c) 3D model of the selected building (d) Measured EUI data for the building in 2019

**Figure 9.** Building information and monthly electricity consumption data.

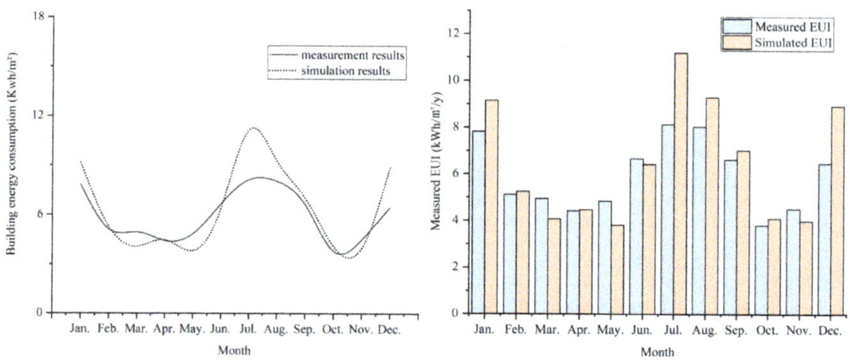


**Figure 10.** Comparison of measured and simulated EUI.

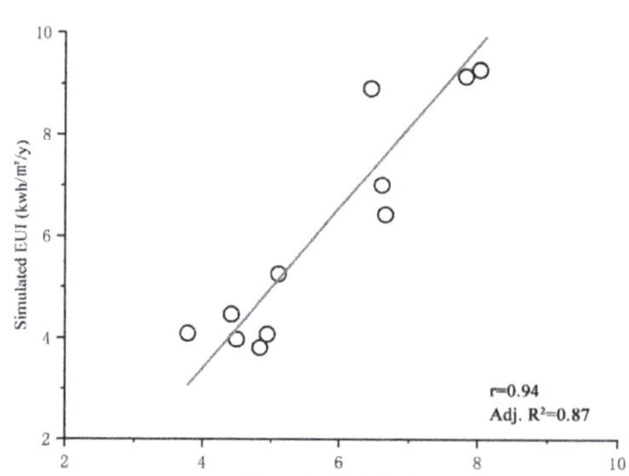


**Figure 11.** Results of linear regression analysis.

### 2.4. Correlation Analysis and Multiple Linear Regression Analysis

Statistical analysis methods such as correlation analysis [51] and multiple linear regression analysis [52] were used to study the extent to which block morphological parameters affected the building energy consumption of office blocks. Firstly, correlation analysis was conducted between the dependent variable (building EUI) and independent variables (block morphological parameters) to examine the intensity of the statistical relationship between variables. The Pearson correlation coefficient in the data analysis software IBM SPSS 24.0 was used to carry out a correlation analysis between block morphology variables and building EUI, and the result was obtained.

Multiple linear regression analysis has been conducted in numerous relevant studies [19,48,53]. This method was chosen to quantitatively analyze the relative significance of eight block morphological parameters on building EUI. Each independent variable (block morphological parameter) is linked to a value of the dependent variable (building EUI). This means that researchers can analyze the impact of a morphological variable on building energy consumption after controlling for other morphological variables through the established equation's result. Thus, multiple linear regression analysis was carried out with SPSS to propose the model for predicting building energy consumption coupled with block morphology in office blocks and then revealed how adequately the block morphological parameters can clarify the differences in building energy consumption among office blocks.

## 3. Results and Discussion

### 3.1. The Effect of Block Typologies on Building EUI

#### 3.1.1. Building EUI for All Office Blocks

The building EUI for the sixty block samples are shown in Figure 12. The results showed that courtyard multi-storey blocks had a maximum EUI of 56.63 kWh/m$^2$/y, while the minimum EUI for high-rise pavilion blocks was 52.80 kWh/m$^2$/y. The average EUI varies among block types by up to 7.0% due to differences in morphological parameters. For the sixty samples, the results of the building energy consumption distribution characteristics revealed that the EUI of the lowest sample C1-4, with 51.88 kWh/m$^2$/y, was 13.82% lower than that of the highest sample A3-5 with 59.05 kWh/m$^2$/y, which was within the range of the existing research threshold that block morphology affects building energy consumption by around 10% to 30% or more [12,15]. Javanroodi et al. studied the effect of block morphology on cooling load and ventilation potential. The results showed that the cooling energy consumption and ventilation potential in Tehran can be optimized by more than 10% and 15%, respectively, due to the difference in block morphology [25]. Similarly, another study which was based on hot and humid climates found a 16–18% reduction in office building cooling energy consumption when the surrounding buildings were taken into account [54]. Mangan et al. evaluated the effect of morphological parameters such as building height to street width, orientation, and building type on the energy consumption of residential buildings. The findings indicated that increasing the building height of rectangular-pavilion, rectangular-slab, and square-pavilion residential buildings can decrease building energy consumption by 14%, 8%, and 18%, respectively [29].

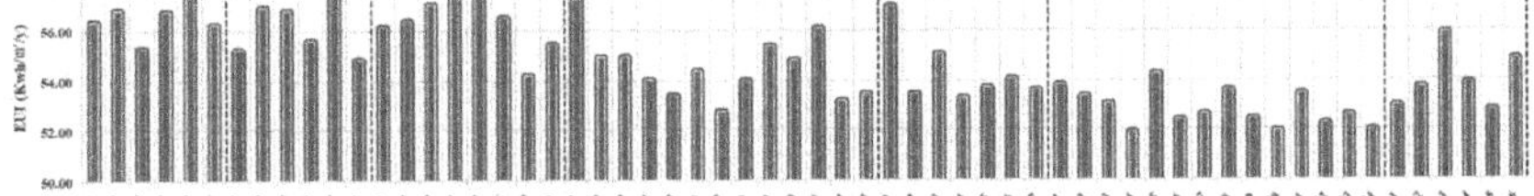


**Figure 12.** Distribution characteristics of total EUI of sixty office block samples.

In the Hot-summer and Cold-winter zone in China, block buildings have a dual demand for cooling in summer and heating in winter, while lighting energy consumption is a non-negligible part of office buildings. The distribution characteristics of the EUI for cooling of the seven urban block typologies are shown in Figure 13. For the overall block samples, the cooling EUI of the lowest block high-rise pavilion (C1-5), with 15.95 kWh/m$^2$/y, was 28.83% lower than that of the highest block slab multi-storey block (A3-5), with 22.41 kWh/m$^2$/y. The distribution characteristics of the EUI for heating of the seven urban block typologies are shown in Figure 14. The heating EUI of the lowest block high-rise pavilion (C1-4), with 10.38 kWh/m$^2$/y, was 28.56% lower than that of the highest block slab multi-storey block (A3-5), with 14.53 kWh/m$^2$/y. The distribution characteristics of the EUI for lighting of the seven urban block typologies are shown in Figure 15. The lighting EUI of the lowest block high-rise pavilion (C1-5), with 22.11 kWh/m$^2$/y, was 23.23% lower than that of the highest block courtyard multi-storey block (A1-3), with 28.80 kWh/m$^2$/y.

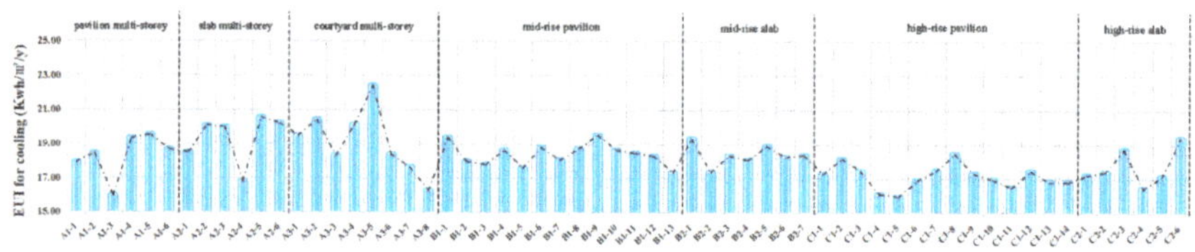


**Figure 13.** Distribution characteristics of EUI for cooling in seven typologies of office blocks.

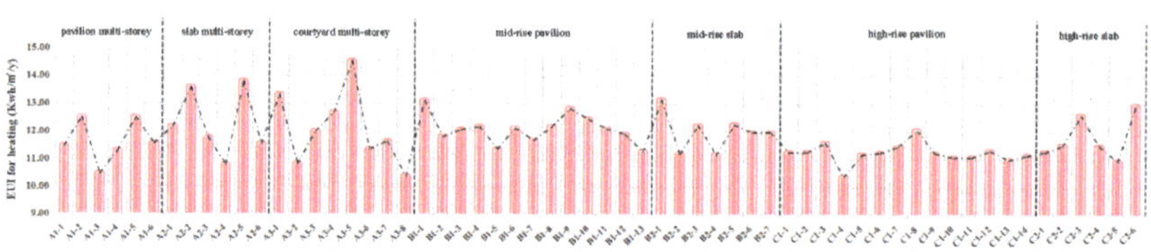


**Figure 14.** Distribution characteristics of EUI for heating in seven typologies of office blocks.

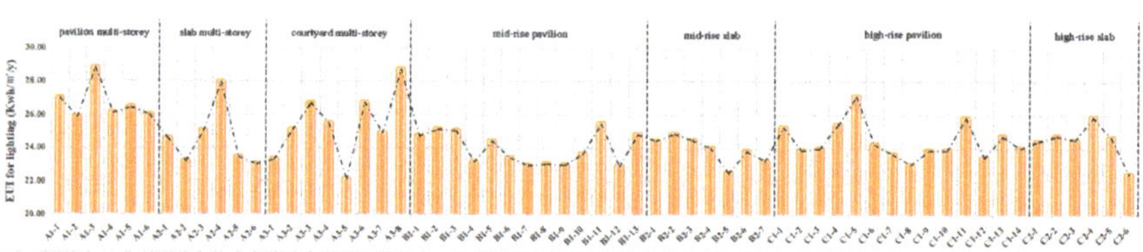


**Figure 15.** Distribution characteristics of EUI for lighting in seven typologies of office blocks.

The relationship between the cooling, heating, and lighting building EUI did not increase and decrease in the same way; there was a discounting phenomenon (Figures 13–15). This is due to the fact that the trend in the EUI for cooling among different office blocks agrees well with the trend in the EUI for heating, while the trend in the EUI for lighting is the opposite of them. Therefore, it is difficult to judge the total building EUI change pattern by the trend of the cooling, heating, and lighting building EUI; in addition to that, the total building energy consumption should be controlled instead of the sub-energy consumption when designing for low-energy office blocks.

### 3.1.2. Building EUI for Different Typologies

The characteristics of the building energy consumption distribution in different typologies of office blocks are shown in Figure 14, from which it can be seen that the building energy consumption varies significantly among the different typologies of office blocks. In terms of average building energy consumption, the pavilion high-rise (52.40 kWh/m$^2$/y) and the slab high-rise (53.97 kWh/m$^2$/y) typologies outperformed the slab mid-rise (54.28 kWh/m$^2$/y) and the pavilion mid-rise (54.45 kWh/m$^2$/y), while the pavilion multi-storey (56.17 kWh/m$^2$/y), slab multi-storey (56.60 kWh/m$^2$/y) and courtyard multi-storey (56.63 kWh/m$^2$/y) had a higher EUI level.

The EUI for cooling, heating, and lighting in the seven typologies of blocks shows a different distribution feature from the EUI for the total (Figures 16–19). The courtyard multi-storey had the higher EUI level, but the cooling, heating, and lighting EUI for the courtyard multi-storey was not the higher. The courtyard multi-storey block had the highest total EUI due to several factors. The L-shaped office buildings are mutually shaded, and the ventilation corridors underneath are poorly ventilated. The high wind resistance of this block typology makes it challenging to remove heat from building surfaces in winter. Conversely, building energy consumption is higher during summer. Additionally, mutual shading affects natural lighting availability, contributing to the highest total EUI value for this block typology.

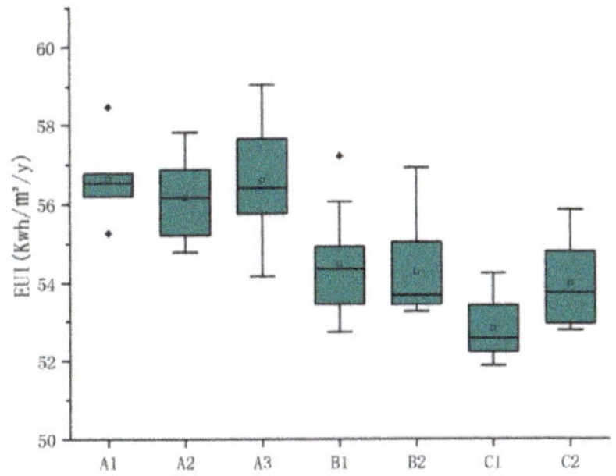


**Figure 16.** Distribution characteristics of total EUI in different typologies of office blocks.

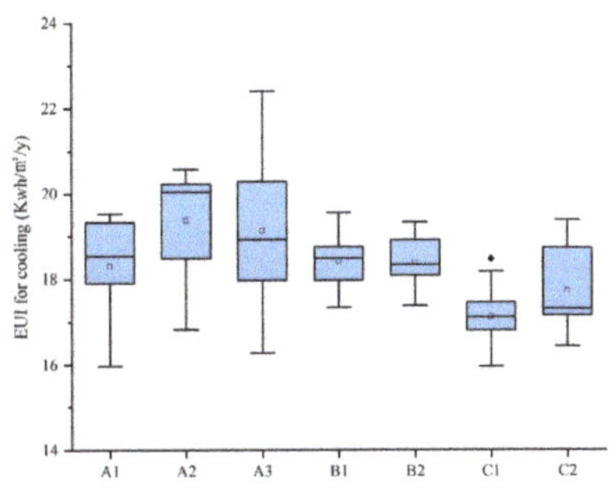


**Figure 17.** Distribution characteristics of EUI for cooling in different typologies of office blocks.

Taking into account the building layout and building height, the office block adopts the high-rise pavilion and high-rise slab typologies, avoiding the multi-storey pavilion and multi-storey slab ones, which can achieve the purpose of reducing the building energy consumption of office blocks in the Hot-summer and Cold-winter zone in China.

Block typology is the external expression of block morphological parameters. In order to analyze the differences in building energy consumption distribution characteristics of different block typologies, it is necessary to research the quantitative relationship between block morphological parameters and building energy consumption.

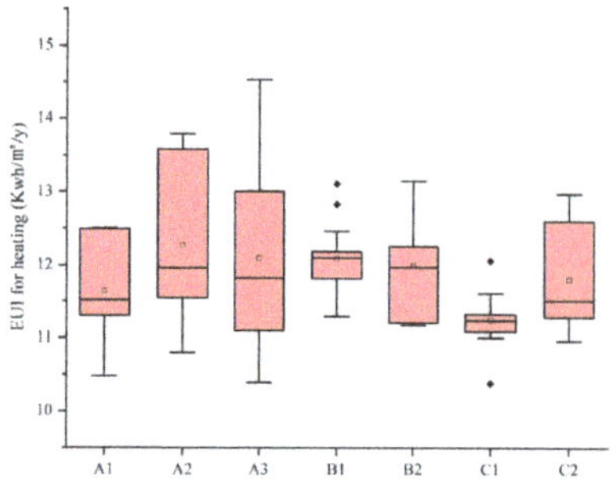


**Figure 18.** Distribution characteristics of EUI for heating in different typologies of office blocks.

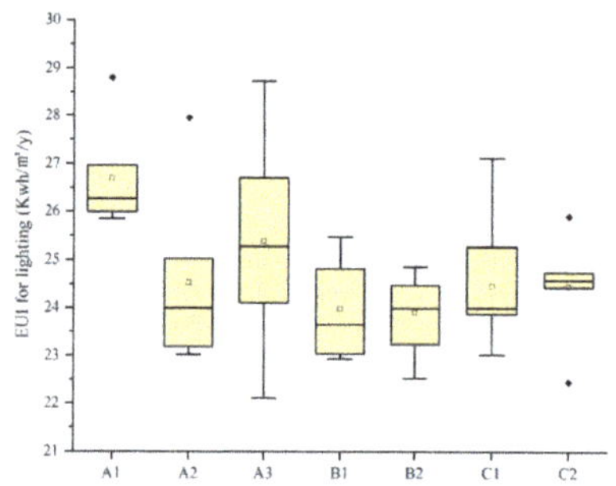


**Figure 19.** Distribution characteristics of EUI for lighting in different typologies of office blocks.

## 3.2. The Effect of Block Morphology on Building EUI

### 3.2.1. Correlation Analysis between Block Morphology and Building EUI

Table 4 shows the results of the correlation analysis between the eight block morphological parameters and building EUI of office blocks. If the significance value (Sig.) is less than 0.05, then the independent variable and dependent variable are significantly correlated. In this paper, the building EUI showed a significant correlation with all independent variables, except BW. Figure 20 depicts the thermal matrix of correlations between block morphological parameters and building EUI. Based on a linear regression analysis, BSF was the most significant factor regarding building EUI, which accounted for 0.833 of its Pearson correlation coefficient, and this was followed by FAR (PCC = −0.810), BH (PCC = −0.644), BD (PCC = −0.623), W/D (PCC = 0.411), and H/D (PCC = −0.316). In addition, BCR (PCC = −0.277) also had a significant impact, although to a lesser extent.

**Table 4.** Pearson correlation coefficient and $p$ value in the linear regression relationship between block morphological parameters and building EUI.

| Block Morphological Parameters | PCC | $p$ Value |
|---|---|---|
| Average Building Width of Block (BW) | −0.125 | 0.343 |
| Average Building Depth of Block (BD) | −0.0623 ** | 0.000 |
| Average Building Height of Block (BH) | −0.644 ** | 0.000 |
| Width-to-Depth Ratio of Block (W/D) | 0.411 ** | 0.001 |
| Height-to-Depth Ratio of Block (H/D) | −0.316 * | 0.014 |
| Building Shape Factor (BS) | 0.833 ** | 0.000 |
| Building Coverage Ratio (BCR) | −0.277 * | 0.032 |
| Floor Area Ratio (FAR) | −0.810 ** | 0.000 |

* means the correlation was significant at level 0.05; ** means the correlation was significant at level 0.01.

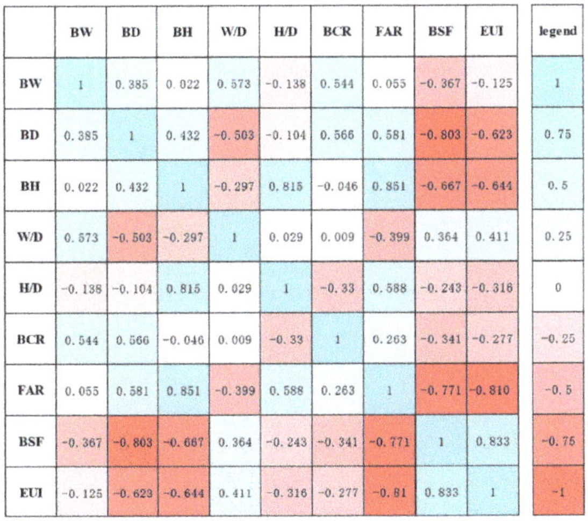


**Figure 20.** Thermal matrix of correlations between block morphological parameters and EUI.

Among these significant independent variables, there is a positive correlation between EUI, and BSF and W/D, which suggests that building EUI is advanced with the increase of BSF and W/D, whereas there is a negative correlation between EUI, and FAR, BH, BD, H/D, and BCR, which indicates that building EUI is reduced with the increase of these five variables. In addition, the larger the absolute value of the Pearson correlation coefficient, the closer the relationship is. Among them, the representative morphological parameters were BSF, FAR, BH, and BCR. Leng et al. [48] researched the influence of block morphology on building heating energy consumption, which demonstrated that there is a high correlation between BD, FAR, and BH, and heating energy consumption in office buildings, and a negative correlation between these morphological parameters and heating energy consumption. The results support the findings of this paper. In addition, the results of another study about the analysis of building energy demand based on block morphology in Maceió showed that there is a positive correlation between BSF and building energy consumption, while a negative correlation between BH, BD, and FAR, and building energy consumption [55], which are consistent with the findings of this paper. Similarly, studies have shown that there is a negative correlation between FAR and energy consumption in office buildings [46,56]. Due to the multitude of urban morphological parameters, there is variation in the morphological parameters selected by different research scholars when studying different issues, which in turn generates some interesting research findings. For example, this study found a correlation between BD, W/D, H/D, and building energy consumption, which is an interesting finding.

### 3.2.2. Predictive Model for Building Energy Consumption with Coupled Block Morphology

(1) Predictive model

Based on the correlation analysis results, this study removed the non-significant correlation variables BW, BD, BH, W/D, H/D, and BCR. The remaining significantly correlated block morphological variables were retained as independent variables, while building EUI was considered the dependent variable for multiple regression analysis. To mitigate the impact of collinearity variables, the study chose a step-by-step regression as the regression method, with the following results obtained:

From Table 5 above, the results of the regression analysis showed that the building energy consumption of office blocks was mainly affected by the combined effect of BSF and

FAR, and the predictive model for building energy consumption with the coupled block morphology of office blocks was as follows:

$$\mathrm{EUI} = 52.047 + 26.682 \times \mathrm{BSF} - 0.7\mathrm{FAR}, \mathrm{R}^2 = 0.755 \quad (2)$$

**Table 5.** Regression analysis results.

| Dependent Variables | Independent Variables | Unstandardized Coefficients | | Standardized Coefficients | Sig. | VIF | R Square ($R^2$) |
|---|---|---|---|---|---|---|---|
| | | B | Standard Error | Beta | | | |
| EUI | (Constants) | 52.047 | 1.229 | | 0.000 | | 0.755 |
| | BSF | 26.682 | 5.244 | 0.514 | 0.000 | 2.464 | |
| | FAR | −0.700 | 0.171 | −0.414 | 0.000 | 2.464 | |

Concurrently, the regression analysis results also produce standardized coefficients (Table 5). These coefficients allow for the comparison of the impact of block morphological parameter variables on the dependent variable, despite having different magnitudes and units. The standardization factor Beta is a quantitative indicator used to unify the different quantitative units, which represents the efficiency of the morphological parameters on the building energy consumption. The standardized coefficient for BSF was 0.514, while the standardized coefficient for FAR was −0.414. The multiple correlation coefficient $R^2$ had a value of 0.755. The determination coefficient $R^2$ is a statistical measure that indicates the fitting effect of the model. When $R^2$ is closer to 1, the better the fitting effect of the model. Therefore, the equation showed a relatively high degree of fit, indicating satisfactory and linearly related block morphological parameters with the dependent variable EUI. In other words, the prediction model of building energy consumption with the coupled block morphology in office blocks had a high accuracy. Additionally, the result of the significant value (Sig.) was 0.000 indicating that the prediction model has high statistical significance and reliability.

From the energy flux perspective, the energy gain and loss of an office building is determined by the solar heat gain from the building surface, the heat exchange between the building and the outdoor environment, and the heat loss from the building. On the one hand, the mutual shading of the building group in the block reduces the solar heat gain of the buildings, which in turn reduces the building energy consumption for cooling in summer and increases the building energy consumption for heating in winter. This is supported by the study of Wong et al. [57]. On the other hand, the building fabric geometry impacts the thermal conduction and convection; the larger the building shape factor of the block building groups, the more thermal conduction and convection there is between the building and the outdoor environment, resulting in increased building energy consumption for cooling in summer and reduced building energy consumption for heating in winter. The study of Martilli provides evidence for this view [58].

The three-dimensional spatial relationship of the equation is shown in Figure 21. With the per unit decrease in BSF, EUI can be saved by 2.67 kWh/m$^2$/y (4.88% of average EUI), while with the per unit increase in FAR, EUI can be saved by 0.7 kWh/m$^2$/y (1.23% of average EUI). The beta in this equation was BSF (0.514) > FAR (−0.414). From an energy-efficiency perspective, both reducing the building BSF and increasing the FAR are effective measures in reducing building energy consumption. Moreover, the former is a higher priority than the latter.

The results of the multiple regression equation showed that the key block morphological parameters which combined affected the building EUI of office blocks are the BSF and FAR. Leng et al. [48] investigated the relationship between block morphology and the building heating energy consumption of office buildings; the analysis results showed that the heating energy consumption of office buildings is affected by both BSF and FAR. This finding is consistent with our research conclusions. The impact of urban block typology on building energy use was explored by Taleghani et al. [59] who found that BSF is a significant geometric factor, and the three-storey courtyard layout uses 22% less energy compared to the one-zone layout.

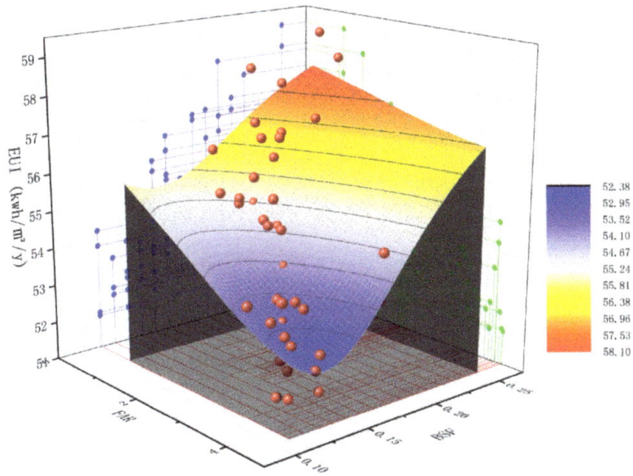


**Figure 21.** 3D nonlinear surface for multiple regression equation.

(2) The predictive model validation

The proposed predictive model was validated through a comparison of the estimated building energy consumption with the simulated values of 10 office blocks that were not used in the model development (Table 6). Validation is measured using the coefficient of variation of *CV(RMSE)*. This approach provides a reliable indicator of model accuracy and ensures that the model is capable of accurately predicting energy consumption for buildings beyond the initial sample [60]. The equation of *CV(RMSE)* is:

$$CV(RMSE) = \frac{1}{\bar{y}} \sqrt{\frac{\sum_{i=1}^{n} (y_i - \bar{y}_i)^2}{n-1}} \tag{3}$$

where $y_i$ is the building energy consumption predictive value of block i, $\overline{y_i}$ is the mean of the building energy consumption simulated values, and n is the total number of blocks. ASHRAE Guideline 14 [61] suggests that a *CV(RMSE)* value lower than 30% indicates a reasonable estimation for predictive models. In this case, the calculated *CV(RMSE)* was 3.91%, indicating that the proposed predictive model for building energy consumption with the coupled block morphology was highly accurate. This model can serve as a valuable reference for exploring the relationship between block morphology and building energy consumption in the Hot-summer and Cold-winter zone in China.

### 3.3. Limitations and Future Research

The building energy assessment tool for office blocks built in this study is constructed based on the Hot-summer and Cold-winter zone in China, and it can be directly applied to similar climate zones worldwide. For different climatic zones, the assessment tool requires an adjustment of the meteorological parameters, thermal performance parameters, window-to-wall ratios, and other simulation parameters before it can be applied. The workflow proposed in this study can provide urban planners and designers with guiding design indicators at the pre-design stage and policymakers with a reference to the range of building energy use intensity indicators.

**Table 6.** Block morphological parameters and EUI of block samples for model validation.

| No. | Block Typology | BW (m) | BD (m) | BH (m) | W/D | H/D | BSF | BCR | FAR | Predictive EUI ($kWh/m^2/y$) | Simulated EUI ($kWh/m^2/y$) |
|---|---|---|---|---|---|---|---|---|---|---|---|
| 1 | Multi-storey | 82.60 | 36.26 | 12.00 | 2.28 | 0.40 | 0.18 | 40.00% | 1.20 | 56.01 | 57.50 |
| 2 | Multi-storey | 28.23 | 10.55 | 20.00 | 2.68 | 2.21 | 0.36 | 36.06% | 1.80 | 61.11 | 62.53 |
| 3 | Multi-storey | 54.59 | 16.98 | 20.00 | 3.24 | 1.19 | 0.21 | 30.63% | 1.53 | 56.58 | 57.86 |
| 4 | Multi-storey | 51.74 | 21.66 | 20.00 | 2.39 | 0.74 | 0.16 | 38.53% | 1.93 | 54.97 | 56.38 |
| 5 | Multi-storey | 85.15 | 23.55 | 16.00 | 3.62 | 0.71 | 0.19 | 34.28% | 1.37 | 56.16 | 57.73 |
| 6 | Mid-rise | 54.96 | 14.82 | 24.00 | 3.71 | 1.62 | 0.22 | 31.73% | 1.90 | 56.59 | 57.88 |
| 7 | Mid-rise | 49.41 | 15.35 | 24.00 | 3.22 | 1.58 | 0.22 | 39.40% | 2.36 | 56.27 | 57.75 |
| 8 | Mid-rise | 51.40 | 17.06 | 24.00 | 3.01 | 1.48 | 0.21 | 21.99% | 1.32 | 56.73 | 58.27 |
| 9 | Mid-rise | 39.34 | 16.45 | 24.00 | 2.39 | 1.46 | 0.22 | 23.86% | 1.43 | 56.92 | 57.68 |
| 10 | Mid-rise | 55.95 | 15.88 | 24.00 | 3.52 | 1.23 | 0.18 | 28.75% | 1.72 | 55.65 | 58.89 |

There are some limitations in this study that need to be addressed in future research. In terms of research assumptions, this study focuses on the combined effect mechanism of block morphology on building energy consumption. The simulation parameters need to be set uniformly in the building energy simulation model. Therefore, the typical values of the simulation parameters were obtained through field research and literature research for setting, such as the building envelope, occupancy rate, operation rate of lighting and equipment, and window-to-wall ratio. Future research should obtain more accurate simulation parameters for each office block, such as the occupancy rate and operation rate of lighting and equipment, and set them separately to improve the accuracy of the building energy simulation.

The research methodology and workflow proposed in this paper offer new possibilities for studying the correlation between block morphological parameters and building energy consumption with limited research samples and relevant data, which met the statistical requirements.

## 4. Conclusions

This paper quantitatively investigated the combined effect of block morphology parameters on the building energy consumption of office blocks in the Hot-summer and Cold-winter zone in China, with Wuhan as a case study. Seventy office block building energy consumption data and eight block morphology parameters of blocks are studied in detail through a combination approach of simulation and statistical analysis. A model for predicting the energy consumption of office blocks was proposed. Several conclusions can be addressed here:

1. Block morphology impacted the total EUI by 13.82%.
2. The effect of block morphology on the building cooling, heating, and lighting EUI was 28.83%, 28.56%, and 23.23%, respectively.
3. The results of the correlation analysis demonstrated that BSF is the most significant factor regarding EUI and this is followed by FAR (PCC = −0.810), BH (PCC = −0.644), and BD (PCC = −0.623).
4. The predictive model for building energy consumption with the coupled block morphology for office blocks was as follows Equation (2).
5. The key morphological parameter which combined affect the building energy consumption of office blocks are BSF and FAR, with standardized coefficients of 0.514 and −0.414, respectively. BSF has 1.24 times the effect on building energy consumption than FAR.

The findings of this work are applied to contribute a prospective block-scale energy consumption assessment and energy-efficient design strategies for urban planners, designers, and policymakers. The workflow proposed in this paper can be applied to other cities around the world for promoting sustainable cities.

**Author Contributions:** Conceptualization, S.X. and H.Z.; methodology, S.X.; software, H.Z. and G.L.; validation, G.L., M.X. and S.X.; formal analysis, G.L.; investigation, H.Z. and G.L.; resources, S.X.; data curation, G.L.; writing—original draft preparation, G.L. and H.Z.; writing—review and editing, M.X., H.D. and T.M.; visualization, G.L.; supervision, S.X.; project administration, S.X.; funding acquisition, S.X. All authors have read and agreed to the published version of the manuscript.

**Funding:** This research was funded by the National Natural Science Foundation (No. 51978296) and the program for the HUST Academic Frontier Youth Team (No. 2019QYTD10).

**Data Availability Statement:** The data presented in this study are available in Appendix A.

**Conflicts of Interest:** The authors declare no conflict of interest.

## Nomenclature

| | |
|---|---|
| BW | Average Building Width of Block |
| BD | Average Building Depth of Block |
| BH | Average Building Height of Block |
| W/D | Width-to-Depth Ratio of Block |
| H/D | Height-to-Depth Ratio of Block |
| BSF | Building Shape Factor |
| BCR | Building Coverage Ratio |
| FAR | Floor Area Ratio |
| EUI | Energy Use Intensity |
| PCC | Pearson Correlation Coefficient |
| R-squared ($R^2$) | Goodness-of-Fit |
| VIF | Variance Inflation Factor |
| BES | Building Energy Simulation |

## Appendix A

**Table A1.** Block morphological parameters of 60 block cases.

| Block Typology | Block Samples | BW | BD | BH | W/D | H/D | BSF | BCR | FAR |
|---|---|---|---|---|---|---|---|---|---|
| Courtyard multi-storey | A1-1 | 43.54 | 15.74 | 24.00 | 2.77 | 1.52 | 0.22 | 0.29 | 1.74 |
| | A1-2 | 50.09 | 23.05 | 20.00 | 2.17 | 0.87 | 0.18 | 0.33 | 1.65 |
| | A1-3 | 78.13 | 33.19 | 16.00 | 2.35 | 0.48 | 0.16 | 0.44 | 1.76 |
| | A1-4 | 45.97 | 14.27 | 20.00 | 3.22 | 1.40 | 0.24 | 0.26 | 1.30 |
| | A1-5 | 42.58 | 13.69 | 24.00 | 3.11 | 1.75 | 0.25 | 0.22 | 1.33 |
| | A1-6 | 57.63 | 19.55 | 20.00 | 2.95 | 1.02 | 0.19 | 0.32 | 1.62 |
| Pavilion multi-storey | A2-1 | 46.44 | 24.51 | 23.33 | 1.89 | 0.95 | 0.17 | 0.30 | 1.78 |
| | A2-2 | 26.43 | 23.13 | 24.00 | 1.14 | 1.04 | 0.21 | 0.18 | 1.10 |
| | A2-3 | 34.74 | 24.40 | 20.00 | 1.42 | 0.82 | 0.20 | 0.30 | 1.49 |
| | A2-4 | 68.76 | 40.31 | 20.00 | 1.71 | 0.50 | 0.15 | 0.33 | 1.66 |
| | A2-5 | 40.00 | 20.15 | 20.00 | 1.99 | 0.99 | 0.22 | 0.18 | 0.90 |
| | A2-6 | 46.34 | 23.94 | 16.00 | 1.94 | 0.67 | 0.19 | 0.32 | 1.26 |
| Slab multi-storey | A3-1 | 72.78 | 25.53 | 8.00 | 2.85 | 0.31 | 0.23 | 0.37 | 0.75 |
| | A3-2 | 59.33 | 18.80 | 20.75 | 3.16 | 1.10 | 0.20 | 0.29 | 1.51 |
| | A3-3 | 60.04 | 19.32 | 24.00 | 3.11 | 1.24 | 0.19 | 0.20 | 1.20 |
| | A3-4 | 44.88 | 21.20 | 20.00 | 2.12 | 0.94 | 0.20 | 0.16 | 0.80 |
| | A3-5 | 34.38 | 16.26 | 17.93 | 2.11 | 1.10 | 0.25 | 0.21 | 0.90 |
| | A3-6 | 69.91 | 24.18 | 23.16 | 2.89 | 0.96 | 0.16 | 0.20 | 1.17 |
| | A3-7 | 77.76 | 24.15 | 24.00 | 3.22 | 0.99 | 0.15 | 0.30 | 1.78 |
| | A3-8 | 84.56 | 27.30 | 20.00 | 3.10 | 0.73 | 0.16 | 0.35 | 1.77 |

**Table A1.** *Cont.*

| Block Typology | Block Samples | BW | BD | BH | W/D | H/D | BSF | BCR | FAR |
|---|---|---|---|---|---|---|---|---|---|
| Mid-rise pavilion | B1-1 | 26.79 | 21.18 | 30.67 | 1.26 | 1.45 | 0.21 | 0.28 | 1.96 |
| | B1-2 | 41.78 | 25.36 | 28.29 | 1.65 | 1.12 | 0.16 | 0.28 | 2.40 |
| | B1-3 | 55.35 | 31.03 | 30.00 | 1.78 | 0.97 | 0.16 | 0.52 | 2.68 |
| | B1-4 | 38.98 | 25.44 | 37.50 | 1.53 | 1.47 | 0.16 | 0.31 | 2.73 |
| | B1-5 | 43.56 | 26.10 | 46.67 | 1.67 | 1.79 | 0.15 | 0.19 | 2.49 |
| | B1-6 | 31.45 | 30.84 | 28.00 | 1.02 | 0.91 | 0.16 | 0.32 | 2.26 |
| | B1-7 | 47.51 | 24.48 | 41.33 | 1.94 | 1.69 | 0.15 | 0.24 | 2.59 |
| | B1-8 | 42.95 | 23.97 | 48.00 | 1.79 | 2.00 | 0.16 | 0.20 | 2.36 |
| | B1-9 | 33.44 | 19.10 | 36.00 | 1.75 | 1.89 | 0.20 | 0.26 | 2.28 |
| | B1-10 | 31.38 | 24.16 | 48.36 | 1.30 | 2.00 | 0.17 | 0.16 | 2.30 |
| | B1-11 | 33.56 | 25.82 | 41.33 | 1.30 | 1.60 | 0.16 | 0.19 | 2.10 |
| | B1-12 | 41.24 | 25.50 | 28.00 | 1.62 | 1.10 | 0.16 | 0.21 | 2.14 |
| | B1-13 | 56.04 | 29.38 | 30.40 | 1.91 | 1.03 | 0.15 | 0.37 | 2.42 |
| Mid-rise slab | B2-1 | 52.20 | 21.32 | 30.91 | 2.45 | 1.45 | 0.18 | 0.31 | 2.05 |
| | B2-2 | 72.29 | 30.10 | 32.67 | 2.40 | 1.09 | 0.14 | 0.37 | 2.14 |
| | B2-3 | 49.51 | 23.86 | 40.44 | 2.07 | 1.69 | 0.16 | 0.30 | 2.30 |
| | B2-4 | 61.05 | 28.67 | 37.71 | 2.13 | 1.32 | 0.15 | 0.35 | 2.50 |
| | B2-5 | 53.43 | 19.90 | 42.86 | 2.68 | 2.15 | 0.17 | 0.23 | 2.31 |
| | B2-6 | 49.12 | 22.57 | 34.29 | 2.18 | 1.52 | 0.16 | 0.22 | 2.21 |
| | B2-7 | 51.76 | 25.22 | 38.22 | 2.05 | 1.52 | 0.16 | 0.26 | 2.24 |
| High-rise pavilion | C1-1 | 53.79 | 28.50 | 53.09 | 1.89 | 1.86 | 0.13 | 0.28 | 3.50 |
| | C1-2 | 41.64 | 31.13 | 56.00 | 1.34 | 1.80 | 0.15 | 0.29 | 3.45 |
| | C1-3 | 33.60 | 30.24 | 100.00 | 1.11 | 3.31 | 0.14 | 0.13 | 3.30 |
| | C1-4 | 58.04 | 41.98 | 69.33 | 1.38 | 1.65 | 0.11 | 0.37 | 4.08 |
| | C1-5 | 92.54 | 46.52 | 69.00 | 1.99 | 1.48 | 0.11 | 0.48 | 4.20 |
| | C1-6 | 50.73 | 36.98 | 66.29 | 1.37 | 1.79 | 0.12 | 0.36 | 4.00 |
| | C1-7 | 34.84 | 31.43 | 57.00 | 1.11 | 1.81 | 0.14 | 0.26 | 4.08 |
| | C1-8 | 31.34 | 21.97 | 44.00 | 1.43 | 2.00 | 0.18 | 0.25 | 4.14 |
| | C1-9 | 41.36 | 34.59 | 50.86 | 1.20 | 1.47 | 0.13 | 0.35 | 4.27 |
| | C1-10 | 42.15 | 33.25 | 53.14 | 1.27 | 1.60 | 0.13 | 0.32 | 4.34 |
| | C1-11 | 58.01 | 35.99 | 56.00 | 1.61 | 1.56 | 0.12 | 0.29 | 3.67 |
| | C1-12 | 48.34 | 25.46 | 54.86 | 1.90 | 2.15 | 0.15 | 0.30 | 3.67 |
| | C1-13 | 55.90 | 32.35 | 50.40 | 1.73 | 1.56 | 0.12 | 0.32 | 3.86 |
| | C1-14 | 56.75 | 34.27 | 76.00 | 1.66 | 2.22 | 0.11 | 0.38 | 4.32 |
| High-rise slab | C2-1 | 72.04 | 32.31 | 64.62 | 2.23 | 2.00 | 0.13 | 0.39 | 3.65 |
| | C2-2 | 58.47 | 24.28 | 50.67 | 2.41 | 2.09 | 0.14 | 0.28 | 3.57 |
| | C2-3 | 65.73 | 21.75 | 66.40 | 3.02 | 3.05 | 0.15 | 0.26 | 2.82 |
| | C2-4 | 69.06 | 33.26 | 73.14 | 2.08 | 2.20 | 0.11 | 0.25 | 3.50 |
| | C2-5 | 54.53 | 26.40 | 61.71 | 2.07 | 2.34 | 0.13 | 0.26 | 3.78 |
| | C2-6 | 52.33 | 16.45 | 90.40 | 3.18 | 5.50 | 0.18 | 0.20 | 4.21 |

## References


1. IEA. *World Energy Outlook 2022;* IEA: Paris, France, 2022.
2. IEA. *World Energy Outlook 2021;* IEA: Paris, France, 2021.
3. Central People's Government of the People's Republic of China. Proposal of the CPC Central Committee on Formulating the 14th Five-Year Plan for National Economic and Social Development and the long-range Goals for the next Five-Year Period. 2020. Available online: http://www.gov.cn/zhengce/2020-11/03/content_5556991.htm (accessed on 20 January 2023).
4. The State Council of China. Carbon Peaking Action Programme by 2030. 2021. Available online: http://www.gov.cn/zhengce/content/2021-10/26/content_5644984.htm (accessed on 20 January 2023).
5. China Association of Building Energy Efficiency. *China Building Energy Consumption Research Report 2016;* China Association of Building Energy Efficiency: Shanghai, China, 2016.
6. Hong, T.; Langevin, J.; Sun, K. Building simulation: Ten challenges. *Build. Simul.* **2018**, *11*, 871–898. [CrossRef]
7. Souza, L.; Bueno, C. City Information Modelling as a support decision tool for planning and management of cities: A systematic literature review and bibliometric analysis. *Build. Environ.* **2022**, *207*, 108403. [CrossRef]
8. Omrany, H.; Ghaffarianhoseini, A.; Ghaffarianhoseini, A.; Clements-Croome, D.J. The uptake of City Information Modelling (CIM): A comprehensive review of current implementations, challenges and future outlook. *Smart Sustain. Built Environ.* **2022**, *ahead-of-print*. [CrossRef]
9. Ang, Y.; Berzolla, Z.; Reinhart, C. From concept to application: A review of use cases in urban building energy modeling. *Appl. Energy* **2020**, *279*, 115738. [CrossRef]
10. Reinhart, C.F.; Davila, C.C. Urban building energy modeling—A review of a nascent field. *Build. Environ.* **2016**, *97*, 196–202. [CrossRef]
11. Hong, T.; Chen, Y.; Luo, X.; Luo, N.; Lee, S.H. Ten questions on urban building energy modeling. *Build. Environ.* **2020**, *168*, 106508. [CrossRef]
12. Ratti, C.; Baker, N.; Steemers, K. Energy consumption and urban texture. *Energy Build.* **2005**, *37*, 762–776. [CrossRef]
13. Oh, M.; Jang, K.M.; Kim, Y. Empirical analysis of building energy consumption and urban form in a large city: A case of Seoul, South Korea. *Energy Build.* **2021**, *245*, 111046. [CrossRef]
14. Steemers, K. Energy and the city: Density, buildings and transport. *Energy Build.* **2003**, *35*, 3–14. [CrossRef]
15. Strømann-Andersen, J.; Sattrup, P.A. The urban canyon and building energy use: Urban density versus daylight and passive solar gains. *Energy Build.* **2011**, *43*, 2011–2020. [CrossRef]
16. Shareef, S.; Altan, H. Urban block configuration and the impact on energy consumption: A case study of sinuous morphology. *Renew. Sustain. Energy Rev.* **2022**, *163*, 112507. [CrossRef]
17. El Bat, S.; Romani, Z.; Bozonnet, E.; Draoui, A.; Allard, F. Optimizing urban courtyard form through the coupling of outdoor zonal approach and building energy modeling. *Energy* **2023**, *264*, 126176. [CrossRef]
18. Bansal, P.; Quan, S.J. Relationships between building characteristics, urban form and building energy use in different local climate zone contexts: An empirical study in Seoul. *Energy Build.* **2022**, *272*, 112335. [CrossRef]
19. Deng, Q.; Wang, G.; Wang, Y.; Zhou, H.; Ma, L. A quantitative analysis of the impact of residential cluster layout on building heating energy consumption in cold IIB regions of China. *Energy Build.* **2021**, *253*, 111515. [CrossRef]
20. Li, C.; Song, Y.; Kaza, N. Urban form and household electricity consumption: A multilevel study. *Energy Build.* **2018**, *158*, 181–193. [CrossRef]
21. Huang, J.; Kaewunruen, S. Forecasting Energy Consumption of a Public Building Using Transformer and Support Vector Regression. *Energies* **2023**, *16*, 966. [CrossRef]
22. Li, X.; Ying, Y.; Xu, X.; Wang, Y.; Hussain, S.A.; Hong, T.; Wang, W. Identifying key determinants for building energy analysis from urban building datasets. *Build. Environ.* **2020**, *181*, 107114. [CrossRef]
23. Vartholomaios, A. A parametric sensitivity analysis of the influence of urban form on domestic energy consumption for heating and cooling in a Mediterranean city. *Sustain. Cities Soc.* **2017**, *28*, 135–145. [CrossRef]
24. Shi, Z.; Fonseca, J.A.; Schlueter, A. A parametric method using vernacular urban block typologies for investigating interactions between solar energy use and urban design. *Renew. Energy* **2021**, *165*, 823–841. [CrossRef]
25. Javanroodi, K.; Nik, V.M.; Mahdavinejad, M. A novel design-based optimization framework for enhancing the energy efficiency of high-rise office buildings in urban areas. *Sustain. Cities Soc.* **2019**, *49*, 101597. [CrossRef]
26. Loeffler, R.; Österreicher, D.; Stoeglehner, G. The energy implications of urban morphology from an urban planning perspective – A case study for a new urban development area in the city of Vienna. *Energy Build.* **2021**, *252*, 111453. [CrossRef]
27. Hadavi, M.; Pasdarshahri, H. Investigating effects of urban configuration and density on urban climate and building systems energy consumption. *J. Build. Eng.* **2021**, *44*, 102710. [CrossRef]
28. Shareef, S. The impact of urban morphology and building's height diversity on energy consumption at urban scale. *The case study of Dubai. Build. Environ.* **2021**, *194*, 107675. [CrossRef]
29. SMangan, D.; Oral, G.K.; Kocagil, I.E.; Sozen, I. The impact of urban form on building energy and cost efficiency in temperate-humid zones. *J. Build. Eng.* **2021**, *33*, 101626. [CrossRef]
30. Zhang, J.; Xu, L.; Shabunko, V.; Tay, S.E.R.; Sun, H.; Lau, S.S.Y.; Reindl, T. Impact of urban block typology on building solar potential and energy use efficiency in tropical high-density city. *Appl. Energy* **2019**, *240*, 513–533. [CrossRef]

31. Kamal, A.; Abidi, S.M.H.; Mahfouz, A.; Kadam, S.; Rahman, A.; Hassan, I.G.; Wang, L.L. Impact of urban morphology on urban microclimate and building energy loads. *Energy Build.* **2021**, *253*, 111499. [CrossRef]
32. Boccalatte, A.; Fossa, M.; Gaillard, L.; Menezo, C. Microclimate and urban morphology effects on building energy demand in different European cities. *Energy Build.* **2020**, *224*, 110129. [CrossRef]
33. Mirzabeigi, S.; Razkenari, M. Design optimization of urban typologies: A framework for evaluating building energy performance and outdoor thermal comfort. *Sustain. Cities Soc.* **2022**, *76*, 103515. [CrossRef]
34. Quan, S.J.; Li, C. Urban form and building energy use: A systematic review of measures, mechanisms, and methodologies. *Renew. Sustain. Energy Rev.* **2021**, *139*, 110662. [CrossRef]
35. Wang, C.; Ferrando, M.; Causone, F.; Jin, X.; Zhou, X.; Shi, X. Data acquisition for urban building energy modeling: A review. *Build. Environ.* **2022**, *217*, 109056. [CrossRef]
36. East China Construction Group Co., Ltd. *Architectural Design Sourcebook, Volume 3*; China Architecture & Building Press: Beijing, China, 2017; p. 2.
37. Ministry of Public Security of the People's Republic of China. *Code for Fire Protection Design of Buildings, GB 50016-2014*; China Planning Press: Beijing, China, 2014; pp. 61–62.
38. Zhejiang Building Design & Research Institute. *Office Building Design Standards, JGJ /T 67-2019*; China Architecture Publishing Media Co., Ltd.: Beijing, China, 2019; pp. 15–16.
39. Wang, W.; Liu, K.; Zhang, M.; Shen, Y.; Jing, R.; Xu, X. From simulation to data-driven approach: A framework of integrating urban morphology to low-energy urban design. *Renew. Energy* **2021**, *179*, 2016–2035. [CrossRef]
40. Ahmadian, E.; Sodagar, B.; Bingham, C.; Elnokaly, A.; Mills, G. Effect of urban built form and density on building energy performance in temperate climates. *Energy Build.* **2021**, *236*, 110762. [CrossRef]
41. Ahmadian, E.; Sodagar, B.; Mills, G.; Byrd, H.; Bingham, C.; Zolotas, A. Sustainable cities: The relationships between urban built forms and density indicators. *Cities* **2019**, *95*, 102382. [CrossRef]
42. Tian, J.; Xu, S. A morphology-based evaluation on block-scale solar potential for residential area in central China. *Sol. Energy* **2021**, *221*, 332–347. [CrossRef]
43. Urquizo, J.; Calderón, C.; James, P. Understanding the complexities of domestic energy reductions in cities: Integrating data sets generally available in the United Kingdom's local authorities. *Cities* **2018**, *74*, 292–309. [CrossRef]
44. Fumo, N.; Mago, P.; Luck, R. Methodology to estimate building energy consumption using EnergyPlus Benchmark Models. *Energy Build.* **2010**, *42*, 2331–2337. [CrossRef]
45. Boyano, A.; Hernandez, P.; Wolf, O. Energy demands and potential savings in European office buildings: Case studies based on EnergyPlus simulations. *Energy Build.* **2013**, *65*, 19–28. [CrossRef]
46. Natanian, J.; Aleksandrowicz, O.; Auer, T. A parametric approach to optimizing urban form, energy balance and environmental quality: The case of Mediterranean districts. *Appl. Energy* **2019**, *254*, 113637. [CrossRef]
47. Kavgic, M.; Mavrogianni, A.; Mumovic, D.; Summerfield, A.; Stevanovic, Z.; Djurovic-Petrovic, M. A review of bottom-up building stock models for energy consumption in the residential sector. *Build. Environ.* **2010**, *45*, 1683–1697. [CrossRef]
48. Leng, H.; Chen, X.; Ma, Y.; Wong, N.H.; Ming, T. Urban morphology and building heating energy consumption: Evidence from Harbin, a severe cold region city. *Energy Build.* **2020**, *224*, 110143. [CrossRef]
49. Maamari, F.; Andersen, M.; de Boer, J.; Carroll, W.L.; Dumortier, D.; Greenup, P. Experimental validation of simulation methods for bi-directional transmission properties at the daylighting performance level. *Energy Build.* **2006**, *38*, 878–889. [CrossRef]
50. Reeves, S.O.T. Validation of Building Energy Modeling Tools: EcotectTM, Green Building StudioTM and IES<VE>TM. In Proceedings of the 2012 Winter Simulation Conference, Berlin, Germany, 9–12 December 2012; pp. 582–593.
51. Tian, W. A review of sensitivity analysis methods in building energy analysis. *Renew. Sustain. Energy Rev.* **2013**, *20*, 411–419. [CrossRef]
52. Kristensen, M.H.; Petersen, S. Choosing the appropriate sensitivity analysis method for building energy model-based investigations. *Energy Build.* **2016**, *130*, 166–176. [CrossRef]
53. Tso, G.K.F.; Guan, J. A multilevel regression approach to understand effects of environment indicators and household features on residential energy consumption. *Energy* **2014**, *66*, 722–731. [CrossRef]
54. Lima, I.; Scalco, V.; Lamberts, R. Estimating the impact of urban densification on high-rise office building cooling loads in a hot and humid climate. *Energy Build.* **2019**, *182*, 30–44. [CrossRef]
55. Martins, T.; Adolphe, L.; Bonhomme, M. Building Energy Demand based on urban morphology analysis. In Proceedings of the PLEA2013—29th Conference, Sustainable Architecture for a Renewable Future, Munich, Germany, 10–12 September 2013.
56. Pan, Y.; Pan, Y.; Yang, Y.; Liu, H. A Parametric Study on the Community Form and Its Influences on Energy Consumption of Office Buildings in Shanghai. *Procedia Eng.* **2017**, *205*, 548–555. [CrossRef]
57. Wong, N.H.; Jusuf, S.K.; Syafii, N.I.; Chen, Y.; Hajadi, N.; Sathyanarayanan, Y.V. Manickavasagam, Evaluation of the impact of the surrounding urban morphology on building energy consumption. *Sol. Energy* **2011**, *85*, 57–71. [CrossRef]
58. Martilli, A. An idealized study of city structure, urban climate, energy consumption, and air quality. *Urban Clim.* **2014**, *10*, 430–446. [CrossRef]
59. Taleghani, M.; Tenpierik, M.; van den Dobbelsteen, A.; de Dear, R. Energy use impact of and thermal comfort in different urban block types in the Netherlands. *Energy Build.* **2013**, *67*, 166–175. [CrossRef]

60. Ruiz, G.; Bandera, C. Validation of Calibrated Energy Models: Common Errors. *Energies* **2017**, *10*, 1587. [CrossRef]
61. American Society of Heating, Ventilation, and Air Conditioning Engineers (ASHRAE). *Guideline 14-2014, Measurement of Energy, Demand and Water Savings*; American Society of Heating, Ventilating, and Air Conditioning Engineers: Atlanta, GA, USA, 2014.

buildings

MDPI

Article

# An Electricity Consumption Disaggregation Method for HVAC Terminal Units in Sub-Metered Buildings Based on CART Algorithm

**Xinyu Yang [1], Ying Ji [1,*], Jiefan Gu [2] and Menghan Niu [1]**

1 Beijing Key Laboratory of Green Built Environment and Energy Efficient Technology, Beijing University of Technology, Beijing 100124, China
2 College of Architecture and Urban Planning, Tongji University, Shanghai 201804, China
* Correspondence: jyesse2015@163.com

**Abstract:** Obtaining reliable and detailed energy consumption information about building service (BS) systems is an essential prerequisite for identifying energy-saving potential and improving energy efficiency of a building. Therefore, in recent years, energy sub-metering systems have been widely implemented in public buildings in China. A majority of electrical systems and equipment can be directly metered. However, in actual sub-metering systems, the terminal units of heating, ventilation and air conditioning (HVAC) systems, such as fan coils, air handling units and so on, are often mixed with the lighting-plug circuit. This mismatch between theoretical sub-metering systems and actual electricity supply circuits constitutes a lot of challenges in BS system management and control optimization. This study proposed an indirect method to disaggregate the energy consumption of HVAC terminal units from mixed sub-metering data based on the CART algorithm. This method was demonstrated in two buildings in Shanghai. The case study results show that the weighted mean absolute percentage errors (WMAPE) are within 5% and 15% during working hours in the cooling and heating seasons, respectively.

**Keywords:** electricity sub-metering; HVAC terminal units; hourly energy use; classification and regression tree

## 1. Introduction

Buildings are one of the world's largest energy consumers and carbon emission producers. According to the tracking report from the International Energy Agency (IEA), the operation of buildings accounted for about 30% of global final energy consumption and 27% of total energy system emission in 2021. About 19% of energy-related carbon emission came from the generation of electricity and heat used in buildings. Compared to 2010, electricity consumption increased by 30% in 2021 [1]. With the widespread use of central heating, ventilation and air conditioning (HVAC) systems, the energy consumption per unit area of public buildings in China increased from 17 kgce/m$^2$ in 2001 to more than 24.7 kgce/m$^2$ in 2020 [2]. Various energy conservation measures need to be adopted to improve building energy efficiency and reduce carbon emission, which increases the possibility of achieving carbon neutrality in China by 2060.

At the building operation stage, accurate building system operation data are indispensable for optimizing HVAC system control strategy and analyzing building energy-saving potential. Therefore, it is very important to obtain accurate building energy consumption data, especially data of HVAC units.

### 1.1. The Status Quo of Sub-Metering Systems

In recent years, energy sub-metering systems have been widely implemented in various buildings. A majority of electrical systems and equipment can be directly monitored and metered. Electricity sub-metering was first proposed by Dr. Hart in 1992 [3]. In 2008, the California Public Utilities Commission (CPUC) adopted a resolution to carry out electricity sub-metering in multi-story commercial buildings [4]. In 2007, the Chinese government issued document No. 558, declaring that a special compensation would be used for building energy monitoring platforms [5]. Large-scale investigations on public building energy consumption have been conducted in Shanghai since 1995 and have then extended to other cities in China, including Beijing, Tianjin, Changsha and Wuhan [6–8]. Statistical data from Shanghai show that more than 1400 large public buildings (the total area exceeding 60 million $m^2$) have installed a sub-metering system [9].

The technical guidelines for electricity sub-metering data collection in China's document No. 114 clearly stipulate that an entire building's electricity consumption should be recorded by four primary sub-metering circuits, including the lighting-plug circuit, HVAC circuit, power circuit and other circuit [10]. This electricity sub-metering classification diagram is illustrated in Figure 1. The technical guidelines also point out that the electricity consumption of HVAC terminal units (e.g., fan coils and variable air volume (VAV) boxes) can be recorded in the lighting-plug circuit because, based on existing electricity-supply circuit design, HVAC terminals are usually plugged in the lighting-plug circuit [10]. Therefore, there is a grey area between an actual sub-metering model (black lines) and the theoretical (red lines) sub-metering classification model (shown in Figure 1). Additionally, the HVAC electricity consumption obtained from a building sub-metering system would be lower than the actual HVAC electricity consumption. In China, this is a common phenomenon. We conducted a survey of about 1400 buildings with a sub-metering system in Shanghai and found that only two buildings had direct sub-meters of HVAC terminal units. In most buildings, it is still impossible to obtain the electricity consumption of HVAC terminal units directly.

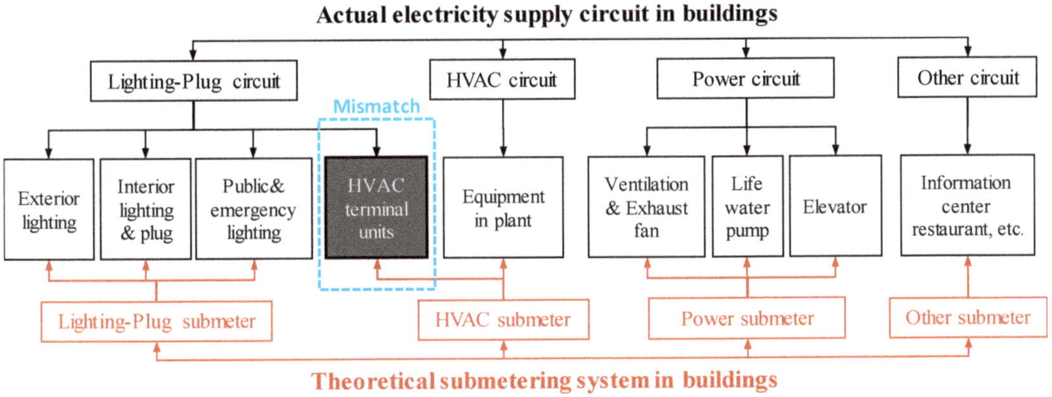


**Figure 1.** Structure of a sub-metering system and the gap between an actual model and the theoretical classification model.

However, accurate sub-metering data are extremely significant in many fields, including evaluating the energy-saving potential of buildings [11], selecting suitable energy efficiency technology [12], reducing carbon emissions of building operation [13], and securing the optimal control of HVAC systems [14]. Therefore, it is necessary to establish an indirect method to acquire the electricity consumption of HVAC terminal units.

### *1.2. Non-Intrusive Load Monitoring (NILM) Method*

In the early 1990s, researchers introduced a method called non-intrusive load monitoring (NILM) to measure the electricity consumption of facilities in buildings [15–18]. This method can also be used to distinguish equipment's on/off signals (e.g., current, voltage, active power and reactive power) and record the power and operational duration. This method is eminently suitable for residential buildings with low-power appliances and simple electricity supply systems [19,20]. Norford et al. applied a NILM method to small-scale commercial buildings [21]. They proposed a load disaggregation model to distinguish the load of air-conditioning systems from other electrical appliances with similar power levels or operation time [22].

NILM methods can be divided into four categories: machine learning method, sparse coding, dynamic time warping (DTW) and Fourier analysis. Machine learning is one of the most popular approaches because it is easy to use and has high accuracy. Rafsanjani et al. combined Density-based Spatial Clustering of Applications with Noise (DBSCAN) algorithm and quadratic discriminant analysis (QDA) to disaggregate total building energy loads into individual energy consumption [23,24]. Kaselimi et al. proposed a Bayesian-optimized Bidirectional LSTM regression model to disaggregate the energy consumption of residential buildings [25]. Schirmer et al. used feed-forward Deep Neural Networks (DNNs), k-Nearest Neighbors (KNNs), Random Forests (RFs) and Support Vector Machines (SVMs) to disaggregate the electricity consumption of residential buildings separately [26]. Kaselimi M et al. proposed a Convolutional Neural Network (CNN)-based structure to estimate the current state of appliances in residential buildings [27]. Yang D et al. proposed an event-driven NILM method based on CNN to extract the energy consumption behavior of occupants [28]. Faustine A et al. proposed a UNet-NILM method based on multi-label learning strategy and multi-target quantile regression to detect the state and power of electrical appliances [29]. Xia et al. presented a deep Long Short-Term Memory (LSTM) model for load disaggregation. In this model, an encoder–decoder structure was designed to solve the problem of time dependency in multi-state devices [30]. Guo et al. presented a load disaggregation method based on combinatorial optimization to disaggregate the electricity load of residential buildings [31]. Monteiro et al. disaggregated residential buildings' electricity load by using Multi-layer Perceptron (MLP), LSTM and Convolutional Neural Network (CNN) [32]. Samadi and Fattahi used K-means method to cluster the daily load profiles of buildings and calculated the common load and occupancy load in different day types [33]. Xiao et al. applied RF for the cooling load disaggregation of office buildings [34]. Athanasiadis C et al. proposed a real-time NILM method, including an event detection algorithm, a CNN classifier and a power estimation algorithm, to estimate the energy consumption of household appliances [35]. Shao et al. proposed a load disaggregation approach based on temporal motif mining. This approach used a clustering method to detect the power of household devices [36]. Burak Gunay et al. proposed an electricity end-use disaggregation method based on regression models for commercial buildings [37]. Zaeri et al. applied multiple linear regression models to disaggregate the electricity and heating consumption in commercial buildings. These models were estimated by a least-square solver with a genetic algorithm [38].

Sparse coding and DTW algorithm are widely used in NILM methods. Elafoudi et al. proposed a NILM method based on DTW algorithm to disaggregate the electricity consumption of residential building devices [39]. Kolter et al. used a sparse coding algorithm to train models for each device in a residential building with electricity consumption and applied these models to predict the electricity consumption of each device [40]. Matsui et al. applied a 0–1 sparse coding method to disaggregate the energy consumption of electric appliances from the total energy consumption of residential buildings [41].

Fourier analysis can be used to disaggregate different loads in the frequency domain. Dhar et al. took advantage of a Fourier series model to simulate real-time energy consumption of some campus buildings in Texas and obtained satisfactory results [42–44]. Ji et al. applied Fourier series models to disaggregate hourly electricity consumption of HVAC system terminal units from mixed lighting-plug sub-meters of commercial buildings [45]. The accuracy of the calculation results is high, although, in this study, the division of the date type is slightly rough, and a standardized and programmed calculation process is not achieved. Table 1 summarizes the details of above studies.

**Table 1.** Summary of existing NILM methods.

| No. | Author | Method | Building Type | Inputs | Output | Reference |
|---|---|---|---|---|---|---|
| 1 | Rafsanjani et al. | Machine learning | Commercial building | Occupant, minimum numbers of data points, entry events, and departure events | Occupants' power changes | [23,24] |
| 2 | Kaselimi et al. | Machine learning | Residential building | Aggregate energy signals over a time window | The consumption of electric appliances | [25] |
| 3 | Schirmer et al. | Machine learning | Residential building | Statistical features, line currents, line voltages, and load angles | The consumption of electric appliances | [26] |
| 4 | Kaselimi M | Machine learning | Residential building | Current, active power, reactive power, and apparent power | The current state of appliances | [27] |
| 5 | Yang D | Machine learning | Residential building | Characteristics of residential appliance | The consumption of electric appliances | [28] |
| 6 | Faustine A | Machine learning | Residential building | Aggregate power signal and power signals | The consumption of electric appliances | [29] |
| 7 | Xia et al. | Machine learning | Residential building | Original main power data and differential processing data | The consumption of electric appliances | [30] |
| 8 | Guo et al. | Machine learning | Residential building | Two-phase current cycle changes, single-phase voltage cycle changes, and apparent power changes | The consumption of electric appliances | [31] |
| 9 | Monteiro et al. | Machine learning | Residential building | Number of connected devices, number of device combinations, and labelled current waveforms | The consumption of electric appliances | [32] |
| 10 | Samadi | Machine learning | Institutional building | Ambient temperature, workday, time of day, daylight length, intensity of sunlight, number of occupants, and humidity | Common and occupancy loads, and exterior and interior lighting | [33] |
| 11 | Xiao et al. | Machine learning | Office building | Total load, outdoor weather or coefficients of Fourier for total load | Occupant load, building envelope load, fresh air load, and equipment load. | [34] |
| 12 | Athanasiadis C | Machine learning | Residential building | Active power transient response | Active power of appliances | [35] |
| 13 | Shao et al. | Temporal motif mining | Residential building | Aggregated power observation time series | The disaggregated time series for each device | [36] |
| 14 | Burak Gunay et al. | Regression model | Commercial building | AHU schedule, outdoor temperature, AHU fan state, chiller pump state, AHU supply air pressure, AHU schedule | Occupant-controlled loads, scheduled distribution loads, and cooling loads | [37] |
| 15 | Zaeri et al. | Regression model | Commercial building | Discharge airflow rate for AHUs and occupancy data | Occupant-controlled loads and distribution loads | [38] |
| 16 | Elafoudi et al. | Dynamic time warping | Residential building | Aggregate active power data and a library of signatures for expert classification | The name of appliances, and the timestamp of the event | [39] |
| 17 | Kolter et al. | Sparse coding | Residential building | Whole home signal and models of each device's electricity consumption | The consumption of electric appliances | [40] |
| 18 | Matsui et al. | Sparse coding | Residential building | Model of each electric appliance's electricity consumption over a week | The consumption of electric appliances | [41] |
| 19 | Ji et al. | Fourier series | Commercial building | Sub-metering data | The energy consumption of HVAC terminals | [45] |

Comprehensive information and detailed monitoring data of HVAC systems are the basis for building energy analysis and system control [46]. However, most previous studies have disaggregated the total electricity consumption of residential buildings into individual devices or the total electricity consumption of public buildings into system levels only. It is difficult for most existing commercial buildings to obtain the electricity consumption of HVAC terminal units directly. Therefore, an effective method is still needed to disaggregate the electricity consumption of HVAC terminal units from mixed sub-metering data.

Previous studies on the characteristics of building electricity consumption have provided a theoretical basis for the disaggregation of mixed sub-metering data. The research conducted by Pandit et al. indicated that the electricity consumption of most commercial buildings follows an obviously periodical pattern [47]. The research by Braun et al. showed that sinusoidal functions could simulate the electricity demand in buildings [48]. They used a trigonometric formula to simulate the electricity consumption of commercial buildings. Afterward, the research by Claridge et al. indicated that the electricity consumption of lights and office equipment varied periodically and was not sensitive to outdoor weather conditions [49]. The investigations by Dhar et al. showed that for commercial buildings, electricity consumption had clear differences between weekdays and non-workdays [42–44]. Additionally, electricity consumption in different meteorological conditions should be treated differently. In our own previous research work, we used EnergyPlus to simulate the energy consumption of HVAC terminal units of a whole building. However, this method is time consuming because a detailed architectural model needs to be established and calibrated. When applied in another building, the model needs to be rebuilt [50].

It is a common practical problem that the electricity consumption of HVAC terminal units is mixed with lighting-plug sub-metering systems in most public buildings in China. This is one of the reasons why sub-metering data quality is poor in China. Many previous research studies focused on electricity consumption disaggregation for residential buildings, while only few research studies discussed the problem for public buildings. Some existing research studies adopted a Fourier series model in sub-metering data disaggregation. However, this algorithm is not suitable for categorical data. Different building types may require different date-type categorization methods. Therefore, in this study, we adopted a CART algorithm to disaggregate the real-time electricity consumption of the terminal units in a HVAC system, and we proposed a general calculation process to make it easier for other users, which would improve the data quality of sub-metering data in building energy management. The main novelties of this study are listed as follows:

(1) This study adopts a CART algorithm to disaggregate the real-time electricity consumption of HVAC terminal units. It makes up for the deficiency that the Fourier series model is not suitable for categorical data.
(2) A general disaggregation framework is proposed. It can be easily extended to different cases without constructing physical building energy models.

In this article, the methodology and model establishment processes are described in Section 2. In Section 3, two case studies in Shanghai are used to illustrate how to apply the proposed method. Discussions and limitations are presented in Section 4, and the conclusions are outlined in Section 5.

## 2. Methodology

### 2.1. Principle of CART Algorithm

Decision tree is a widely used data mining method because of its high interpretability [51]. A decision tree model consists of root, internal and leaf nodes. The indicators of a decision tree model for picking branches are information gain, entropy and Gini impurity. Figure 2 shows a schematic diagram of a decision tree model. It uses a top-down recursive method at each node on the sample set to select the branch attribute according to the given criteria from a root node to a leaf node. A leaf node represents an objective function value, which is determined by the input variables along the path from the root node to the leaf node.

Compared to other data mining methods, the decision tree algorithm has some strong points that make it suitable for both classification and regression problems [52]: (1) it is simple to understand and interpret; (2) it is able to process both numerical and categorical data; (3) it requires little data preparation; (4) it can validate a model using statistical tests; (5) it uses a non-statistical approach that makes no assumptions of the training data or prediction residuals; and (6) it performs well with large datasets.

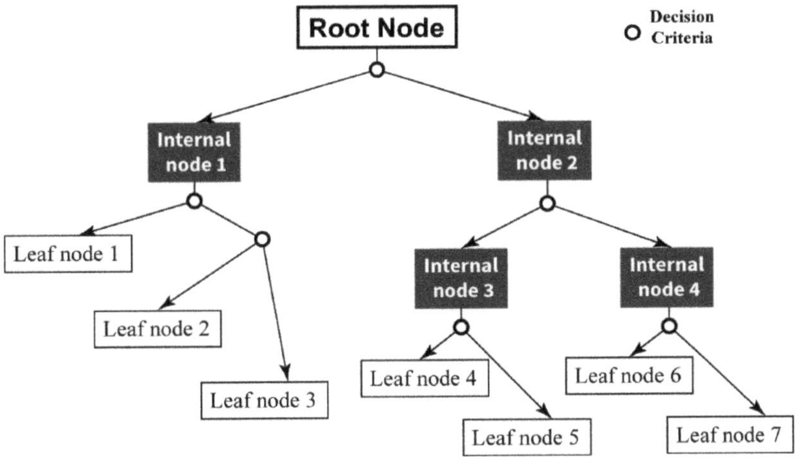


**Figure 2.** Schematic diagram of a decision tree model.

CART algorithms, first proposed by Breiman in 1984 [53,54], is a specific decision tree algorithm that is currently widely used. For classification problems, "Gini impurity" is used as the criteria for choosing branches, while for regression problems, mean squared error (MSE) is used.

Gini impurity represents the probability that a randomly chosen sample is misclassified in the sample set. Gini impurity can be calculated as follows:

$$Gini(A) = 1 - \sum_{k=1}^{n} P_k^2 \tag{1}$$

where $A$ represents node $A$; $n$ is the number of items in a dataset, $i = 1, 2, 3, \ldots, n$; and $P_k$ is the fraction of item $i$ in the set.

For regression problems, the goal of CART is to minimize MSE. MSE represents the difference between the predicted value and the actual value in a leaf node, and the calculation formula is as follows:

$$\mathrm{MSE} = \frac{1}{n}\sum_{i=1}^{n}(y_i - \hat{y}_i)^2 \tag{2}$$

where $n$ is the number of items in a dataset, $i = 1, 2, 3, \ldots, n$; $y_i$ is the actual value; and $\hat{y}_i$ is the predicted value.

The right part of Figure 3 illustrates the procedure of a CART model, which can be referred to in [53–55] for details. The CART algorithm has the dual functions of classification and regression, which meets the research needs of this paper.

### 2.2. *Establishment of an Extended CART Algorithm for HVAC Sub-Metering Systems*

Previous studies [42–44,49] have indicated that electricity consumption of lighting-plug systems in commercial buildings presents periodical changes and is not related to both outdoor weather conditions and indoor occupancy rate. Hence, electricity consumption of lighting plugs follows the same pattern over a year, while that of a HVAC system is seasonal (only in summer and winter). Therefore, during a transition season when the HVAC system is closed, the electricity consumption measured by lighting-plug circuits is all lighting-plug electricity consumption. For these reasons, real-time lighting-plug sub-meter data in transition seasons can be used to train a CART model, which can then be used to predict the lighting-plug sub-meter electricity consumption in air-conditioning seasons.

Thus, the electricity consumption of HVAC terminal units in air-conditioning seasons can be calculated by the following equation:

$$E_{ter} = E_{mix} - E_{L_{cal}} \tag{3}$$

where $E_{L_{cal}}$ is the calculated lighting-plug electricity consumption during an air-conditioning season (kWh); $E_{mix}$ is the mixed electricity consumption of the lighting plug and HVAC terminal units that is directly collected from the sub-metering system(kWh); and $E_{ter}$ is the electricity consumption of the HVAC terminal units (kWh).

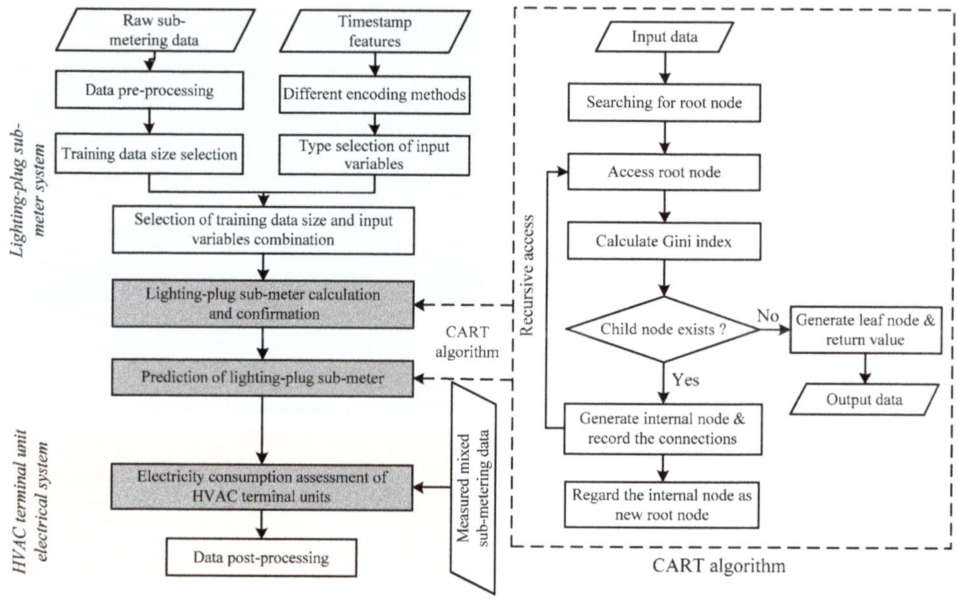


**Figure 3.** The establishment procedure of a CART model.

This procedure is demonstrated on the left side of Figure 3. The model includes two main modules: (1) the lighting-plug electricity consumption prediction module, and (2) the calculation module for the electricity consumption of HVAC terminal units. The main steps are described in the following sections.

### 2.2.1. Data Pre-Processing

Data pre-processing consists of data cleaning and data transformation. Data cleaning is an important part of data pre-processing. Due to various faults in data collection, transmission and storage procedures, there are a lot of abnormal values, such as missing value, duplicates and outliers, in sub-metering data. Therefore, the data from sub-meters should be cleaned before training models. In this study, a box graph is used to find abnormal data, and linear interpolation is used to fill the missing data. For the data cleaning process, see [55,56] for details.

Commonly, the measurement intervals of sub-metering data are 1–15 min, and the recorded data are instantaneous power. Therefore, in order to facilitate the use of the model, instantaneous power should be converted to hourly electricity consumption. The calculation is shown in Formula (4):

$$E = \sum_{i=1}^{\frac{60}{\tau}} \left( P_i \times \frac{\tau}{60} \right) \tag{4}$$

where $E$ is the electricity consumption (kWh); $P_i$ is the instantaneous power (kW); and $\tau$ is the measurement interval (min).

### 2.2.2. Input Variable Selection

Because electricity consumption of lighting plugs changes periodically, timestamp features are selected as the input variables of the CART model. Table 2 demonstrates different encoding methods and corresponding symbols of timestamp features.

**Table 2.** Summary of symbols and encoding methods of timestamp features.

| Feature | Symbol | Code | Explanation |
|---|---|---|---|
| Date | D | 1–365 (or 366) | Days in one year |
| Month | M | 1–12 | Months from January to December |
| Date Type | T1 | 0–8 | Monday–Sunday (1–7), holiday (8), and compensated workday (0). |
| | T2 | 0–2 | Weekday (1), weekend (0), and holiday (2) |
| | T3 | 0–1 | Workday (1) and non-workday (0) |
| Hour | H1 | 0–24 | Hours in one day |
| | H2 | 8–17, 0 | For office building, service hours (8:00–17:00) are labeled as 8–17, other hours (0) |
| | | 10–22, 0 | For shopping mall, service hours (10:00–22:00) are labeled as 10–22, other hours (0) |
| | | 8–22, 0 | For office–shopping complex building service hours (8:00–22:00) are labeled as 8–22, other hours (0) |
| | | —— | The special serve time is adjusted according to the actual building information |

Features in Table 2 can be combined into different input patterns, as summarized in Table 3. For example, No. 1 means to select D, M, T1 and H1 as the model inputs.

**Table 3.** Summary of different schedule input combinations.

| No. | Input Combination |
|---|---|
| 1 | D-M-T1-H1 |
| 2 | D-M-T1-H2 |
| 3 | D-M-T2-H1 |
| 4 | D-M-T2-H2 |
| 5 | D-M-T3-H1 |
| 6 | D-M-T3-H2 |
| 7 | M-T1-H1 |
| 8 | T1-H1 |
| 9 | T2-H1 |
| 10 | T3-H1 |
| 11 | H1 |

In order to select the best input combinations, a CART model was built for each combination. The coefficient of variability (CV, Equation (5)), the weighted mean absolute percentage error (WMAPE, Equation (6)) and the mean time of multiple calculations are chosen as the criteria to select the best input combinations:

$$CV = \frac{\sqrt{\frac{\left[\sum_{i=1}^{N}(E_{Mi}-E_{Pi})^2\right]}{N}}}{\frac{\left(\sum_{i=1}^{N}E_{Mi}\right)}{N}} \quad (5)$$

$$WMAPE = \frac{\sum_{i=1}^{N} \mid E_{Mi} - E_{Pi} \mid}{\sum_{i=1}^{N} E_{Mi}} \quad (6)$$

where $E_{Mi}$ is the metered electricity consumption of the $i$th data point (kWh); $E_{Pi}$ is the predicted electricity consumption of the $i$th data point (kWh); and $N$ is the total point number of the dataset.

2.2.3. Training Period Selection

This section discusses the training period required for model training. Electricity consumption of the lighting-plug system in a transition season is used for training the CART model, and the data in an air-conditioning season are used for testing. In China, cooling seasons are usually from June to September, while heating seasons occur from December to March in the following year. Therefore, data in April, May, October and November can be used in the training process. Table 4 shows the different training periods discussed in this paper.

**Table 4.** Summary table of the training data size selection.

| No. | Training Period | Description |
|---|---|---|
| 1 | 4 months | whole transition season data |
| 2 | 2 months | 1-month data in spring and 1-month data in fall |
| 3 | 4 weeks | 2-week data in spring and 2-week data in fall |
| 4 | 2 weeks | 2-week data in transition season |
| 5 | 1 week | 1-week data in transition season |
| 6 | 2 days | 1-workday data and 1-non-workday data |

2.2.4. Electricity Consumption Prediction of Lighting-Plug System

In this study, the electricity consumption of a lighting-plug system and corresponding timestamp features in a transition season are used to train a CART model. The data in an air-conditioning season are the testing data. CV and WMAPE are chosen to evaluate the prediction results.

2.2.5. Electricity Consumption Calculation of HVAC Terminal Units

The real-time electricity consumption of HVAC terminal units can be calculated by using Equation (3). Similarly, CV and WMAPE are chosen to evaluate the disaggregation results. The relative error is also selected for evaluation, and the calculation formula is as follows:

$$RE = \frac{E_{Mi} - E_{Pi}}{E_{Mi}} \times 100\% \tag{7}$$

## 3. Case Study

The model established in Section 3 is demonstrated and tested in this section. It is applied to two actual buildings in Shanghai where the electricity consumption of HVAC terminal units can be measured directly. Basic facilities and operation information about the two buildings are shown in Table 5.

### *3.1. Data Pre-Processing*

In this part, the raw data are cleaned by removing outliers and divided into a training set and a testing set. Table 6 summarizes the data collected from the two buildings and describes the training data and testing data, respectively. Because the sub-metering system has not been used since 2014, the latest sub-metering data of Building A could not be obtained. Therefore, for Building A, data in 2013 are used in the case study, while for Building B, data in 2021 are used. The training data are then re-organized based on seasons and periods according to Table 4.

**Table 5.** Structural information for the two buildings in Shanghai used in this study.

| Building Name | Building A | Building B |
|---|---|---|
| City | Shanghai | Shanghai |
| Building type | Complex building | Shopping mall |
| Floors | Shopping mall (1–4) and offices (5–34) | 9 |
| Area ($m^2$) | 68,330 | 40,000 |
| Service time | Shopping mall: Full year 10:00–22:00;<br>Office: Weekdays 8:00–18:00 | Full year: 10:00–22:00 |
| HVAC terminal units | AHU (shopping mall) and FCU+FAU (office) | AHU |
| Energy type | Electric | Electric |

**Table 6.** Selected electrical input data from two buildings studied in Shanghai.

| Building Name | Data of Lighting-Plug System | Training Data | Testing Data | |
|---|---|---|---|---|
| | | Data of Transition Season | Data of HVAC Terminal Units | |
| | | | Cooling Season | Heating Season |
| A | 1 January–31 December 2013 | 1 April–31 May<br>1 October–30 November | 1 June–30 September | 1 January–31 March<br>1 December–31 December |
| B | 1 January–31 December 2021 | 1 April–31 May<br>1 October–30 November | 1 June–30 September | 1 January–31 March<br>1 December–31 December |

*3.2. Input Variable Selection*

The prediction accuracy of the model for the electricity consumption of the lighting-plug system during air-conditioning on and off time are considered separately. As shown in Table 5, the working period of Building A are from 8:00 to 22:00, and the working period of Building B are from 10:00 to 22:00.

Tables 7 and 8 list the WMAPE, CV and a multiple-computing-mean time $\overline{T}$ (average time for 1000 runs) for different input combinations. The WMAPE and CV indices are calculated separately for three periods: working hours in the cooling season (WHC), working hours in the heating season (WHH), and non-working hours (NW).

**Table 7.** Test results for lighting-plug system in different schedule input patterns in Case A.

| Model No. | Input Combination | WMAPE (%) | | | CV (%) | | | $\overline{T}$ (s) |
|---|---|---|---|---|---|---|---|---|
| | | WHC | WHH | NW | WHC | WHH | NW | |
| 1 | D-M-T1-H1 | 1.00 | 1.22 | 1.89 | 1.39 | 1.91 | 2.75 | 0.1780 |
| 2 | D-M-T1-H2 | 1.00 | 1.22 | 5.20 | 1.39 | 1.91 | 7.08 | 0.1432 |
| 3 | D-M-T2-H1 | 1.09 | 1.18 | 2.82 | 1.50 | 1.79 | 4.58 | 0.2503 |
| 4 | D-M-T2-H2 | 1.09 | 1.18 | 6.09 | 1.50 | 1.79 | 7.94 | 0.1892 |
| 5 | D-M-T3-H1 | 1.09 | 1.18 | 2.88 | 1.49 | 1.79 | 4.58 | 0.2090 |
| 6 | D-M-T3-H2 | 1.09 | 1.18 | 6.13 | 1.49 | 1.79 | 7.95 | 0.1706 |
| 7 | M-T1-H1 | 1.55 | 1.88 | 3.38 | 2.18 | 2.81 | 5.06 | 0.1394 |
| 8 | T1-H1 | 2.30 | 3.34 | 4.57 | 2.99 | 4.60 | 6.56 | 0.0195 |
| 9 | T2-H1 | 2.38 | 3.48 | 5.96 | 3.16 | 4.84 | 8.30 | 0.0148 |
| 10 | T3-H1 | 2.38 | 3.48 | 6.25 | 3.16 | 4.84 | 8.65 | 0.0140 |
| 11 | H1 | 12.28 | 14.02 | 22.82 | 15.42 | 18.23 | 37.31 | 0.0130 |

In Table 7, Models No. 1 and 2, No. 3 and 4, and No. 5 and 6 are considered as three groups for the comparative analysis. The comparison results show that in each group, the prediction accuracy in working hours is the same, while in the non-working period, the WMAPE and CV of the models with the input H1 are 3.31% and 4.33%, respectively, which are less than those of the models with the input H2. Additionally, when comparing Models No. 8, 9 and 10, the prediction accuracy is very close, while Model No. 10 has the fastest calculation speed. Therefore, T3 is chosen as the date-type encoding form. The calculation speed of Model No. 11 is the highest with 0.0130 s, but the accuracy is significantly lower than the other models. When the model operates during working hours, the WMAPE and CV values of Model No. 11 exceed 12%, while these two criteria of the other models are all within 5%. Model No. 1 is the most accurate; however, its speed is nearly 14 times slower than Model No. 11. Taking model accuracy and running speed into consideration, Model No. 10 is most suitable for Building A. In summary, the electricity consumption prediction model of the lighting-plug system in Building A is mainly influenced by two variables: date type and hour.

**Table 8.** Test results for lighting-plug system in different schedule input patterns in Case B.

| Model No. | Input Combination | WMAPE (%) | | | CV (%) | | | $\bar{T}$ (s) |
|---|---|---|---|---|---|---|---|---|
| | | WHC | WHH | NW | WHC | WHH | NW | |
| 1 | D-M-T1-H1 | 0.82 | 0.74 | 8.15 | 2.90 | 1.07 | 12.41 | 0.0529 |
| 2 | D-M-T1-H2 | 0.82 | 0.74 | 20.79 | 2.90 | 1.07 | 32.20 | 0.0970 |
| 3 | D-M-T2-H1 | 0.82 | 0.75 | 7.93 | 2.90 | 1.09 | 11.99 | 0.0508 |
| 4 | D-M-T2-H2 | 0.82 | 0.75 | 20.80 | 2.90 | 1.09 | 32.20 | 0.0809 |
| 5 | D-M-T3-H1 | 0.83 | 0.75 | 7.88 | 2.91 | 1.08 | 11.95 | 0.0544 |
| 6 | D-M-T3-H2 | 0.83 | 0.75 | 20.79 | 2.91 | 1.08 | 32.20 | 0.0893 |
| 7 | M-T1-H1 | 2.22 | 1.58 | 11.83 | 6.97 | 5.55 | 18.46 | 0.0653 |
| 8 | T1-H1 | 3.26 | 4.07 | 16.47 | 7.54 | 8.75 | 23.31 | 0.0127 |
| 9 | T2-H1 | 3.17 | 4.05 | 16.53 | 8.88 | 7.51 | 23.50 | 0.0098 |
| 10 | T3-H1 | 3.16 | 4.08 | 16.41 | 8.84 | 7.71 | 23.56 | 0.0101 |
| 11 | H1 | 3.14 | 4.09 | 16.48 | 8.84 | 7.77 | 23.59 | 0.0084 |

The test results of Building B are displayed in Table 8. The evaluation criteria of the model during the WHC and WHH periods of the two buildings are similar. Additionally, during non-working hours, the model of Building A performs better than that of Building B.

Similar to Building A, the results of Model No. 1 to Model No. 6 indicate that in working hours, the predictive accuracy is similar, while in non-working period, the WMAPE and CV of the models with the input H1 are 12.64% and 17.79%, respectively, which are less than those of the models with the input H2. The WMAPE and CV of Model No. 8, Model No.9 and Model No.10 are close. Compared to the above three models, the performance of Model No. 11 is also satisfactory, and the calculation speed is the fastest. In working hours, the WMAPE values of all models are less than 5% and the CV values of all models are less than 9%. Similar to the results of Building A, Model No. 1 is the most accurate, but the calculation speed is more than six times slower than that of model No. 11. Model No. 7 has no significant advantage in both accuracy and calculation speed. Therefore, when comparing the test results of the 11 models, Model No. 11 has the fastest calculation speed while ensuring predictive accuracy. That means the electricity consumption prediction model of the lighting-plug system in Building B is mainly influenced by hour.

### 3.3. Training Period Selection

Based on Section 3.2, Model No. 10 is selected for Building A and Model No. 11 is selected for Building B. This section discusses which training period is best for the models in both buildings. The test results of the models with different training periods are shown in Figures 4 and 5. The horizontal axis of the picture refers to the training periods used in the model, and the vertical axis represents the WMAPE and CV of the model with different training periods.

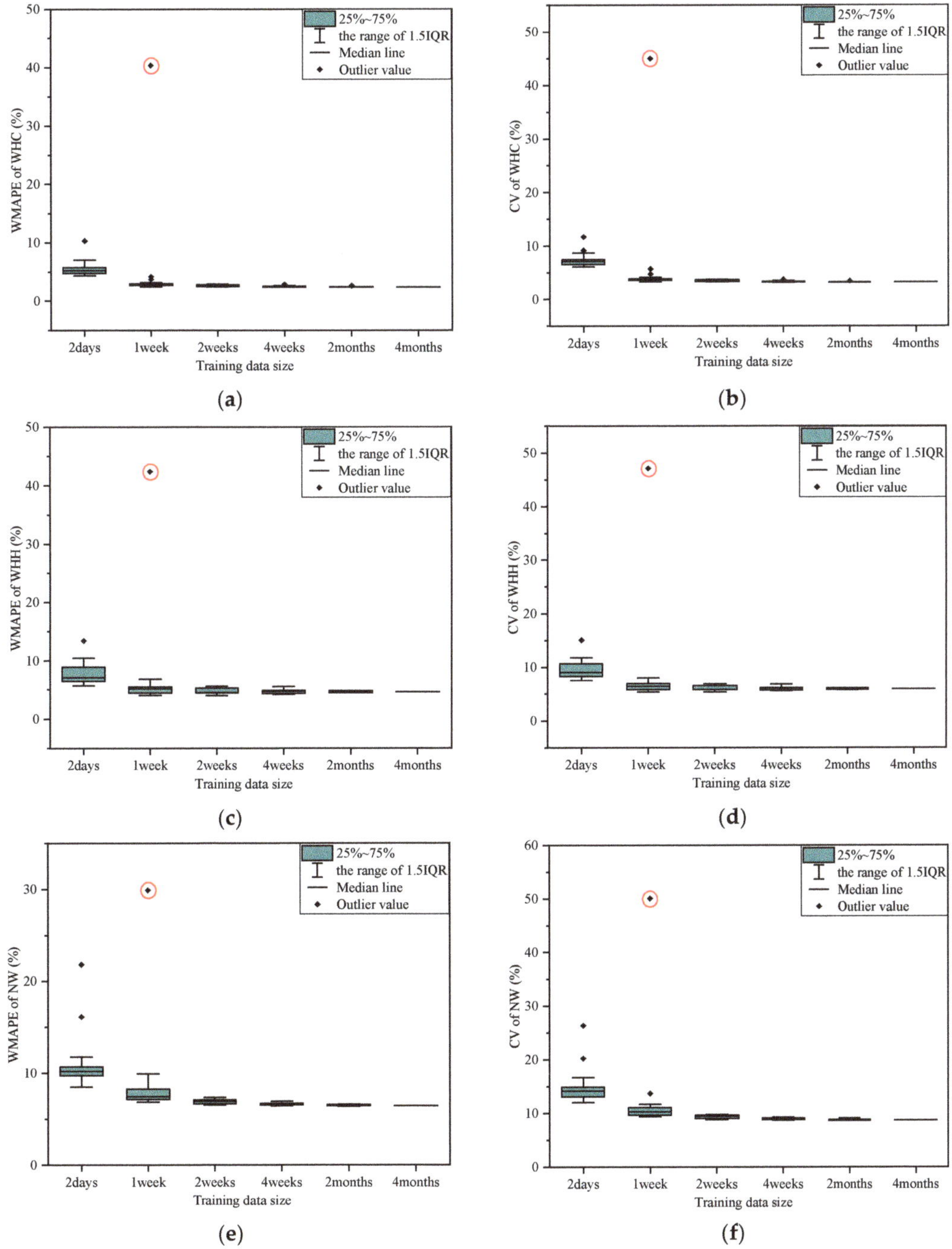


**Figure 4.** Case A test results with different training data sizes: (**a**) the range of WMPE varies during the WHC; (**b**) the range of CV varies during the WHC; (**c**) the range of WMPE varies during the WHH; (**d**) the range of CV varies during the WHH; (**e**) the range of WMPE varies during the NW; and (**f**) the range of CV varies during the NW.

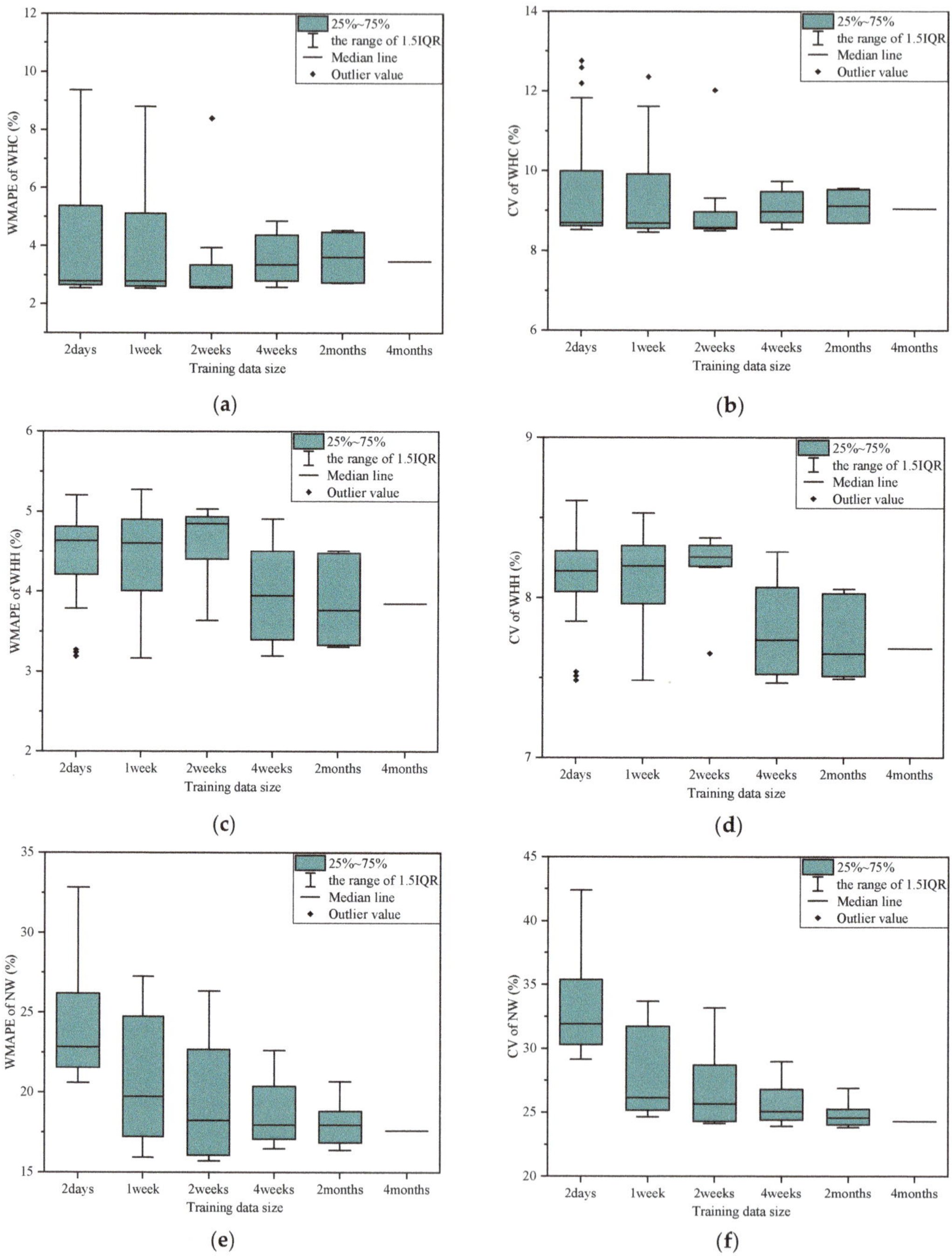


**Figure 5.** Test results of Building B with different training periods: (**a**) the range of WMPE varies during the WHC; (**b**) the range of CV varies during the WHC; (**c**) the range of WMPE various during the WHH; (**d**) the range of CV various during the WHH; (**e**) the range of WMPE various during the NW; and (**f**) the range of CV various during the NW.

The prediction results for Building A displayed in Figure 4 show that training period has an influence on the accuracy of the model, especially during the NW period. The results shown in (a) and (b) indicate that the model prediction effect during the WHC period is the worst when the 2-day data are used to train the model, and the prediction effects with the other five training periods are close. Figure 4c–e illustrate that during the WHH and NW periods, when the model uses the 1-week and 2-day data as the model training sets, the estimation errors are greater than the errors using the other four training periods. The prediction results of models during the WHC and WHH periods are better than those of the NW period.

In the case of the 1-week training period, there is an outlier shown in the red circle. This is caused by using the data during the National Day Holiday (1 October–7 October 2013) as the training set. The input parameters used by Building A are date type and hour. Since the days during this period are all non-working time, the data in working hours cannot be trained. Thus, the model estimation errors are extremely large.

When the model uses the 2-das data as the training set, there are two outliers in the WMAPE of the model during the NW period. This training set corresponds to the combination of one day in the National Day Holiday and a shift day, which is on a weekend but is shifted to a workday (12 October 2013, Saturday). Due to the existence of holidays and shift days, the electricity consumption of normal workdays shows the features of non-workdays, while the regularity of electricity consumption on normal weekends shows the features of workdays, so the estimation errors of the model is relatively high.

Figure 4 shows that the model can achieve good prediction performance when the training period is four months, two months, four weeks and two weeks. There is no significant difference among these four different periods. Therefore, the 2-week data are selected as the training period for Building A.

Figure 5 shows the estimation errors for Building B. Figure 5a,b show that for WHC, the prediction accuracy of the model trained with the 2-week data is better than the other models. For WHH, as shown in Figure 5c,d, the model with the 2-week training period does not performs well. Figure 5e,f show that the model has the worst prediction accuracy during the NW period, especially when it is trained with the 2-day data. The input variable of Model No. 11 used by Building B is only hour, so other input patterns do not have significant impacts on model accuracy. Since Building B is a shopping mall that is open all year, the building's electricity consumption does not show significant differences among different date types. Therefore, when the National Day Holiday data are used as the training set, the accuracy of the model in Building B does not decrease significantly. To sum up, the 2-week data are selected as the model training set.

### 3.4. Electricity Consumption Disaggregation Results of Lighting-Plug System

Based on the results of Sections 3.2 and 3.3, "T3-H1" is selected as the input combination and the 2-week data are used as the training set for the model in Case A. In Case B, "H1" is selected as the input variable and the 2-week data are used as the training set. In order to verify the accuracy of the model established in this study, the data sets with the best performance and the worst performance are selected to predict the hourly electricity consumption of the lighting-plug system separately. In Case A, the training set with the best performance is from 1 November to 14 November 2013, and the training set with the worst performance is from 15 April to 28 April 2013. In Case B, the training set with the best performance is from 18 May to 31 May 2021, and the training set with the worst performance is from 6 April to 19 April 2021.

Table 9 shows the model performance of Building A and Building B. For Case A, the WMAPE values of the best performance model during the WHC, WHH and NW are 2.50%, 4.42% and 6.59%, and the CV values are 3.35%, 5.75 % and 9.11%, respectively. The WMAPE values of the worst performance model during the WHC, WHH and NW are 2.94%, 5.62% and 7.15%, and the CV values are 3.82%, 6.87% and 9.75%, respectively. For Case B, the best WMAPE values during the WHC, WHH and NW are 2.57%, 4.81% and

16.00%, and the best CV values are 8.59%, 8.23% and 24.13%, respectively, while the worst WMAPE values are 5.55%, 4.89% and 21.09%, and the worst CV values are 8.27%, 8.54% and 27.23%, respectively.

**Table 9.** The input parameters, training periods and disaggregation estimation errors of the model in Case A and Case B.

| Case | Input Pattern | Training Data Size and Performance | WMAPE (%) | | | CV (%) | | |
|---|---|---|---|---|---|---|---|---|
| | | | WHC | WHH | NW | WHC | WHH | NW |
| A | T3-H1 | 2 weeks (best) (1 November–14 November 2013) | 2.50 | 4.42 | 6.59 | 3.35 | 5.75 | 9.11 |
| | | 2 weeks (worst) (15 April–28 April 2013) | 2.94 | 5.62 | 7.15 | 3.82 | 6.87 | 9.75 |
| B | H1 | 2 weeks (best) (18 May–31 May 2021) | 2.57 | 4.81 | 16.00 | 8.59 | 8.23 | 24.13 |
| | | 2 weeks (worst) (6 April–19 April 2021) | 5.55 | 4.89 | 21.09 | 8.27 | 8.54 | 27.23 |

### 3.5. Electricity Consumption Calculation of HVAC Terminal Units

This section verifies the accuracy of the method proposed in this research to predict the hourly electricity consumption of HVAC terminal units. The model disaggregation results of the best training set for the two buildings are shown in Figure 6. Additionally, Table 10 summarizes the WMAPE and CV values of the electricity consumption disaggregation results of the HVAC terminal units in the two buildings. Figures 7 and 8 show the probability cumulative curves of the relative error of all data points.

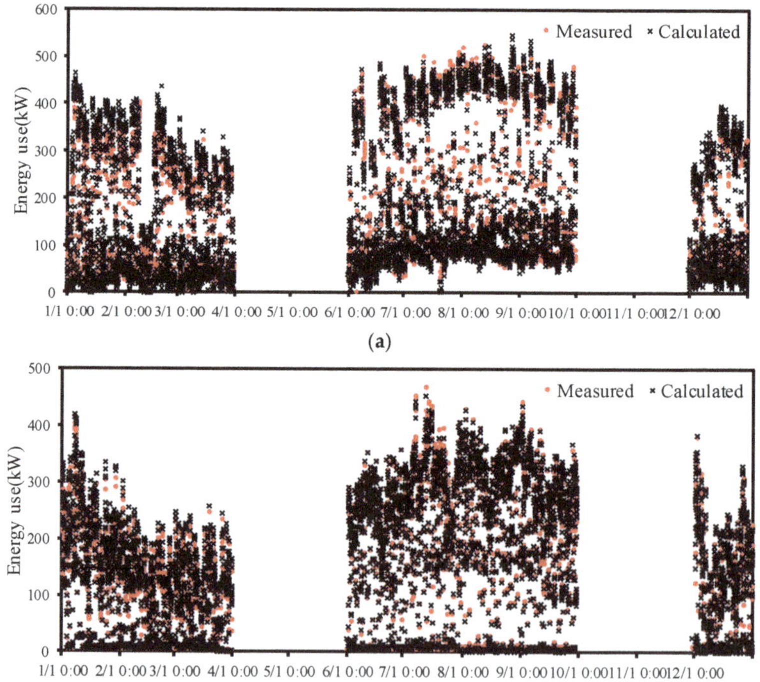


**Figure 6.** Energy consumption disaggregation results of HVAC terminal units: (**a**) Building A with the best training set, and (**b**) Building B with the best training set.

**Table 10.** Electricity consumption disaggregation results of HVAC terminal units.

| Building | Performance | WMAPE (%) | | | CV (%) | | |
|---|---|---|---|---|---|---|---|
| | | WHC | WHH | NW | WHC | WHH | NW |
| A | Best | 3.79 | 10.05 | 21.90 | 5.09 | 13.07 | 30.26 |
| | Worst | 4.47 | 12.76 | 23.75 | 5.80 | 15.60 | 32.39 |
| B | Best | 2.25 | 8.34 | 6.47 | 7.54 | 14.28 | 9.75 |
| | Worst | 2.24 | 8.48 | 5.83 | 7.50 | 14.35 | 11.01 |

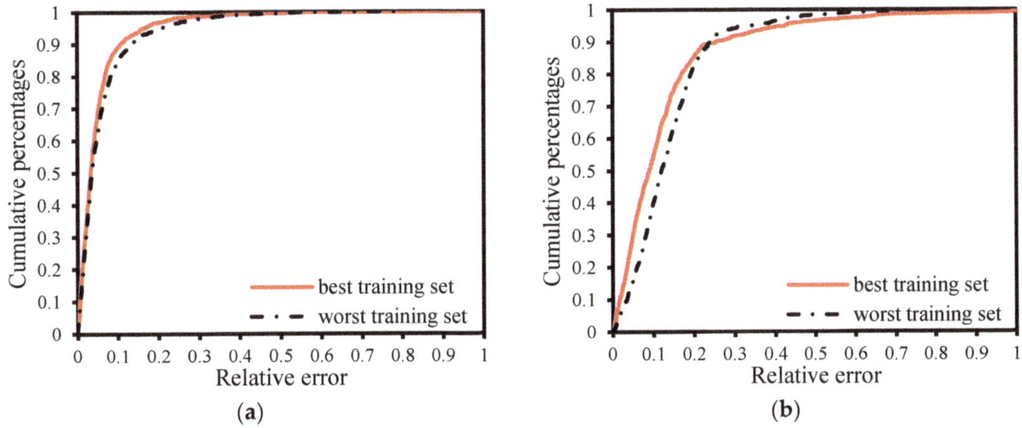


**Figure 7.** Building A: the probability cumulative curve of the samples at different relative error levels. (**a**) Cooling season. (**b**) Heating season.

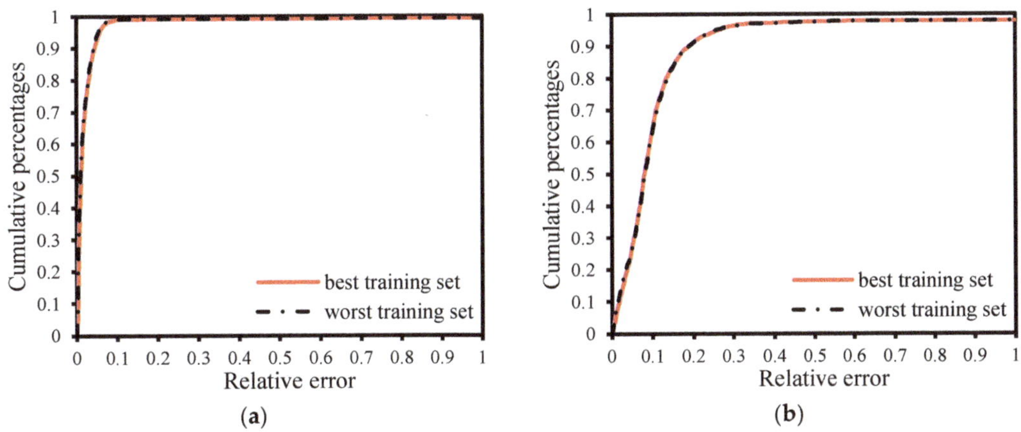


**Figure 8.** Building B: the probability cumulative curve of the samples at different relative error levels. (**a**) Cooling season. (**b**) Heating season.

In Figure 6, the red dots are actual electricity consumption of the HVAC terminal units during working hours in the two buildings, while the black crosses indicate the electricity consumption during working hours predicted by the CART model established in this study. It can be seen that the predicted values change consistently with the actual values.

For Building A, the minimum WMAPE values during the WHC, WHH and NW are 3.79%, 10.05% and 21.90%, and the minimum CV values are 5.09%, 13.07% and 30.26%, respectively, while the maximum WMAPE values during the WHC, WHH and NW are 4.47%, 12.76% and 23.75%, and the maximum CV values are 5.80%, 15.60% and 32.39%, respectively.

The probability cumulative curve of the samples at different relative error levels is shown in Figure 7. There are 1274 testing samples for the cooling season and 1227 testing samples for the heating season. During the WHC, the cumulative percentage of samples with a relative error lower than 15% is 93.7% when using the best training set, while it is 91.7% when using the worst training set. During the WHH, for the best data set, 75.6% of the samples' relative errors are less than 15% and more than 91.8% of the samples' relative errors are less than 30%. For the worst data set, 63.9% of the samples' relative errors are less than 15% and 93.9% of the samples' relative errors are less than 30%. Considering Table 10 and Figure 7 comprehensively, the following conclusions can be drawn for Case A: (A1) models with the best and the worst training sets have little difference; (A2) the disaggregation accuracy for the cooling season is better than the heating season; and (A3) the predication accuracy for working hours is much better than non-working hours.

Similarly, for Building B, the minimum WMAPE values during the WHC, WHH and NW are 2.25%, 8.34% and 6.47%, and the minimum CV values are 7.54%, 14.28% and 9.75%, respectively. The maximum WMAPE values during the WHC, WHH and NW are 2.24%, 8.48% and 5.83%, and the maximum CV values are 7.50%, 14.35% and 11.01%, respectively.

The number of testing samples for Case B is 1586 in the cooling season and 1573 in the heating season. For the model with the best training set, the proportion of the samples with relative errors less than 15% is 99.1% (during WHC) and 83.8% (during WHH). For the model with the worst training set, this proportion is 99.1% (during WHC) and 83.4% (during WHH), respectively. The conclusions for Case B are as follows: (B1) during the WHC and WHH, the disaggregation resluts of these two models are similar; (B2) the model accuracy in the cooling season is better than that in the heating season; and (B3) the disaggregation accuracy of WHC and non-working hours are better than WHH.

## 4. Discussion

Based on the results of the two case studies, the application effects of the proposed disaggregation method are further discussed in this section.

The time schedule input pattern of Building A is optimized as "T3-H1" and that for Building B is "H1". This means that the model of Building A uses "Date Type" and "Hour" as the input variables, and the model of Building B only uses "Hour". Building A is a complex commercial building with both office and shopping areas, while Building B is purely a shopping mall. Thus, for buildings with office area, "Date Type" should be taken into consideration because the occupancy pattern on weekdays, weekends and holidays are quite different. Lighting system is usually off on weekends and holidays. However, for shopping malls, lighting systems follow the same pattern. Only "Hour" is used as the input.

By comparing the changes in model prediction accuracy under different training sets, it is determined that the amount of training data required for the two buildings are same. A 2-week data set can be used to train the models.

Based on the conclusions (A2) and (B2) drawn in Section 3.5, the disaggregation accuracy in the cooling season is better than in the heating season. During the WHC, the disaggregation accuracy of the model in two buildings are similar, while during the WHH, the disaggregation accuracy of the model in Building B is higher than that in Building A. The on-site survey results show that some other types of heating facilities (such as electric heaters) are occasionally used during the heating season in Building A, which affects the energy consumption regularity of the lighting-plug system to some extent. For example, some staff use electric heaters in the office when it is too cold in winter. This phenomenon incidentally affects the accuracy of the hourly electricity use calculation of the lighting-plug system and HVAC terminal units. Therefore, the disaggregation accuracy in the cooling season is better than in the heating season.

Based on the conclusions (A3) drawn in Section 3.5, the result for working hours is much better than that for non-working hours. One reason is that the electricity consumption of HVAC terminal units is very small during non-working hours. A little difference could cause large WMAPE and CV errors. Another reason is that during working hours, occupants have formed a certain behavior pattern and the lighting systems are generally turned on. However, during non-working hours, occupant behaviors are quite random. For example, sometimes some people would work overtime. It is difficult to accurately predict the electricity consumption of the HVAC terminal units caused by these random occupant behaviors. Compared to the electricity consumption of the HVAC terminal units during working hours, the consumption during non-working hours only accounts for a little part of the operating electricity consumption of the HVAC terminal units. Thus, the disaggregation results in working hours are more importance in practice.

Generally, the HVAC system keeps off during non-working hours. However, the hourly electricity consumption of the HVAC terminal units in non-working hours is still considered in this research. Our purpose is to monitor the operating schedule of the HVAC terminals using the calculated energy data. For example, if energy use appears during non-working time, it indicates that some HVAC terminal units may not be closed.

## 5. Conclusions

In order to solve the mismatch problem of sub-metering systems in public buildings, this research proposes a disaggregation method based on a CART algorithm. This method performs well in two buildings in Shanghai.

The main conclusions of this study are as follows: (1) Considering the prediction accuracy and running time of the model, T3-H1 is selected as the input parameter for the office–shopping mall complex building (Building A), and H1 is selected for the shopping mall (Building B). (2) The impact of date type on electricity consumption of office buildings is more significant than shopping malls. (3) The 2-week data set can meet the model accuracy requirements. (4) For the hourly electricity consumption disaggregation result of the HVAC terminal units in Building A, the minimum WMAPE values of WHC and WHH are 3.79% and 10.05%, and the minimum CV values are 5.09% and 13.07%, respectively. For Building B, the minimum WMAPE values of WHC and WHH are 2.25% and 8.34%, and the minimum CV values are 7.54% and 14.28%, respectively.

The advantages of this proposed method are the following: (1) It is simple and convenient to applied. The input variables in the model just need time features, which are all very easy to obtain and sufficient to support the model accuracy. According to the case study results, two-week data are enough for training a model, so it is very convenient in practical application. (2) This method is versatile. Although there are only two case studies in this paper, this method can be used in other buildings to disaggregate seasonal sub-metering data with non-seasonal sub-metering data.

The limitations of this proposed method are the following: (1) This method can only disaggregate seasonal data from non-seasonal data. It cannot disaggregate the electricity consumption of equipment or devices with the same operation patterns. (2) Other features, such as weather and occupant behaviors, can be taken into consideration in future works.

**Author Contributions:** Conceptualization, Y.J.; Software, X.Y. and Y.J.; Validation, X.Y.; Formal analysis, M.N.; Investigation, J.G.; Resources, Y.J.; Writing—original draft, Y.J.; Writing—review & editing, X.Y. and J.G.; Visualization, X.Y.; Funding acquisition, Y.J. All authors have read and agreed to the published version of the manuscript.

**Funding:** This study was funded by the Youth Program of National Natural Science Foundation of China (No. 51908006).

**Data Availability Statement:** Not applicable.

**Conflicts of Interest:** The authors declare no conflict of interest.

## Abbreviations

| | |
|---|---|
| HVAC | Heating, Ventilation and Air Conditioning |
| VAV | Variable Air Volume System |
| CART | Classification and Regression Tree |
| NILM | Non-intrusive Load Monitoring |
| EDA | End-use Disaggregation Algorithm |
| ANN | Artificial Neural Network |
| Gini | Gini Impurity |
| CV | Coefficient of Variability |
| WMAPE | Weighted Mean Absolute Percentage Error |
| WHC | Working Hours in Cooling Season |
| WHH | Working Hours in Heating Season |
| NW | Non-working Hours |
| $E_{L\_cal}$ | Calculated lighting-plug electricity consumption in the air conditioning season (kWh) |
| $E_{mix}$ | Mixed electricity consumption of the lighting plug and HVAC terminal units (kWh) |
| $E_{ter}$ | Electricity consumption of terminal units in the air conditioning system (kWh) |
| E | Electricity consumption (kWh) |
| $P_i$ | Instantaneous power (kW) |
| $\tau$ | Count cycle of instantaneous power (min). |
| D | Days in one year |
| M | Months from January to December |
| T | Date types |
| H | Equipment use time |
| $E_{Mi}$ | Metered electricity consumption of the $i^{th}$ data point (kWh) |
| $E_{pi}$ | Predicted electricity consumption of the $i^{th}$ data point (kWh) |
| N | Total point number of the dataset |
| $\overline{T}$ | Multiple-computing-mean time (s) |

## References


1. IEA. Buildings: IEA, Paris. License: CC BY 4.0. 2022. Available online: https://www.iea.org/reports/buildings (accessed on 10 December 2022.).
2. Building Energy Conservation Research Center of Tsinghua University. *Annual Report on China Building Energy Efficiency 2022*; Building Energy Conservation Research Center of Tsinghua University: Beijing, China, 2022.
3. Hart, G.W. Nonintrusive appliance load monitoring. *Proc. IEEE* **1992**, *80*, 1870–1891.
4. Howard, E. *California. Guide to the Implementation of Tenant Submetering and Billing for Commercial Buildings in Northern California*; SILO Inc.: Brooklyn, NY, USA, 2008.
5. Ministry of Finance of the People's Republic of China. *Interim Measures for the Administration of Special Funds for Energy Saving of State-owned Buildings and Large-scale Public Buildings*; Document No. 5582007; Ministry of Finance of the People's Republic of China: Beijing, China, 2007.
6. Junruo, W. Case analysis on energy consumption of large-scale public buildings in shanghai area based on partial subentry measurement system. *New Build. Mater.* **2010**, *37*, 48–50.
7. Zhang, H.W.Q.; Li, Y.; Wang, X. Research and development on electricity sub-metering system for large public buildings (Part I): System structure. *Heat. Vent. Air Cond.* **2010**, *40*, 10–13+50.
8. Wang, X.W.Q.; Shen, Q.; Zhang, H. Research and development on electricity sub-metering system for large public buildings (Part II): Uniform model and methodology of energy consumption classification. *Heat. Vent. Air Cond.* **2010**, *40*, 14–17.
9. Tongji University BEAD. *Building Energy Consumption Sub-metering Data Status and Application Seminar*; Tongji University BEAD: Shanghai, China, 2016.
10. Ministry of Housing and urban-rural development of the People's Republic of China. *Technical Guideline for Sub-Metering Data Collection of Energy Consumption Monitoring System in State-Owned Buildings and Large-Scale Public Buildings*; Document No. 114; Ministry of Housing and urban-rural development of the People's Republic of China: Beijing, China, 2008.
11. Qu, Y.; Zhang, Z.; Wang, H.; Yang, F. Energy Consumption Analysis of Public Buildings Located in the Severe Cold Region. *Procedia Eng.* **2017**, *205*, 2111–2117.
12. Guo, C.; Bian, C.; Liu, Q.; You, Y.; Li, S.; Wang, L. A new method of evaluating energy efficiency of public buildings in China. *J. Build. Eng.* **2022**, *46*, 103776.
13. Sizirici, B.; Fseha, Y.; Cho, C.-S.; Yildiz, I.; Byon, Y.-J. A Review of Carbon Footprint Reduction in Construction Industry, from Design to Operation. *Materials* **2021**, *14*, 6094. [CrossRef]

14. Granderson, J.; Lin, G.; Singla, R.; Fernandes, S.; Touzani, S. Field evaluation of performance of HVAC optimization system in commercial buildings. *Energy Build.* **2018**, *173*, 577–586. [CrossRef]
15. Marceau, M.; Zmeureanu, R. Nonintrusive load disaggregation computer program to estimate the energy consumption of major end uses in residential buildings. *Energy Convers. Manag.* **2000**, *41*, 1389–1403. [CrossRef]
16. Laughman, C.; Lee, K.; Cox, R.; Shaw, S.; Leeb, S.; Norford, L.; Armstrong, P. Power signature analysis. *IEEE Power Energy Mag.* **2003**, *1*, 56–63.
17. Giri, S.; Bergés, M. An energy estimation framework for event-based methods in Non-Intrusive Load Monitoring. *Energy Convers. Manag.* **2015**, *90*, 488–498. [CrossRef]
18. Murata, H.; Onoda, T. Estimation of power consumption for household electric appliances. In Proceedings of the Conference Estimation of Power Consumption for Household Electric Appliances, Singapore, 18–22 November 2002; Volume 5, pp. 2299–2303.
19. Liu, Q.; Nakoty, F.M.; Wu, X.; Anaadumba, R.; Liu, X.; Zhang, Y.; Qi, L. A secure edge monitoring approach to unsupervised energy disaggregation using mean shift algorithm in residential buildings. *Comput. Commun.* **2020**, *162*, 187–195. [CrossRef]
20. Dash, S.; Sahoo, N. Electric energy disaggregation via non-intrusive load monitoring: A state-of-the-art systematic review. *Electr. Power Syst. Res.* **2022**, *213*, 108673. [CrossRef]
21. Norford, L.K.; Mabey, N. *Non-Intrusive Electric Load Monitoring in Commercial Buildings. Symposium on Improving Building Systems in Hot and Humid Climates*; Texas A&M University: College Station, TX, USA, 1992.
22. Norford, L.K.; Leeb, S.B. Non-intrusive electrical load monitoring in commercial buildings based on steady-state and transient load-detection algorithms. *Energy Build.* **1996**, *24*, 51–64. [CrossRef]
23. Rafsanjani, H.N.; Ahn, C. Linking Building Energy-Load Variations with Occupants' Energy-Use Behaviors in Commercial Buildings: Non-Intrusive Occupant Load Monitoring (NIOLM). *Procedia Eng.* **2016**, *145*, 532–539. [CrossRef]
24. Rafsanjani, H.N.; Ahn, C.R.; Chen, J. Linking building energy consumption with occupants' energy-consuming behaviors in commercial buildings: Non-intrusive occupant load monitoring (NIOLM). *Energy Build.* **2018**, *172*, 317–327. [CrossRef]
25. Kaselimi, M.; Doulamis, N.; Doulamis, A.; Voulodimos, A.; Protopapadakis, E. Bayesian-optimized Bidirectional LSTM Regression Model for Non-intrusive Load Monitoring. In Proceedings of the ICASSP 2019–2019 IEEE International Conference on Acoustics, Speech and Signal Processing (ICASSP), Brighton, UK, 12–17 May 2019; pp. 2747–2751.
26. Schirmer, P.A.; Mporas, I.; Paraskevas, M. Evaluation of Regression Algorithms and Features on the Energy Disaggregation Task. In Proceedings of the 2019 10th International Conference on Information, Intelligence, Systems and Applications (IISA), Patras, Greece, 15–17 July 2019; pp. 1–4.
27. Kaselimi, M.; Protopapadakis, E.; Voulodimos, A.; Doulamis, N.; Doulamis, A. Multi-Channel Recurrent Convolutional Neural Networks for Energy Disaggregation. *IEEE Access* **2019**, *7*, 81047–81056. [CrossRef]
28. Yang, D.; Gao, X.; Kong, L.; Pang, Y.; Zhou, B. An Event-Driven Convolutional Neural Architecture for Non-Intrusive Load Monitoring of Residential Appliance. *IEEE Trans. Consum. Electron.* **2020**, *66*, 173–182. [CrossRef]
29. Faustine, A.; Pereira, L.; Bousbiat, H.; Kulkarni, S. UNet-NILM: A Deep Neural Network for Multi-Tasks Appliances State Detection and Power Estimation in NILM. In Proceedings of the 5th International Workshop on Non-Intrusive Load Monitoring NILM'20 Association for Computing Machinery, New York, NY, USA, 18 November 2020; pp. 84–88.
30. Xia, M.; Liu, W.; Wang, K.; Song, W.; Chen, C.; Li, Y. Non-intrusive load disaggregation based on composite deep long short-term memory network. *Expert Syst. Appl.* **2020**, *160*, 113669. [CrossRef]
31. Guo, Y.; Xiong, X.; Fu, Q.; Xu, L.; Jing, S. Research on non-intrusive load disaggregation method based on multi-model combination. *Electr. Power Syst. Res.* **2021**, *200*, 107472. [CrossRef]
32. Monteiro, R.; de Santana, J.; Teixeira, R.; Bretas, A.; Aguiar, R.; Poma, C. Non-intrusive load monitoring using artificial intelligence classifiers: Performance analysis of machine learning techniques. *Electr. Power Syst. Res.* **2021**, *198*, 107347. [CrossRef]
33. Samadi, M.; Fattahi, J. Energy use intensity disaggregation in institutional buildings—A data analytics approach. *Energy Build.* **2021**, *235*, 110730. [CrossRef]
34. Xiao, Z.; Gang, W.; Yuan, J.; Zhang, Y.; Fan, C. Cooling load disaggregation using a NILM method based on random forest for smart buildings. *Sustain. Cities Soc.* **2021**, *74*, 103202. [CrossRef]
35. Athanasiadis, C.; Doukas, D.; Papadopoulos, T.; Chrysopoulos, A. A Scalable Real-Time Non-Intrusive Load Monitoring System for the Estimation of Household Appliance Power Consumption. *Energies* **2021**, *14*, 767. [CrossRef]
36. Shao, H.; Marwah, M.; Ramakrishnan, N. A Temporal Motif Mining Approach to Unsupervised Energy Disaggregation: Applications to Residential and Commercial Buildings. *Proc. Conf. AAAI Artif Intell* **2013**, *27*, 1327–1333. [CrossRef]
37. Burak, G.H.; Shi, Z.; Wilton, I.; Bursill, J. Disaggregation of commercial building end-uses with automation system data. *Energy Build.* **2020**, *223*, 110222.
38. Zaeri, N.; Ashouri, A.; Gunay, H.B.; Abuimara, T. Disaggregation of electricity and heating consumption in commercial buildings with building automation system data. *Energy Build.* **2021**, *258*, 111791. [CrossRef]
39. Elafoudi, G.; Stankovic, L.; Stankovic, V. Power disaggregation of domestic smart meter readings using dynamic time warping. In Proceedings of the Conference Power Disaggregation of Domestic Smart Meter Readings Using Dynamic Time Warping, Athens, Greece, 21–23 May 2014; pp. 36–39.
40. Kolter, J.Z.; Batra, S.; Ng, A.Y. Energy Disaggregation via Discriminative Sparse Coding. In Proceedings of the 23rd International Conference on Neural Information Processing Systems, Vancouver, Canada, 6–9 December 2010; Volume 1, pp. 1153–1161.

41. Matsui, K.; Yamagata, Y.; Nishi, H. Disaggregation of Electric Appliance's Consumption Using Collected Data by Smart Metering System. *Energy Procedia* **2015**, *75*, 2940–2945.
42. Dhar, A.; Reddy, T.A.; Claridge, D.E. A Fourier Series Model to Predict Hourly Heating and Cooling Energy Use in Commercial Buildings with Outdoor Temperature as the Only Weather Variable. *J. Sol. Energy Eng.* **1999**, *121*, 47–53. [CrossRef]
43. Dhar, A.; Reddy, T.A.; Claridge, D.E. Modeling Hourly Energy Use in Commercial Buildings with Fourier Series Functional Forms. *J. Sol. Energy Eng.* **1998**, *120*, 217–223. [CrossRef]
44. Dhar, A.; Reddy, T.A.; Claridge, D.E. Generalization of the Fourier Series Approach to Model Hourly Energy Use in Commercial Buildings. *J. Sol. Energy Eng.* **1999**, *121*, 54–62. [CrossRef]
45. Ji, Y.; Xu, P.; Ye, Y. HVAC terminal hourly end-use disaggregation in commercial buildings with Fourier series model. *Energy Build.* **2015**, *97*, 33–46. [CrossRef]
46. Fan, C.; Wang, J.; Gang, W.; Li, S. Assessment of deep recurrent neural network-based strategies for short-term building energy predictions. *Appl. Energy* **2018**, *236*, 700–710. [CrossRef]
47. Parzen, E.; Pandit, S.M.; Wu, S.-M. Time Series and System Analysis with Applications. *J. Am. Stat. Assoc.* **1985**, *80*, 251. [CrossRef]
48. Braun, J.E. Performance and control characteristics of a large cooling system. *ASHRAE Trans.* **1987**, *93*, 1830–1852.
49. Claridge, D.E.; Haberl, J.S.; Turner, D.; O'Neal, D.L.; Jaeger, S. Improving energy conservation retrofits with measured savings. *ASHRAE J.* **1991**, *33*, 10–14.
50. Ji, Y.; Xu, P. A bottom-up and procedural calibration method for building energy simulation models based on hourly electricity submetering data. *Energy* **2015**, *93*, 2337–2350. [CrossRef]
51. Lior, R.; Maimon, O. *Data Mining with Decision Trees: Theory and Applications*; World Scientific Publishing Co. Inc.: Singapore, 2007.
52. James, G.; Witten, D.T.H. *An Introduction to Statistical Learning*; Springer: New York, NY, USA, 2013; Volume 103, pp. 78–129.
53. Breiman, L.; Friedman, J.R.O.; Stone, C. *Classification and Regression Trees*; Wadsworth International Group: Belmont, CA, USA, 1984; Volume 40, pp. 17–23.
54. Breiman, L.; Friedman, J.H.R.O. Classification and Regression Trees. Wadsworth & Brooks. *J. Am. Stat. Association.* **1984**, *18*, 358.
55. Wang, H.; Xu, P.; Lu, X.; Yuan, D. Methodology of comprehensive building energy performance diagnosis for large commercial buildings at multiple levels. *Appl. Energy* **2016**, *169*, 14–27. [CrossRef]
56. Fu, Y.; Li, Z.; Feng, F.; Xu, P. Data-quality detection and recovery for building energy management and control systems: Case study on submetering. *Sci. Technol. Built Environ.* **2016**, *22*, 798–809. [CrossRef]

buildings

MDPI

*Article*

# Research on Multiple Energy-Saving Strategies for Existing Coach Stations: A Case of the Xi'an Area, China

**Xueping Li, Luo Qin * and Jingjing Li**

School of Architecture and Civil Engineering, Xi'an University of Science and Technology, Xi'an 710054, China
* Correspondence: qinluo@stu.xust.edu.cn

**Abstract:** In the context of China's dual-carbon goals, energy efficiency in public buildings has become a focal point of public concern. As large-scale public transportation buildings, the indoor thermal comfort and the current state of energy consumption of coach stations are increasingly being emphasized. This research used existing coach stations in the Xi'an region as the object; through on-site investigations and field tests of indoor thermal environments in winter and summer seasons, it was found that the coach stations had energy waste and high energy consumption; the enclosure structures had poor thermal performance; and the stations lacked effective energy-saving measures. Energy-saving transformation strategies were proposed from two aspects: enclosure structures and renewable energy utilization. Using DeST-C for energy consumption, the external walls, roofs, insulation materials, and glass materials were simulated, and nine different combinations of energy-saving schemes were simulated using orthogonal experiments. The optimal scheme was selected based on the comprehensive energy-saving rate and economic analysis results, which included using 80 mm XPS external insulation for the external walls, low-e hollow glass for the windows (low transmittance type), and an 80 mm PUR board for the roof insulation. The energy-saving rate of this scheme was 26.84%. The use of rooftop solar photovoltaic power generation and fresh air heat recovery devices can effectively reduce building energy consumption, and the investment payback period is less than 5 years. The research applications have practical significance for improving the indoor environment of existing coach stations and saving energy consumption.

**Keywords:** coach stations; energy consumption simulation; renewable energy utilization; energy-saving transformation strategy

## 1. Introduction

With the rapid development of the global economy, it is inevitable that there has been overexploitation of natural resources and excessive energy consumption. However, with the introduction of sustainable development goals (SDG), there is a growing emphasis worldwide on reducing various forms of energy consumption and increasing the utilization of renewable energy [1–3]. Energy consumption is mainly concentrated in three areas: buildings, transportation, and industry [4,5]. Various types of transportation buildings have been identified as the focus of energy conservation and emission reduction work. These public transportation buildings generally have higher energy consumption levels than ordinary public buildings [6]. The highway transportation industry occupies a dominant position in the transportation system [7]. By 2020, the total mileage of highways in China had reached 5.1981 million kilometers, and the annual passenger traffic of commercial passenger transportation had reached 6.894 billion people, accounting for 71.3% of the total passenger transportation traffic [8]. The continuous development of the highway transportation industry has also led to a continuous increase in the number of coach stations. However, while the number and building area of coach stations continue to increase, the energy consumption of these buildings remains high. Most of coach stations were built earlier and designed based on early energy-saving design specifications [9], with little consideration for building energy conservation, resulting in poor thermal performance of the enclosure structure, poor quality of the indoor thermal environment, and high energy consumption of some coach buildings.

The research on energy consumption of coach stations mainly focuses on establishing evaluation indicators and analyzing the important factors affecting energy consumption, while there is less research on the analysis of indoor thermal environment and the energy-saving strategies of coach station buildings. Fan et al. [10] calculated and ranked 18 factors affecting the energy consumption of coach stations in four aspects: site selection, hardware configuration, building facilities, and layout technology. The study was based on the group decision-making hierarchical analysis method (GAHP) and provided a basis for determining evaluation indicators. Yang [11] conducted a survey of bus stations in five climate zones in China and used five factors that affect energy efficiency to determine seven highway coach station energy efficiency evaluation indicators. The energy efficiency classification of the bus station was determined by scoring the seven indicators. The research on energy consumption of other types of public transportation buildings mainly focuses on the indoor thermal environment and energy consumption status. Huang et al. [12] conducted indoor thermal environment testing and thermal sensation voting surveys at railway stations in cold regions; they found that the thermal comfort evaluation of waiting halls was significantly correlated with passenger's age and the length of residence in the local area. After analyzing the indoor thermal environment data of airports, Mambo et al. [13] found that energy-saving measures could be implemented by considering different comfort set points while ensuring temperature and humidity control, which not only improved passenger comfort but also reduced energy consumption by 20% to 25%. Alkis et al. [14] analyzed the measured indoor temperatures of three airports and proposed the maximum use of natural light to save lighting energy based on the acceptable temperature range for passengers.

Due to the promotion of the Sustainable Development Goals worldwide, there are higher requirements for the energy efficiency of large public buildings, such as green buildings [15,16]. Not only are energy-saving and environmentally friendly building materials necessary, but also the use of renewable energy sources such as solar and geothermal energy is required to reduce building energy consumption [17–19]. Therefore, for the energy-saving renovation of existing coach stations, the thermal performance of the envelope structure can be improved by using high-efficiency insulation materials, and renewable energy technologies such as solar photovoltaic power generation and heat recovery can be utilized. The research found that energy-saving renovation of building envelopes and the use of renewable energy sources are also common ways to achieve energy efficiency in public buildings [20–24]. The envelope structure has a significant impact on the energy consumption of buildings [25]. By using orthogonal experimental design (OED) to optimize the types of envelope structures, such as external walls and roofs, as well as the thickness of insulation layers [26,27], the optimal combination is determined to reduce the operational energy consumption of buildings. Feng et al. [28] established a thermal performance function model for the enclosure structure of residential buildings in cold regions based on energy-saving standards, and they used the Lagrange's method of multipliers to calculate the optimal insulation thickness for external walls and roofs. Gindi et al. [29] simulated the transformation of building roofs, south-facing facades, and west-facing facades using photovoltaic components and obtained the optimal tilt angles and component types for each direction, effectively reducing energy consumption. Fouad et al. [30] established a regression prediction model for the power of photovoltaic panels, indoor lighting, and indoor temperature with the tilt angle of the photovoltaic system based on experimental data and determined the optimal tilt angle of the photovoltaic shading system in Egypt. Angeliki et al. [31] used Matlab to calculate the electricity generation from applying photovoltaic technology on buildings in six European cities and analyzed the impact of different building orientations on the internal rate of return of building-integrated photovoltaics (BIPV). Lu et al. [32] conducted a study on the advantages and challenges of applying Building Integrated Photovoltaics (BIPV) technology in Singapore. Although BIPV can bring economic benefits and reduce carbon emissions, its high cost and uncertainty in energy conversion efficiency are still issues that must be considered before applying this technology. Faranda et al. [33] have applied photovoltaic power generation to train stations in Italy, achieving the station's sustainable development goals. They have discussed the feasibility of the plan in terms of specific parameters such as the converter and distributor of photovoltaic generators. Zhang et al. [34] compared the energy-saving effects of using new air heat recovery in public buildings in seven typical cities and found that full heat recovery is the most energy-saving and cost-effective method in cold regions. Heat recovery technology is widely used in high latitude countries such as Germany and Sweden [35,36]. In cold climates, using heat recovery technology can reduce building energy consumption by about 20%. Liu et al. [37] used an optimization design method to determine the optimal shape of the heat recovery device from the perspectives of energy-saving and reducing material costs. The aforementioned studies have explored the application of energy-saving strategies that utilize renewable energy in public buildings. However, renewable energy utilization technology has not been widely applied in China's coach stations, which presents enormous energy-saving potential. The application of energy-saving retrofit strategies must consider not only their energy-saving effects but also their economic feasibility. Therefore, when analyzing these strategies, it is necessary to calculate the cost of each retrofitting plan as well as the payback period.

As a major urban transportation hub, the coach station has increasingly high requirements for its indoor thermal comfort and thermal performance of the envelope structure. Taking the coach stations in Xi'an as an example, this research collected building information, passenger thermal comfort feedback, and indoor thermal environment testing through field investigations to analyze the existing problems and the potentials of energy saving. Energy-saving renovation strategies were proposed, including the transformation of the building envelope structure such as the external walls, windows, and roof, the use of solar photovoltaic systems for power generation, and the installation of new air supply and heat recovery devices to reduce the building's operational energy consumption. In addition, the feasibility of these energy-saving strategies is estimated through economic analysis to ensure that they are both energy-efficient and economically viable. The energy-saving renovation strategies can provide reference value for the energy-saving renovation of existing coach stations or the energy-saving design of newly built coach stations in cold regions, with the aim of achieving sustainable development goals for buildings [38].

## 2. Survey and Analysis on Coach Stations

The following subchapter first conducted on-site visits and research to the automobile passenger transportation stations in the Xi'an area, obtaining basic information and energy consumption status of each station. Secondly, the indoor thermal environment of the buildings was tested and surveyed in both winter and summer seasons, and the results were analyzed to determine the indoor comfort level. Finally, based on the research information, the Xi'an Chengxi Passenger Transportation Station was selected as the benchmark model for modeling and analysis in the DsST-C software.

### *2.1. Overview of Survey Area*

Six first-class coach stations in the Xi'an area were selected for field research to collect basic information about the buildings. The Xi'an area is located in the central part of the Guanzhong Plain in northwest China, with hot and rainy summers and cold and dry winters. According to the architectural thermal design partition, the Xi'an area belongs to the cold region. The meteorological data of Xi'an are shown in Figure 1, with an annual average temperature of 13.0~13.7 °C and an annual precipitation of 522.4~719.5 mm. The highest precipitation occurs from July to October, and the relative humidity is above 70% throughout the year, with about 70% of the time falling within the range of 40~70% of the human body humidity comfort zone. The annual sunshine duration is more than 2000 h, with peak solar radiation of 647.23 $W/m^2$ in August, and the solar radiation is particularly high from May to August.

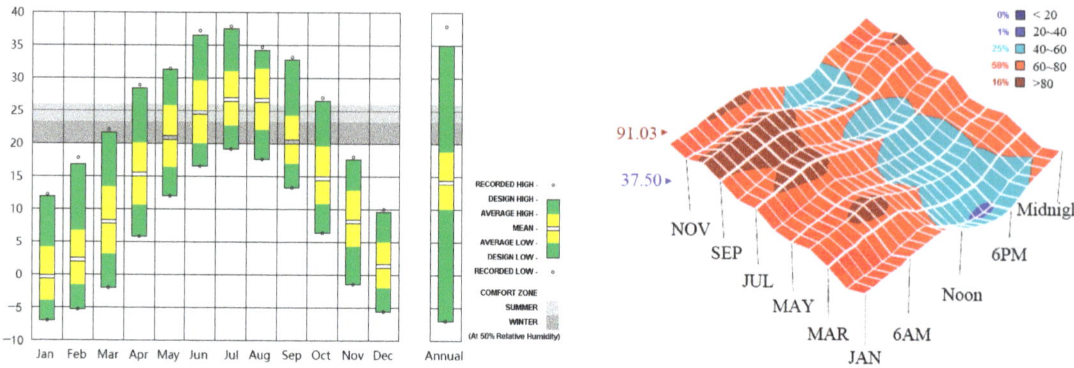


(**a**) (**b**)

**Figure 1.** Annual weather data map of Xi'an. (**a**) Monthly temperature range in the Xi'an area; (**b**) Xi'an area's relative humidity range throughout the year.

### *2.2. The Current Status of the Buildings and Energy Consumption Situation*

The surveyed coach stations in Xi'an area all face the north–south direction. The waiting halls and ticket halls of the complexes are mostly 1–2 stories high and are open for more than 12 h a day with an annual passenger volume over 500,000 people. The selection of heating and cooling equipment varies among each station, but the main energy consumption comes from electricity and water usage, with the majority of the air conditioning systems being fan coil unit and fresh air system. The construction year of the researched coach stations ranged from 1983 to 2013—relatively early construction years in which there was less consideration for energy conservation. The thermal performance of the building envelope structures no longer meets the requirements of current public building energy-saving design standards [39]. Although the envelope structures and air conditioning systems have been renovated in subsequent operation processes, there are still problems with high energy consumption per unit area, poor indoor thermal comfort, and low energy efficiency of the air conditioning and lighting systems. The survey information regarding each coach station is shown in Table 1.

The annual electricity consumption and comprehensive energy consumption per unit area of coach stations A to F are shown in Figure 2. The energy consumption of the coach station buildings ranges from 23.68 to 89.58 $kW{\cdot}h/(m^2{\cdot}a)$. The electricity consumption peak mainly occurred in the air conditioning and heating period during the summer months of July and August and the winter months of December and January, when the heating and cooling equipment were operated for longer periods of time, consuming a large amount of electricity and natural gas. During the remaining transition periods, electricity consumption mainly came from the lighting systems and electrical equipment. Station E was jointly built with a large shopping mall and subway station, and no separate energy consumption measurement system was installed, so it was not included in the statistical analysis.

### *2.3. Indoor Thermal Environment Overview*

The quality of indoor thermal environment will have a significant impact on the physical sensations of waiting passengers. Therefore, crowded waiting halls and ticketing halls were chosen as test locations for thermal environment analysis. The following section analyzes the indoor thermal environment from three aspects: indoor thermal environment testing, evaluation, and questionnaire survey.

**Table 1.** Basic information on coach station buildings.

| Building Sign | A | B | C | D | E | F |
|---|---|---|---|---|---|---|
| Construction year | 1983 | 2012 | 2005 | 1999 | 2013 | 2002 |
| Construction area ($m^2$) | 12,000 | 29,000 | 9200 | 18,000 | 8946.2 | 8000 |
| Building height (m) | 10 | 22.5 | 22 | 24 | 7 | 15 |
| Annual passenger volume | 730 k | 1640 k | 820 k | 750 k | 2600 k | 510 k |
| Cold source | Direct-fired lithium bromide cold and hot water unit | Ground source heat pump unit | Multi-connected air conditioning unit | Multi-connected air conditioning unit | Water-cooled chiller | Air-cooled chiller |
| Heat source | Direct-fired lithium bromide cold and hot water unit | Ground source heat pump unit | Multi-connected air conditioning unit | Multi-connected air conditioning unit | Municipal heating network | Air energy heat pump unit |

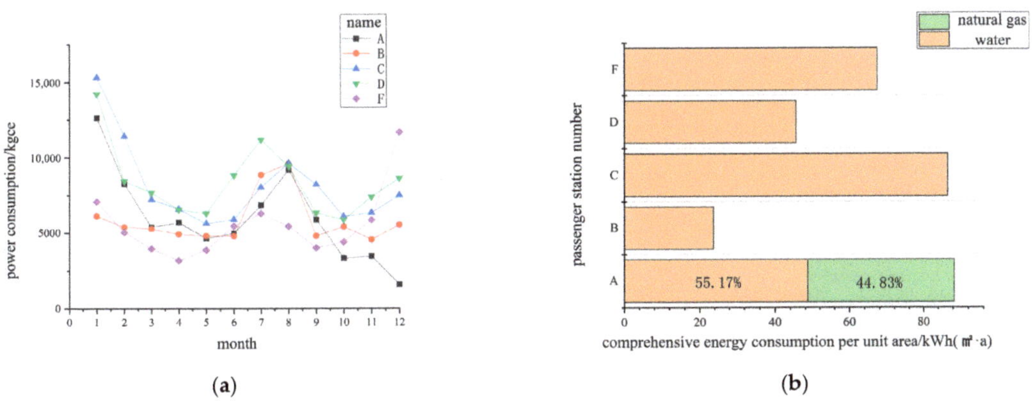


**Figure 2.** Energy consumption situation of each coach station: (**a**) monthly power consumption; (**b**) energy Intensity.

### 2.3.1. Questionnaire Survey Analysis

In the summer, a questionnaire survey was conducted among passengers waiting in the waiting hall to understand the passengers' intuitive perception about the indoor thermal environment of the waiting hall. A total of 208 questionnaires were distributed. Among the surveyed passengers, passengers aged 20 to 30 accounted for the largest proportion, with a total of 134 male passengers, accounting for 64.42% of the total. Most of the passengers traveled for short distances within the province.

It can be seen in Tables 2 and 3 that 63.02% of the passengers felt that the waiting area was relatively comfortable; 10.57% of the passengers felt that the indoor environment of the waiting room was relatively cool; and 19.23% of the passengers thought that the waiting room was a bit hot. Largely, the passengers did not feel obviously cold when they first entered the waiting hall from the outside, but after sitting for a long time, they felt slightly cold after a period of heat loss. Additionally, 81.25% of the passengers felt that the humidity in the waiting area was moderate, and 11.05% of the passengers felt a bit damp. Since Xi'an is rainy in summer, the evaporation of rainwater increases the water vapor content in the air, resulting in an increase in the humidity in the waiting hall.

Table 2. Questionnaire on hot and cold feelings of respondent passengers.

| Hot and Cold Feeling | Cold | Slightly Cooler | Comfortable | Slightly Hotter |
|---|---|---|---|---|
| Number of people | 17 | 22 | 129 | 40 |
| Percentage taken by | 8.17% | 10.57% | 62.02% | 19.23% |

Table 3. Questionnaire on Air Humidity Feelings of Respondent Passengers.

| Wet and Dry Feeling | More Humid | Slightly Damp | Moderate Humidity | Slightly Dry | Drier |
|---|---|---|---|---|---|
| Number of people | 4 | 19 | 169 | 14 | 2 |
| Percentage taken by | 1.92% | 9.13% | 81.25% | 6.73% | 0.96% |

2.3.2. Indoor Thermal Environment Test

The Xi'an Chengxi coach station was selected for indoor thermal environment testing in both winter and summer seasons. Figure 3 shows the distribution of the measuring points. A total of four measuring points were set up, which were outdoor measuring point A, measuring point B in the ticket office, and measuring points C and D in the waiting area on both sides of the east and west sides. The height of the indoor testing equipment was placed on the resting level of the waiting passengers at 0.8 m above the ground. Outdoors, it was placed away from direct sunlight, 1.5 m above the ground.

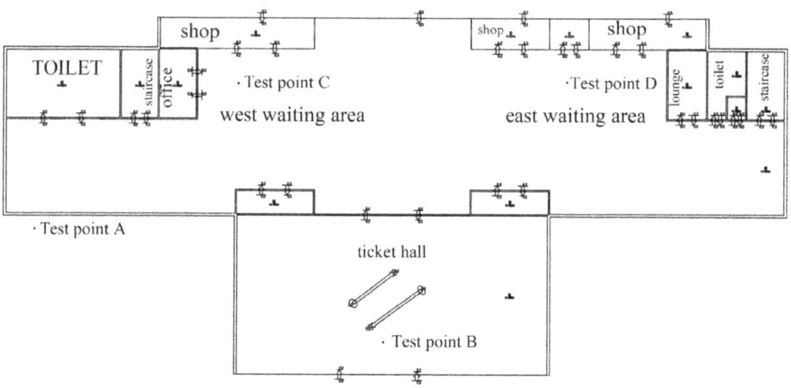


Figure 3. First floor plan in DeST software.

As shown in Figure 4, during the winter testing period, the average outdoor temperature was 4.4 °C; the lowest temperature in the indoor waiting area was 7.2 °C, while the highest temperature was 13.7 °C; the average temperature was 10.9 °C, and the average temperature in the ticketing area was 10.8 °C. Due to the proximity of the waiting area to the entrance and the station shops and the low energy efficiency of the heating system, the temperature in the waiting area was lower than 16 °C, and the thermal comfort of the human body was low. If their waiting time was longer than half an hour, the passengers felt cold if they sat for a long time. The average relative humidity of the outdoor air was 45.0%; the relative humidity of the indoor ticket lobby was 37.7%; and the average relative humidity of the waiting hall area was 34.6%. The overall relative humidity level was low, resulting in dry indoor air.

As shown in Figure 5, during the summer testing period, the average outdoor temperature was 32.3 °C, and the average temperatures in the east and west passenger waiting areas and ticketing areas were 28.3 °C, 27 °C, and 27.7 °C. Although the average temperature was within the specified range of 22 °C to 28 °C, the temperature in the waiting hall and ticket lobby was higher than 28 °C due to the increase in the afternoon. Since the waiting area is close to the outside, the outdoor heat can enter through the ticket gate and store passage, and the temperature varies with the outdoor temperature. The average relative humidity of the outdoor air during the summer was 50.4%. The average relative humidity in the ticket hall and waiting rooms on the east and west sides was 60.5%, 63.0%, and 58.4%, respectively. The relative humidity of the air does not fluctuate significantly, remaining within 10%.

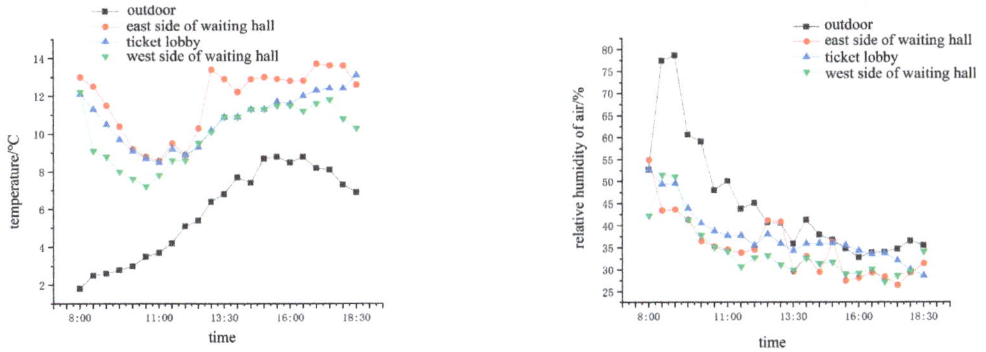


(**a**) (**b**)

**Figure 4.** Variations of indoor and outdoor air temperature and relative humidity in winter sea: (**a**) Indoor and outdoor temperature (**b**) relative humidity.

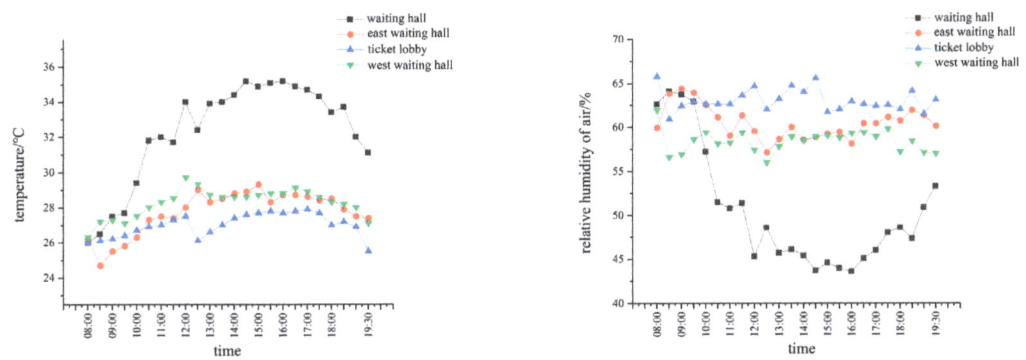


(**a**) (**b**)

**Figure 5.** Variations of indoor and outdoor air temperature and relative humidity in summer season. (**a**) Indoor and outdoor temperature; (**b**) relative humidity.

With the changes and fluctuations, the passengers felt slightly hotter during this time. According to the "Indoor Air Quality Standard" (GB/T18883-2022) [40], the standard value of air-conditioned room temperature in summer is 16~24 °C, and that of heating in winter is 22~28 °C. The comparison shows that the indoor temperatures in winter and summer do not meet the requirements.

### 2.3.3. Indoor Thermal Environment Evaluation

The waiting hall where passengers spend the longest time in a day was selected to analyze the indoor thermal environment, using the International Organization for Standardization ISO7730 as the evaluation standard, with an indoor air velocity of 0.1 m/s. Based on the survey questionnaire results, the thermal resistance of passengers' winter clothing was taken as 1.5 clo, and the thermal resistance of summer clothing was taken as 0.5 clo. Considering that waiting passengers are in a sedentary state, the metabolic rate of the human body was taken as 1.0 met. According to the indoor thermal environment data obtained from the test, the predicted average thermal sensation index PMV and predicted percent dissatisfied (PPD) were calculated for the same time period.

As shown in Figure 6, the predicted average thermal sensation index PMV value during winter season was between −1.75 and 0.75, with an average value of −1.23; the predicted unsatisfactory percentage PPD value was between 30% and 70%, with an average value of 39.13%. According to the "Evaluation Standards for Indoor Humidity and Heat Environment of Civil Buildings" (GB/T 50785-2012), the indoor environment in winter met the overall evaluation standard of level III, indicating poor thermal comfort, with more than half of the passengers waiting for the bus feeling relatively cold.

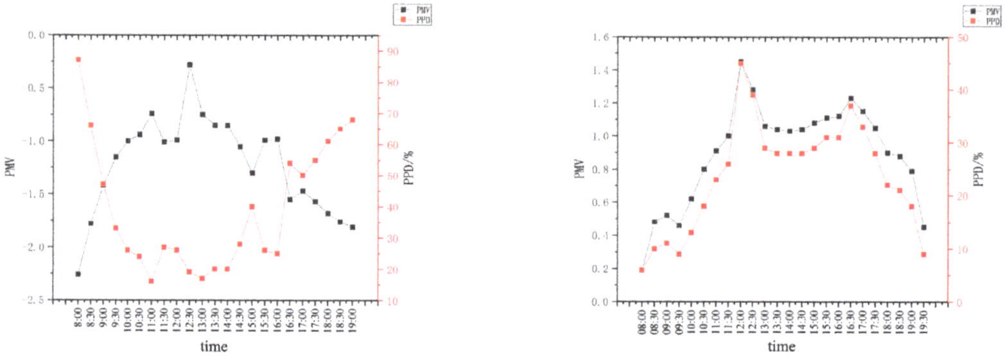


(**a**) (**b**)

**Figure 6.** Thermal comfort change in the waiting hall of Chengxi coach station between winter and summer seasons. (**a**) winter; (**b**) summer.

The predicted average thermal sensation index PMV value during summer season was between 0.4 and 1.3, with an average value of 0.90; the predicted unsatisfactory percentage PPD value was between 5% and 35%, with an average value of 23.83%. The indoor PMV value was slightly higher in summer, meeting the overall evaluation standard of level II, indicating that over 75% of passengers felt comfortable. However, during some periods from 1:00 p.m. to 5:00 p.m., the PMV value was high, and the indoor environment was slightly hot.

### 2.4. *Influencing Factors of Energy Consumption and Benchmark Models*

The following section combines research findings to analyze the existing problems and energy-saving potential of the Xi'an region's automobile passenger transport stations. The Xi'an Chengxi Passenger Transport Station is selected as the benchmark model in the DeST-C software to simulate and establish a foundation for various energy-saving renovation schemes.

#### 2.4.1. Influencing Factors of Energy Consumption

Through the analysis of the results from the on-site research and testing mentioned above, it was found that there are the following energy-saving problems in the coach station in Xi'an: the low energy efficiency of the air-conditioning system results in energy waste and high energy consumption and poor thermal insulation performance of the external wall and roof, leading to excessively high or low indoor temperatures. Additionally, no effective energy-saving measures have been taken, and the utilization rate of renewable energy is not enough. For example, the bus station has a large roof area, but solar energy is not fully utilized for power generation, etc.

The factors that affect the energy consumption of coach station buildings mainly include climate zone, construction age, enclosure structure condition, building orientation, cooling and heating equipment, indoor personnel density, etc., while the reconstruction of existing coach station buildings is mostly envelope structure retrofitting [41–44], adding energy-saving measures [45,46], etc. In response to these problems and with the aim of reducing building energy consumption, combined with the geographical and meteorological environment of Xi'an area, the energy-saving strategy of coach stations was analyzed by taking Xi'an Chengxi Coach Station as an example, including the use of high-efficiency heat insulation materials for the building envelope to reduce heat loss and the use of solar photovoltaic power generation and fresh air heat recovery to reduce building power consumption.

### 2.4.2. Benchmark Models

A Xi'an Chengxi Coach Station was selected as the benchmark model, and energy consumption simulation software, DeST-C, developed by Tsinghua University, was used to develop the model, as shown in Figure 7. DeST-C is developed based on AutoCAD, and its operation and interface are similar to CAD software, making it very user-friendly. The software simulation results can be exported through Excel spreadsheets and visualized for further analysis.

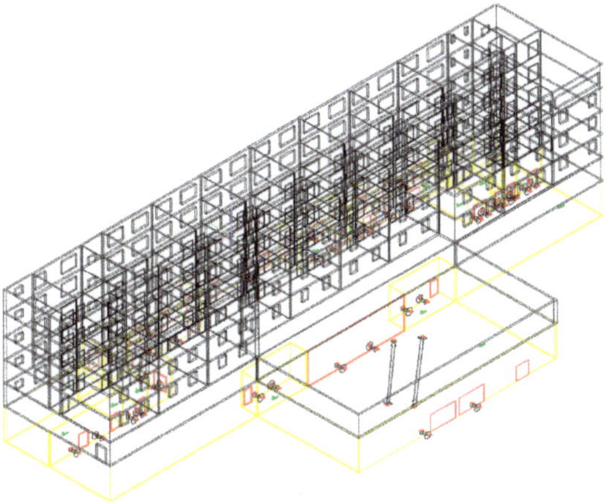

**Figure 7.** Three-dimensional building model drawing in DeST software.

Firstly, the actual information of the architecture of the coach station is used to build a model. After the modeling is completed, specific model materials are set based on the parameters of the building enclosure structure, including meteorological parameters, room thermal disturbance, indoor and outdoor ventilation, etc. Then, building energy consumption calculations can be performed, and the calculation results can be exported and viewed in report form. Finally, based on the benchmark model, quantitative analysis of the impact of different building materials or material thicknesses on building energy consumption can be achieved by conducting building energy simulations after adjusting the global settings of the building or replacing local building components with different materials.

The passenger station is located in Lianhu District, Xi'an City, Shaanxi Province. It was built in 2001. The construction area is 9200 square meters, and the building faces south. There are six floors in total; the first floor is the waiting hall and ticket hall area, including lounges, offices, shops and foyers, etc.; the second floor is the pedestrian corridor for passengers leaving the station; the third to sixth floors are hotel rooms and offices. The building envelope parameters are shown in Table 4.

**Table 4.** Envelope parameters.

| Envelope | Structure |
|---|---|
| External wall | 20 mm cement mortar + 240 mm hollow brick + 20 mm cement lime mortar |
| Outside window | Single-layer plastic steel window |
| Roof | 20 mm cement mortar + 120 mm reinforced concrete precast panel + 20 mm cement mortar + 50 mm cement expanded perlite + 20 mm cement mortar |

The air conditioning and lighting equipment in the building are turned on during the bus station business hours, that is, from 8:00 to 20:00. The cooling period is set from May to October, and the heating period is set from 15 November to 15 March of the following year. Parameters such as indoor heat disturbance and the temperature settings are shown in Table 5.

**Table 5.** Indoor thermal disturbance parameters.

| Type | Lounge | Office | Shop | Corridor/Foyer | Hotel Room |
|---|---|---|---|---|---|
| Calorific value per capita/(W/m$^2$) | 62 | 66 | 58 | 58 | 61 |
| Humidity production per capita/(g/(h·person)) | 68 | 102 | 184 | 184 | 109 |
| Fresh air volume/(m$^3$/(h·p)) | 30 | 30 | 20 | 20 | 20 |
| Maximum light heat disturbance/(W/m$^2$) | 11 | 18 | 12 | 15 | 12 |
| Maximum equipment heat disturbance/(W/m$^2$) | - | 13 | 13 | - | 20 |
| Indoor temperature in summer/°C | 24–26 | 23–25 | 22–27 | 18–28 | 24–26 |
| Indoor temperature in winter/°C | 20–22 | 21–23 | 20–22 | 18–28 | 18–26 |
| Relative humidity in summer/% | 50–60 | | | | |
| Relative humidity in winter/% | 30–40 | | | | |

The hourly air conditioning load calculation results for the whole year are shown in Figure 8.

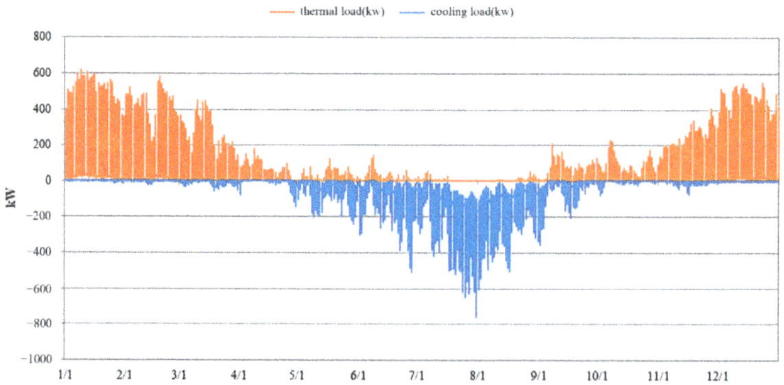


**Figure 8.** Annual calculation results of the building's heating and cooling load.

## 3. Energy-Saving Renovation of Enclosure Structure

The enclosure structure is an important component of the main body of a building, and the material selection and performance of the enclosure structure directly affect the indoor thermal environment and energy consumption of the building. The following subchapter discusses the energy-saving renovation of the enclosure structure from three aspects, external walls, windows, and roofs, and uses orthogonal experiments and economic evaluation to analyze the optimal energy-saving solution.

### 3.1. External Wall Insulation Improvement

External wall insulation is the main way to achieve energy-saving for the external wall. Common forms of thermal insulation include the external thermal insulation of external walls and internal thermal insulation of external walls. Considering the renovation of existing buildings, in order to ensure that the indoor space is not reduced, external thermal insulation of the external walls was selected. External wall insulation materials include expanded polystyrene board (EPS), extruded polystyrene board (XPS), rock wool board, polyurethane board, glass wool, etc. Among them, EPS has high compressive strength, good stability, and a low price; XPS has good water resistance and low thermal conductivity and is lightweight; rock wool board has strong fire resistance and low thermal conductivity. Therefore, these three materials were selected as the research objects. The thickness of the insulation board was used as a variable, and the selected range was from 20 to 120 mm.

As shown in Figure 9a, with the increase in the thickness of the insulation board, the annual cumulative total load of the building decreased significantly. Meanwhile, the calculation results show that the building's annual cumulative heat load was reduced by 10.8%, 10.6%, and 11.5%, respectively, when using EPS board, XPS board, and rock wool board as insulation materials. However, the building's annual cumulative cooling load increased by 2.13%, 2.11%, and 2.34%, respectively. The annual cumulative heating load of the building decreased, while the annual cumulative cooling load increased. The rate gradually decreased as the thickness of the insulation board increased. This was due to the increase in the thermal insulation effect of the external wall, which led to the slowing down of the accumulated heat loss in the building after the outdoor temperature decreased at night in summer, which is similar to increasing the indoor heat disturbance, resulting in an increase in the cooling load of the building the next day.

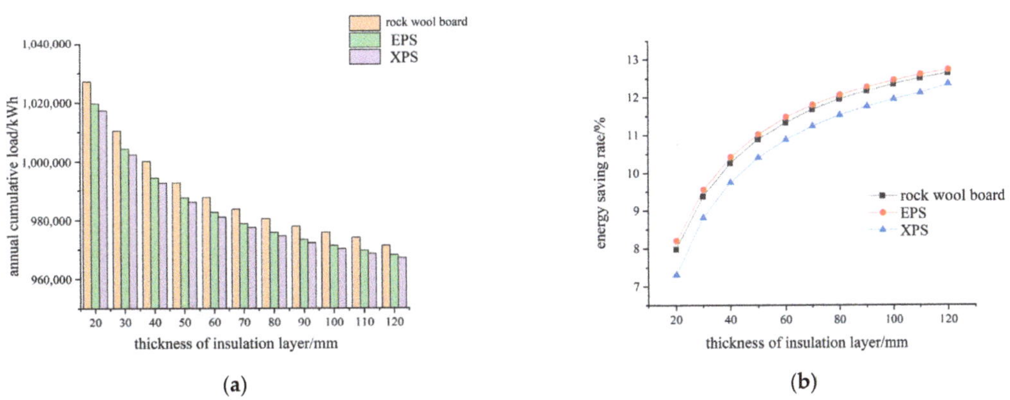


**Figure 9.** Influence of different thickness of insulation materials on building energy consumption (1): (**a**) Changes in energy consumption; (**b**) changes in energy-saving rate.

According to the results in Figure 9b, it can be seen that the three insulation materials had good energy-saving effects. When the thickness was in the range of 0–90 mm, the annual cumulative total load decreased significantly. The energy-saving rates for three types of insulation materials used were 12.17%, 12.27%, and 11.76%, respectively. When the thickness increased from 90 mm to 120 mm, the energy-saving rates only increased by 0.47%, 0.45%, and 0.59%. After the thickness of the insulation material reached 90 mm, there was limited space for the energy-saving rate to rise, and the material cost continued to increase. Therefore, it is advisable to choose 80 mm-thick EPS board and XPS board and 90 mm-thick rock wool board as the external wall optimization scheme.

### 3.2. External Window Material

As the weak link of the thermal insulation of the external enclosure structure, external windows have great energy-saving potential. By selecting different glass materials for the outer windows and conducting energy consumption simulations for each type of glass material, the annual cumulative load calculation results for the building are shown in Table 6. Low-e glass can reduce the indoor and outdoor heat exchange while ensuring indoor lighting due to its far-infrared filtering characteristics. Low-e glass costs more but has better energy-saving effects. Low-transmittance, low-e film-coated glass windows had the lowest energy consumption: the energy-saving rate reached 6.40%. The energy-saving rate of high-transparency low-e insulating glass reached 5.82%. Considering the economy of energy-saving optimization, low-transmittance, low-e film-coated glass windows and 6 mm + 12 mm air + 6 mm insulating glass windows were selected as the optimal solutions.

**Table 6.** Influence of window type on building load.

| Type | Annual Accumulative Total Load/kW·h | Energy-Saving Rate |
|---|---|---|
| Low-e hollow 6 + 9 + 6 (low permeability type) | 1,037,282 | 6.40% |
| Low-e hollow 6 + 9 + 6 (high transparency type) | 1,043,617 | 5.82% |
| Vacuum + Low-e glass (3 + 0.1 + 3) | 1,046,767 | 5.54% |
| Vacuum glass 3 + 0.1 + 3 | 1,060,542 | 4.30% |
| 6 mm + 12 mm air + 6 mm | 1,059,880 | 4.36% |
| 6 mm + 9 mm air + 6 mm | 1,062,122 | 4.15% |

### 3.3. Roof Insulation Improvement

The roof receives the longest and strongest exposure to solar radiation, and it is the main channel for indoor–outdoor heat exchange in the top-level room. Therefore, the thickness of the roof insulation layer should be greater than that of the external wall insulation layer. However, the fire resistance of the insulation material is lower than that required for the external wall. The essence of the roof insulation method is to add or replace a new insulation layer on the basis of the original roof. After investigation, it was found that the roof insulation layer in the Xi'an area did not suffer major damage and still had a certain insulation effect. Therefore, a new insulation layer was chosen to be added above the original roof. XPS board, rock wool board, and rigid polyurethane foam board (PUR) were used as the roof insulation materials. The PUR board had good thermal insulation and waterproof performance and a long service life. The material thickness range was 20~120 mm. The simulation results of each thickness are shown in Figure 10.

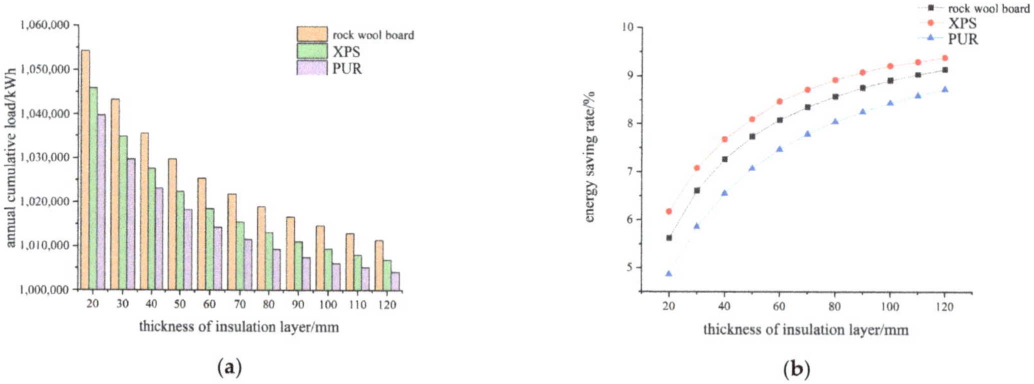


**Figure 10.** Influence of different thickness of insulation materials on building energy consumption (2): (**a**) Changes in energy consumption; (**b**) changes in energy-saving rate.

As shown in Figure 10a, it can be seen that with the increase in the thickness of the insulation layer, the cumulative total load of the building for the whole year was significantly reduced. According to the results in Figure 9b, when the thickness of the insulation board increased from 0 mm to 80 mm, the energy-saving rates of buildings using PUR board, XPS board, and rock wool board increased by 8.92%, 8.58%, and 8.05%. PUR board had the highest energy-saving rate. After the thickness of the material reached more than 80 mm, the rising curve of the energy-saving rate tended to be flat, and the energy-saving effect was limited, but the cost continued to increase. Thus, the energy-saving effect of the XPS and PUR boards was better than that of the rock wool boards. Therefore, considering the economy and energy-saving effects, 80 mm thick XPS board and 70 mm and 80 mm thick PUR boards are the optimal solutions for the roof.

### *3.4. Orthogonal Experimental Design*

In order to analyze the energy-saving effects under the joint action of different envelope structure optimization schemes, orthogonal experiments were conducted on the external walls, external windows, and roof, as shown in Table 7. Each factor has three levels, and the energy consumption in the single-factor optimization is the lower basis for selection. A total of nine energy-saving optimization schemes are designed, and the energy consumption simulation of each scheme is simulated, as shown in Table 8. It can be seen in the results that the building's cooling and heating loads of each scheme decreased compared to the benchmark model, and the average annual cumulative load energy-saving rate of different schemes was 25.00%. Among them, scheme 12 has the best energy-saving effect, with an energy-saving rate of 26.84%.

**Table 7.** Factor level settings for orthogonal test.

| Level | External Wall Insulation Improvement | External Window Material | Roof Insulation Improvement |
|---|---|---|---|
| 1 | 80 mm XPS board<br>$X_1$ | Low-e hollow 6 + 9 + 6 (low permeability type)<br>$Y_1$ | 80 mm PUR board<br>$Z_1$ |
| 2 | 80 mm EPS board<br>$X_2$ | Vacuum + Low-e glass (3 + 0.1 + 3)<br>$Y_2$ | 70 mm PUR board<br>$Z_2$ |
| 3 | 90 mm rock wool board<br>$X_3$ | 6 mm + 12 air + 6 mm<br>$Y_3$ | 80 mm XPS board<br>$Z_3$ |

**Table 8.** Orthogonal test simulation results.

| Scheme | External Wall Insulation Improvement | External Window Material | Roof Insulation Improvement | Annual Accumulative Heat Load/kW·h | Annual Accumulative Cooling Load/kW·h | Annual Accumulative Total Load/kW·h | Energy-Saving Rate |
|---|---|---|---|---|---|---|---|
| 1 | $X_1$ | $Y_1$ | $Z_1$ | 376,780.55 | 433,944.79 | 810,725.34 | 26.84% |
| 2 | $X_1$ | $Y_2$ | $Z_3$ | 374,201.42 | 458,422.60 | 832,624.02 | 24.86% |
| 3 | $X_1$ | $Y_3$ | $Z_2$ | 379,959.12 | 466,495.38 | 846,454.50 | 23.62% |
| 4 | $X_2$ | $Y_1$ | $Z_3$ | 381,954.06 | 433,428.19 | 815,382.25 | 26.42% |
| 5 | $X_2$ | $Y_2$ | $Z_2$ | 374,337.12 | 457,924.82 | 832,261.94 | 24.90% |
| 6 | $X_2$ | $Y_3$ | $Z_1$ | 379,544.65 | 466,034.03 | 845,578.68 | 23.70% |
| 7 | $X_3$ | $Y_1$ | $Z_2$ | 382,429.24 | 432,710.07 | 815,139.31 | 26.44% |
| 8 | $X_3$ | $Y_2$ | $Z_1$ | 374,235.05 | 457,219.02 | 831,454.07 | 24.97% |
| 9 | $X_3$ | $Y_3$ | $Z_3$ | 384,946.61 | 465,143.74 | 850,090.35 | 23.29% |

### *3.5. Economic Analysis*

In order to select a scheme with both economical and energy-saving effects, an economic analysis of the above nine energy-saving schemes was carried out, and the net present value (*NPV*) and static investment payback period of each scheme were calculated. The formula for calculating the net present value is as follows:

$$NPV = \sum_{t=1}^{n} (CI - CO) \cdot (1+i)^{-t} \quad (1)$$

where *NPV*—net present value, CNY; *CI*—annual savings, CNY; *i*—discount rate, %; *t*—service life, years.

By conducting research, the discount rate was set at 5%. The electricity fee in Xi'an was 0.62 CNY/kW·h, and the material prices of the EPS, XPS, PUR, and rock wool insulation boards were 420 CNY/$m^3$, 600 CNY/$m^3$, 680 CNY/$m^3$, and 750 CNY/$m^3$, respectively. The prices of the ordinary insulating glass (12 mm), low-transmittance, low-e glass, and vacuum + low-e coated glass were 80 CNY/$m^2$, 190 CNY/$m^2$, and 240 CNY/$m^2$, respectively. The service life is 15 years, and the annual income is assumed to be consistent.

The net present value and static payback period of each energy-saving scheme were calculated, and $P_t$ is the initial investment amount/annual income. The results of the net present value and static payback period of each scheme are shown in Figure 11 and Table 9. According to the results, it can be concluded that schemes 1, 4, 5, and 7 had large net present values and short static payback periods, and the energy-saving rates among the four schemes ere not much different, so schemes 1, 4, and 7 are all optimal schemes, among which the energy-saving rate of scheme 1 was the highest at 26.84%.

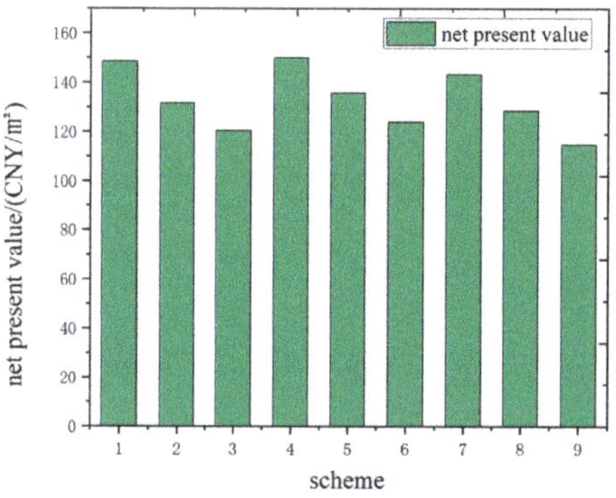


**Figure 11.** Net present value of nine energy-saving schemes.

**Table 9.** Static payback period of energy-saving scheme.

| Energy-Saving Scheme | Energy-Saving Rate | Net Present Value/CNY | Static Return on Investment Period/Year |
|---|---|---|---|
| 1 | 26.84% | 1,365,672 | 2.97 |
| 2 | 24.86% | 1,210,859 | 3.29 |
| 3 | 23.62% | 1,107,574 | 3.55 |
| 4 | 26.42% | 1,378,575 | 2.79 |
| 5 | 24.90% | 1,248,268 | 3.08 |
| 6 | 23.70% | 1,138,926 | 3.38 |
| 7 | 26.44% | 1,317,305 | 3.13 |
| 8 | 24.97% | 1,181,268 | 3.49 |
| 9 | 23.29% | 1,054,848 | 3.79 |

## 4. Renewable Energy Utilization

Due to the strict environmental and operational requirements of new energy-saving technologies in practical applications, as well as the high cost of using renewable energy equipment and technology, the utilization of renewable energy in automobile passenger transportation stations has not yet matured, become widespread, or been universally applied. Improving the energy form and increasing the use of clean energy is an important way to improve the energy consumption structure of existing automobile passenger transportation stations. The following subchapter focuses on the study of two aspects: the use of roof solar photovoltaic power generation and new air heat recovery.

### *4.1. Roof Photovoltaic Energy-Saving Design*

According to the survey findings, it was discovered that the roofs of the buildings at the coach stations in Xi'an area have large areas of idle space, aside from the placement of outdoor air conditioning units. The unsheltered rooftop space is the optimal choice for installing solar photovoltaic equipment that the building can support. The annual sunshine duration in Xi'an is 2200~3000 h, and the annual solar radiation is 5016~5852 MJ/($m^2$·a). The separated photovoltaic roof was selected for the energy-saving design of the existing coach station, and the photovoltaic system was integrated into the public grid for operation. Autodesk Ecotect software was used to simulate the annual solar radiation on the outer surface of the area where the target building is located, and the obtained 3D color map of solar radiation is shown in Figure 12. Based on the distribution of solar radiation shown in the figure, it can be seen that the roof receives the most annual solar radiation per unit area on the outer surface of the building. A polysilicon power generation system with higher power generation efficiency is selected for the photovoltaic system on the roof of the building, and the photoelectric conversion efficiency is 15%.

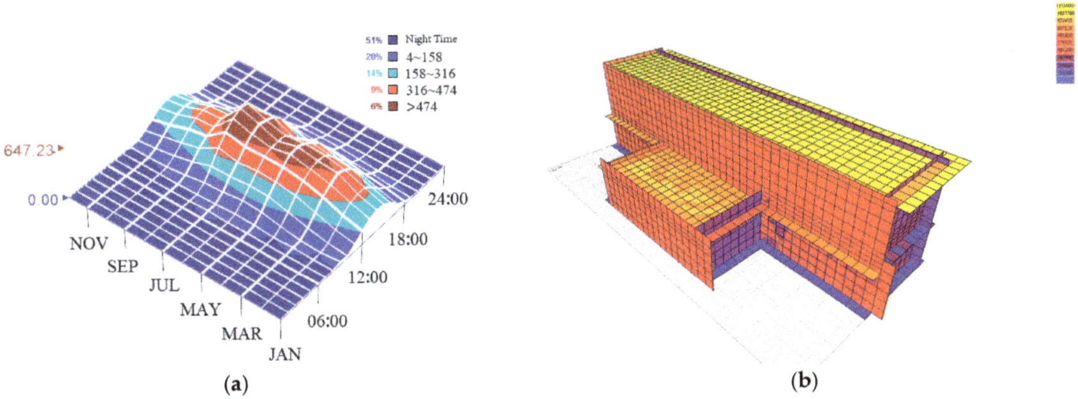


**Figure 12.** (**a**) Total solar radiation range at all times throughout the year in Xi'an region; (**b**) Three-dimensional annual solar radiation map of building surface.

Using Pvsyst software to simulate and calculate the polysilicon photovoltaic system, set the azimuth of photovoltaic array as 0°, and install it in the due south direction, the optimal installation inclination of photovoltaic panel is 25°; the annual power generation of the photovoltaic system was 169,057 kW·h; and the range of solar light available for photovoltaic modules under different solar elevation angle and solar azimuth angle are shown in Figures 13 and 14. The power generation of photovoltaic systems varies with different regions, temperatures, and installation inclinations. The comparison shows that the peak periods of power generation are concentrated in April, May, June, July, and August, and the power generation in July is the largest. Considering the attenuation of photovoltaic modules, the attenuation rate was set as 2% for the first year; the average attenuation rate from 2nd to 25th year was 0.72%; the total attenuation rate for 25 years was 20%; and the total power generation of photovoltaic system for 25 years was 3,826,130 kW·h. The annual power generation is shown in Table 10.

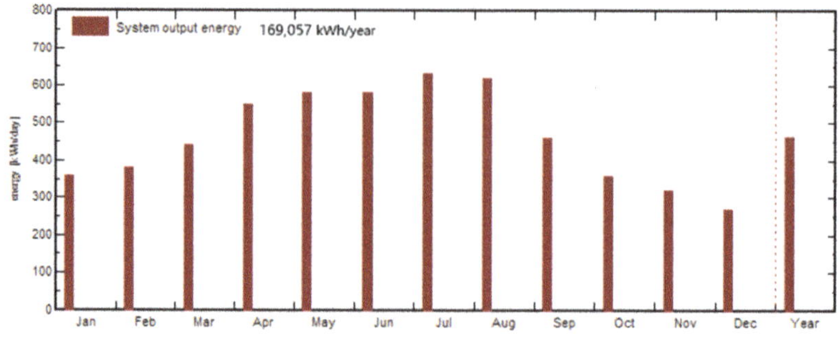


**Figure 13.** Monthly power generation of roof photovoltaic system.

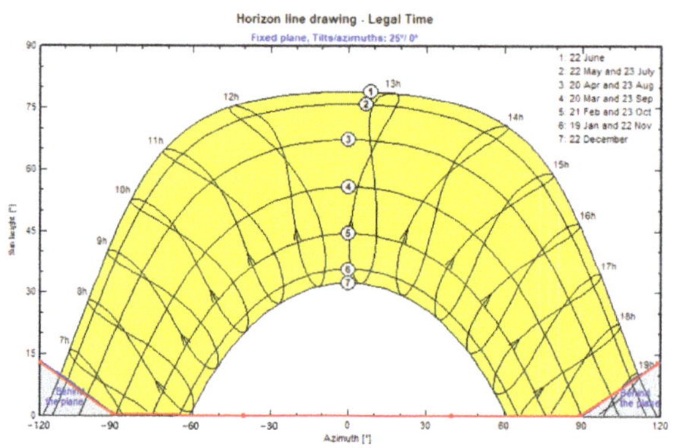


**Figure 14.** Range of solar light available for photovoltaic modules.

**Table 10.** Annual power generation of photovoltaic system over 25 years.

| Year | Power Generation/(kW·h) | Year | Power Generation/(kW·h) | Year | Power Generation/(kW·h) |
|---|---|---|---|---|---|
| 1 | 169,057 | 10 | 156,463 | 19 | 145,665 |
| 2 | 166,701 | 11 | 155,215 | 20 | 144,473 |
| 3 | 165,424 | 12 | 154,023 | 21 | 143,267 |
| 4 | 164,148 | 13 | 152,831 | 22 | 142,075 |
| 5 | 162,858 | 14 | 151,639 | 23 | 140,883 |
| 6 | 161,582 | 15 | 150,447 | 24 | 139,677 |
| 7 | 160,306 | 16 | 149,241 | 25 | 138,485 |
| 8 | 159,015 | 17 | 148,049 | | |
| 9 | 157,739 | 18 | 146,857 | | |

The net income in the life cycle of the photovoltaic system is the difference between the total power generation income of the photovoltaic system and the total investment cost in 25 years.

The power generation income $W$ in the 25-year life cycle is

$$W = [(B + D) \times Q \times 5 + D \times Q \times 20] \tag{2}$$

The total investment cost $Z$ is

$$Z = Z_1 + Z_2 + Z_3 \tag{3}$$

The static payback period $T$ is

$$T = \frac{Z_1}{W - Z} \tag{4}$$

where $B$ is the photovoltaic subsidy price of Xi'an, which is 0.1 CNY/kW·h for 5 consecutive years; $D$ is the electricity price of Xi'an; and $Q$ is the annual photovoltaic power generation. $Z_1$ is the initial investment cost of the photovoltaic system; $Z_2$ is the maintenance cost during system operation; and $Z_3$ is other costs, such as inverter maintenance and replacement. The calculation results of the main economic parameters are shown in Table 11. Combining the various costs, the payback period of investment was 5.61 years, and the total power generation income of the photovoltaic system is

$$W = 169{,}057 \times (0.62 + 0.10) \times 5 + 169{,}057 \times 0.62 \times 20 = 270.49 \text{ million CNY}$$

**Table 11.** Main economic parameters of photovoltaic system.

| Annual Power Generation/kW·h | Life Cycle/year | Initial Investment/ 10,000 CNY | Annual Maintenance Cost/10,000 CNY | Other Expenses/ 10,000 CNY |
|---|---|---|---|---|
| 169,057 | 25 | 203.53 | 2.60 | 23.85 |

## 4.2. Fresh Air Heat Recovery

Due to the large number of passengers transported daily at the coach station and the high density of people in the indoor waiting halls and ticket halls, mechanical ventilation is required to ensure fresh air volume for the indoor personnel. According to on-site investigations, it was found that the fresh air load of the coach station building can account for 40% to 70% of the total load. The use of a fresh air heat recovery system can preheat or precool the fresh air in summer and winter, as well as prehumidify or dehumidify it, thus reducing the energy consumption of the air conditioning system when handling fresh air. Both total heat and sensible heat recovery devices were used for the target buildings, and the two types of energy savings were calculated using the annual hourly dynamic calculation method. The year-round outdoor meteorological data for Xi'an were retrieved from the "China Building Thermal Environment Analysis Meteorological Dataset", including outdoor air temperature and enthalpy, as shown in Figure 15.

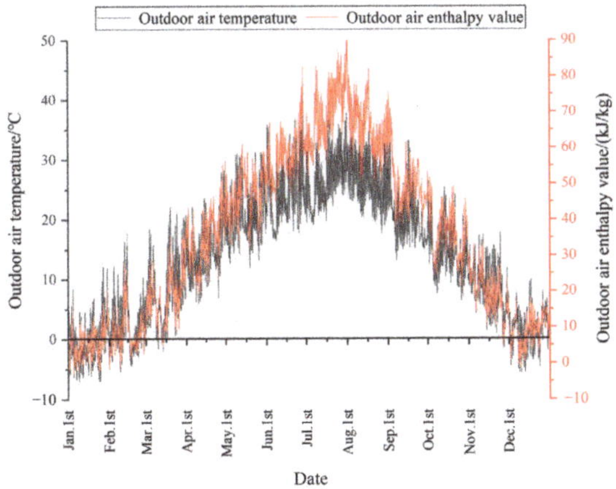


**Figure 15.** Annual parameters of outdoor air temperature and outdoor air enthalpy in Xi'an.

Due to the additional fan energy consumption caused by the resistance of the heat recovery module after the fresh air heat recovery system is started, this part of the energy consumption increment should be subtracted from the energy saving. When the net energy saving is greater than 0, the energy saving of the fresh air heat recovery system will be calculated. Using the calculation formula, the conditions when the net energy-saving rate is greater than 0 are shown in Table 12.

**Table 12.** Effective energy-saving conditions for fresh air heat recovery system.

| Type | Summer | Winter |
|---|---|---|
| Sensible heat recovery | tw ≥ 26.85 °C | tw ≤ 18.15 °C |
| Total heat recovery | hw ≥ 57.36 kJ/kg | hw ≤ 35.72 kJ/kg |

The formula for calculating the net energy saving is

$$W_t = \frac{G_E \rho \eta_t}{3600 \times COP} \sum_{i=1}^{h_t} |h_w - h_n| - W_f h_t \tag{5}$$

$$W_s = \frac{G_E \rho \eta_s C_P}{3600 \times COP} \sum_{i=1}^{h_s} |t_w - t_n| - W_f h_s \tag{6}$$

$$W_f = \frac{(G_F + G_P) \cdot \Delta P}{3600 \times \varnothing \times 1000} \tag{7}$$

where $W_t$ and $W_s$ are the net energy saving of total heat and the sensible heat recovery, kW·h; $W_f$ is the increase in fan energy consumption due to the resistance from the heat recovery module; $h_t$ and $h_s$ are the opening time of the heat recovery device, $h$; $h_n$ and $h_w$ are the enthalpy values of the indoor and outdoor air, KJ/kg; $t_n$ and $t_w$ are the temperatures of the indoor and outdoor air, °C; $\rho$ is the air density, and the value was 1.29 kg/m$^3$; $C_p$ is the mass specific heat capacity of the air at constant pressure, and the value was 1.005 kJ/(kg·°C); $G_E$ and $G_F$ are the exhaust air volume and fresh air volume, m$^3$/h; $\eta_t$ is the heat recovery efficiency of the total heat recovery device, which was 70% according to the standard; $\eta_s$ is the sensible heat recovery device efficiency, which was 75%; $\Delta P$ is the resistance of the heat recovery section, and had a value of 250 Pa; $\phi$ is the efficiency of the fan and had a value of 0.75; the COP of the air conditioning system had a value of 3.0.

The indoor temperature and humidity parameters were 20 °C, 40% in winter and 25 °C, 50% in summer. According to the calculation formula, the net energy saving in winter and summer is 37,273.77 kW·h and 10,525.15 kW·h, respectively, when using the total heat recovery device. When recycling the device, the net energy saving in winter and summer is 27,697.52 kW·h and 1635.73 kW·h. The economic efficiency of the heat recovery device is judged according to the static investment payback period, and its calculation formula is

$$C = \frac{I}{MW} \tag{8}$$

where C is the static payback period, a, and I is the initial investment cost of the fresh air heat recovery system. The fresh air full heat recovery module was 11 CNY/(m$^3$/h), and the sensible heat recovery was 8 CNY/(m$^3$/h). W is the annual energy saving, kW·h, and M is the electricity price in Xi'an, which was 0.62 CNY/kW·h.

The calculation result shows that the payback period for using the total heat recovery device was 2.09 years, while the payback period for the sensible heat recovery device was 2.97 years. Xi'an is located in a cold region where the winter time is longer than the summer time. The energy-saving effect of the heat recovery device in winter is significantly better than that in summer. As the total heat recovery device can simultaneously recover both the sensible heat caused by the temperature difference between the indoor and outdoor air and the latent heat caused by the moisture content, its energy-saving effect of the total heat recovery device is better than that of the sensible heat recovery device.

## 5. Conclusions

Through an on-the-spot investigation and indoor thermal environment tests of a highway bus station in the Xi'an area, the existing problems of the bus station in this area were analyzed. Based on simulation models, the energy-saving potential of the passenger bus station was studied from two aspects of the envelope structure and the utilization of renewable resources, and the following conclusions are drawn.

(1) The coach stations in the Xi'an area were built earlier as a whole; the majority of passenger stations were built before the year 2005; and the thermal performance of the building envelope structure does not meet the requirements of the new energy-saving design standards for public buildings. The annual comprehensive energy consumption is 46.68~89.58 kW·h/(m$^2$·a). The average indoor temperature of the indoor waiting hall area in winter and summer is 10.9 °C and 28.3 °C, both of which do not meet the standards for indoor environmental quality, with a requirement of 22–28 °C during summer and 16–24 °C during winter. The results show that there are energy waste and high energy consumption problems in the coach stations in the Xi'an area; while energy consumption measurements for various systems are lacking, the potential for energy efficiency in buildings is significant.

(2) Through the simulation of different thicknesses of external wall and roof insulation materials and building energy consumption under external window glass materials, it was found that with an increase in the insulation material thickness and the improvement of insulation materials, the energy-saving rate increased significantly, but the upward of the energy-saving rate change curve gradually tended to be flat. DeST-C software was used to calculate the building energy consumption of different material combinations. The energy-saving rate and net present value were combined, and the best enclosure structure combination was selected, that is, the external wall adopting 80 mm XPS external insulation, the external window glass adopting low-e film-coated insulating glass (6 + 9 + 6) (low permeability type), and using 80 mm PUR board for the roof insulation. The energy-saving rate of this scheme is 26.84%.

(3) The solar photovoltaic power generation system can effectively utilize the large-area roof space of the passenger station. Taking Chengxi Coach Stations as an example, the annual power generation is 169,057 kW·h, and the investment payback period is 5.61 years.

(4) Using a heat recovery device to pre-treat fresh air is also an effective measure to reduce energy consumption. The energy-saving effect of using total heat recovery is better than sensible heat recovery; the annual energy savings can reach 47,798.92 kW·h; and its investment payback period is less than 2.97 years.

**Author Contributions:** X.L.: conceptualization, writing—review and editing, funding acquisition; L.Q.: investigation, writing—original draft preparation, methodology, validation; J.L.: software, data curation, formal analysis; All authors have read and agreed to the published version of the manuscript.

**Funding:** This research was funded by the Shaanxi Provincial Social Science Fund Project (Grant No. 2022J052).

**Institutional Review Board Statement:** Not applicable.

**Informed Consent Statement:** Not applicable.

**Data Availability Statement:** Some data or models used during the study are available from the corresponding author by request.

**Conflicts of Interest:** The authors declare no conflict of interest.

## References


1. Sanchez Rodriguez, R.; Ürge-Vorsatz, D.; Barau, A.S. Sustainable Development Goals and climate change adaptation in cities. *Nat. Clim. Chang.* **2018**, *8*, 181–183. [CrossRef]
2. Pogge, T.; Sengupta, M. The Sustainable Development Goals: A plan for building a better world? *J. Glob. Ethics* **2015**, *11*, 56–64. [CrossRef]
3. Di Foggia, G. Energy efficiency measures in buildings for achieving sustainable development goals. *Heliyon* **2018**, *4*, e00953. [CrossRef]
4. Goubran, S.; Cucuzzella, C. Integrating the sustainable development goals in building projects. *J. Sustain. Res.* **2019**, *1*, e190010.
5. Huo, T.F.; Cai, W.G.; Ren, H.; Feng, W.; Zhu, M.L.; Lang, N.N.; Gao, J.X. China's building stock estimation and energy intensity analysis. *J. Clean. Prod.* **2019**, *207*, 801–813. [CrossRef]
6. Lin, L.; Liu, X.H.; Zhang, T.; Liu, X.T. Energy consumption index and evaluation method of public traffic buildings in China. *Sustain. Cities Soc.* **2020**, *57*, 102132. [CrossRef]
7. Han, Y.; Long, Z.J.; Wen, J.P.; Sun, Z.; Li, D.; Li, K. Energy Consumption Simulation and Economic Study on the Insulation Layer Thickness of Desert Passenger Station Building. *Eng. Adv.* **2023**, *3*, 1–10. [CrossRef]
8. National Bureau of Statistics of China. China Statistical Yearbook-2021. Available online: http://www.stats.gov.cn/tjsj/ndsj/2021/indexch.htm (accessed on 22 July 2022).
9. Zheng, X.J.; Xu, B.X.; You, S.J.; Zhang, H.; Wang, Y.R.; Sun, L.Z. Energy consumption and $CO_2$ emissions of coach stations in China. *Energies* **2020**, *13*, 3600. [CrossRef]
10. Fan, L.; Tian, X.; Wang, S. Study on the Factors of Energy Consumption of Highway Transport Passenger Station. In Proceedings of the 2013 the International Conference on Remote Sensing, Environment and Transportation Engineering (RSETE 2013), Nanjing, China, 26–28 July 2013; Atlantis Press: Amsterdam, The Netherlands, 2013; pp. 224–226.
11. Yang, T.J. Study on the energy efficiency classification method of highway passenger stations based on climatic divisions. In *AIP Conference Proceedings*; AIP Publishing LLC: College Park, MD, USA, 2018; Volume 1971.
12. Huang, M.; Lin, Y.J. Thermal comfort of railway station's waiting room in severe cold regions of China. *Energy Procedia* **2017**, *134*, 749–756. [CrossRef]
13. Mambo, A.D.; Eftekhari, M.; Steffen, T. Evaluation of indoor environment system performance for airport buildings. *Int. J. Sustain. Green Energy* **2015**, *4*, 73–84.
14. Kotopouleas, A.; Marialena, N. Evaluation of comfort conditions in airport terminal buildings. *Build. Environ.* **2018**, *130*, 162–178. [CrossRef]
15. Sinha, A.; Gupta, R.; Kutnar, A. Sustainable development and green buildings. *Drv. Ind.* **2013**, *64*, 45–53. [CrossRef]
16. Balali, A.; Valipour, A. Identification and selection of building façade's smart materials according to sustainable development goals. *Sustain. Mater. Technol.* **2020**, *26*, e00213. [CrossRef]
17. Glaser, G. Base sustainable development goals on science. *Nature* **2012**, *491*, 35. [CrossRef]
18. Ragheb, A.; El-Shimy, H.; Ragheb, G. Green architecture: A concept of sustainability. *Procedia-Soc. Behav. Sci.* **2016**, *216*, 778–787. [CrossRef]
19. Wen, B.H.; Musa, S.N.; Onn, C.C.; Ramesh, S.; Liang, L.H.; Wang, W.; Ma, K. The role and contribution of green buildings on sustainable development goals. *Build. Environ.* **2020**, *185*, 107091. [CrossRef]
20. Mora, R.; Girma, B.; Miljana, H. Integrated life-cycle design of building enclosures. *Build. Environ.* **2011**, *46*, 1469–1479. [CrossRef]
21. Zhou, Y. Evaluation of renewable energy utilization efficiency in buildings with exergy analysis. *Appl. Therm. Eng.* **2018**, *137*, 430–439. [CrossRef]
22. Li, D.Y.; Jiang, H.; Lin, L. A review of renewable energy applications in buildings in the hot-summer and warm-winter region of China. *Renew. Sustain. Energy Rev.* **2016**, *57*, 327–336. [CrossRef]
23. Chel, A.; Kaushik, G. Renewable energy technologies for sustainable development of energy efficient building. *Alex. Eng. J.* **2018**, *57*, 655–669. [CrossRef]
24. Tong, Z.N. Building Enclosure Structure Energy Saving Discussion. In *Advanced Materials Research*; Trans Tech Publications Ltd.: Stafa-Zurich, Switzerland, 2013; Volume 671.
25. Kheiri, F. A review on optimization methods applied in energy-efficient building geometry and envelope design. *Renew. Sustain. Energy Rev.* **2018**, *92*, 897–920. [CrossRef]
26. Huang, J.N.; Lv, H.L.; Gao, T.; Feng, W.; Chen, Y.X.; Zhou, T. Thermal properties optimization of envelope in energy-saving renovation of existing public buildings. *Energy Build.* **2014**, *75*, 504–510. [CrossRef]
27. Zhu, J.J.; Chew, D.A.S.; Lv, S.N.; Wu, W.W. Optimization method for building envelope design to minimize carbon emissions of building operational energy consumption using orthogonal experimental design (OED). *Habitat Int.* **2013**, *37*, 148–154. [CrossRef]
28. El Gindi, S.; Abdin, A.R.; Hassan, A. Building integrated Photovoltaic Retrofitting in office buildings. *Energy Procedia* **2017**, *115*, 239–252. [CrossRef]
29. Feng, W.; Huang, J.N.; Lv, H.L.; Guo, D.P.; Huang, Z.C. Determination of the economical insulation thickness of building envelopes simultaneously in energy-saving renovation of existing residential buildings. *Energy Sources Part A Recovery Util. Environ. Eff.* **2019**, *41*, 665–676. [CrossRef]
30. Fouad, M.M.; Lamia, A.S.; Mohamed, A.H. Modeling and analysis of Building Attached Photovoltaic Integrated Shading Systems (BAPVIS) aiming for zero energy buildings in hot regions. *J. Build. Eng.* **2019**, *21*, 18–27. [CrossRef]

31. Kylili, A.; Paris, A.F. Investigation of building integrated photovoltaics potential in achieving the zero energy building target. *Indoor Built Environ.* **2014**, *23*, 92–106. [CrossRef]
32. Lu, Y.J.; Chang, R.D.; Shabunko, V.; Amy, T.L.Y. The implementation of building-integrated photovoltaics in Singapore: Drivers versus barriers. *Energy* **2019**, *168*, 400–408. [CrossRef]
33. Faranda, R.; Sonia, L. Energetic sustainable development of railway stations. In Proceedings of the 2007 IEEE Power Engineering Society General Meeting, Tampa, FL, USA, 24–28 June 2007; IEEE: Piscataway, NJ, USA, 2007.
34. Zhang, W.; Nie, J.Z. *Study on the Applicability of Fresh Air Heat Recovery System in Typical Buildings in Different Climatic Regions in China*; Architecture Science: Beijing, China, 2022.
35. Mardiana-Idayu, A.; Riffat, S.B. Review on heat recovery technologies for building applications. *Renew. Sustain. Energy Rev.* **2012**, *16*, 1241–1255. [CrossRef]
36. Cuce, P.M.; Riffat, S. A comprehensive review of heat recovery systems for building applications. *Renew. Sustain. Energy Rev.* **2015**, *47*, 665–682. [CrossRef]
37. Liu, X.P.; Niu, J.L. An optimal design analysis method for heat recovery devices in building applications. *Appl. Energy* **2014**, *129*, 364–372. [CrossRef]
38. Scrucca, F.; Ingrao, C.; Barberio, G.; Matarazzo, A.; Lagioia, G. On the role of sustainable buildings in achieving the 2030 UN sustainable development goals. *Environ. Impact Assess. Rev.* **2023**, *100*, 107069. [CrossRef]
39. *GB 50189-2015*; Design Standard for Energy Efficiency of Public Buildings. China Architecture & Building Press: Beijing, China, 2015.
40. *GB/T 18883-2022*; Indoor Air Quality Standard. State Environmental Protection Administration: Beijing, China, 2022.
41. *GB/T 50785-2012*; Evaluation Standard for Indoor Thermal Environment in Civil Buildings. China Architecture & Building Press: Beijing, China, 2012.
42. Sozer, H. Improving energy efficiency through the design of the building envelope. *Build. Environ.* **2010**, *45*, 2581–2593. [CrossRef]
43. Pacheco, R.; Javier, O.; Germán, M. Energy efficient design of building: A review. *Renew. Sustain. Energy Rev.* **2012**, *16*, 3559–3573. [CrossRef]
44. Huo, H.; Jing, C.J.; Li, K.; Huo, H.B. Synergic relationships between thermophysical properties of wall materials in energy-saving building design. *Int. J. Heat Mass Transf.* **2015**, *90*, 246–253. [CrossRef]
45. Zhu, J.; Li, D. Current situation of energy consumption and energy saving analysis of large public building. *Procedia Eng.* **2015**, *121*, 1208–1214. [CrossRef]
46. Feng, Q. Energy-Saving System Analysis about the Exterior Building Enclosures of Wenyuan Building. In *Advanced Materials Research*; Trans Tech Publications Ltd.: Stafa-Zurich, Switzerland, 2011; Volume 163.

buildings

MDPI

*Article*

# Development, Calibration, and Validation of a Simulation Model for Indoor Temperature Prediction and HVAC System Fault Detection

**Darko Palaić [1], Ivan Štajduhar [1,2], Sandi Ljubic [1,2] and Igor Wolf [1,*]**

1 Faculty of Engineering, University of Rijeka, Vukovarska 58, HR-51000 Rijeka, Croatia; darko.palaic@riteh.hr (D.P.); ivan.stajduhar@riteh.hr (I.Š.); sandi.ljubic@riteh.hr (S.L.)
2 Center for Artificial Intelligence and Cybersecurity, University of Rijeka, R. Matejcic 2, HR-51000 Rijeka, Croatia
* Correspondence: igor.wolf@riteh.hr

**Abstract:** An effective approach to increasing energy efficiency in buildings without compromising thermal comfort is to optimize heating, ventilation, and air conditioning (HVAC) systems through the use of advanced building-management system features, such as fault detection and diagnosis. Such functions are usually developed based on simulation models that must be calibrated and validated to achieve an appropriate level of accuracy and reliability. The objective of this study was to develop and calibrate a room-level simulation model of a hotel building and its HVAC system using TRNSYS 18 software and real data collected from the smart room system installed in the building. The calibration process was performed with 100 rooms using 5-min samples of room temperatures in selected 1-month periods during the summer and winter seasons by minimizing the root mean squared error *(RMSE)* in the average of the observed rooms using a genetic algorithm. The calibrated model was able to predict room temperatures with an *RMSE* of 0.79 ± 0.14 °C and a coefficient of variation in the root mean squared error (*cvRMSE*) of 3.58 ± 0.7%, which is well below the limits prescribed by international guidelines. The model was then applied to detect faults in the operation of fan coil units in the rooms based on the residual analysis and defined if–then rules. The results obtained show that the model can track the trends of temperature changes in real conditions and successfully detect major anomalies in a system.

**Keywords:** building energy performance; HVAC system; smart-room system; building modeling and simulation; building thermal dynamics; model validation and calibration; genetic algorithm

## 1. Introduction

Climate change, caused by increasing greenhouse gas (GHG) emissions due to human activities, is of major concern worldwide. The buildings sector contributes significantly to this problem, both directly through GHG emissions and indirectly through energy consumption. Buildings account for around 34% of global final energy consumption and 33% of energy- and process-related GHG emissions, with heating, ventilation, and air-conditioning (HVAC) systems playing a significant role [1]. Increasing energy efficiency in buildings without compromising occupant comfort is critical to reducing their environmental impact.

To reduce both energy consumption and GHG emissions in this sector, the development of energy-efficient, environmentally friendly and sustainable buildings, such as net-zero-energy/emission buildings, and the refurbishment of existing buildings are required [2]. The mainly renewable-energy-based technical systems in these buildings must be equipped with efficient, properly installed and commissioned components controlled by advanced building automation and control systems. Significant improvements in reducing energy consumption and GHG emissions in existing buildings can already be achieved in a simple, cost-effective and sustainable manner by implementing smart technologies and control algorithms in HVAC systems that optimize their performance and avoid unnecessary energy waste [3].

The control of HVAC systems in most buildings currently relies on rule-based methods [4]. However, a promising approach for improvement is model predictive control, which aims to save energy by predicting future events and controlling system operation based on these predictions [5,6]. In addition, detection and diagnosis of faults in HVAC systems can also contribute to energy savings, as faulty operation can lead to increased energy demand and affect occupant thermal comfort [7]. The two strategies for improving control and troubleshooting are based on developing a model of the building and its systems that can simulate specific features such as indoor temperatures.

Models for predicting building energy consumption and indoor temperatures can be developed using different techniques—white-box, black-box, and grey-box—with each technique having its own advantages and disadvantages [8–10]. In white-box models, the thermal behavior of the building and its HVAC system is described by physical laws. The advantages are interpretable results, high accuracy in cases where all required information is known, universality of modeling tools, and the fact that white-box models can be created without historical data on building operation. However, model creation requires detailed specification of physical parameters, expertise, and some effort, which can lead to less accurate results than other modeling techniques when model simplifications and constraints must be applied. In contrast, black-box models—also called data-driven models—use only data collected in the field for modeling. The models are generally more accurate than white-box models when detailed information about building physics is lacking, and they can be adjusted and extended with newly collected data. On the other hand, the results obtained cannot be physically explained, the accuracy of the models under previously unknown working conditions is unreliable, and high-quality and informative data sets are needed. Finally, grey-box models—also called hybrid models—combine and balance the advantages of both modeling techniques. Grey-box models, resistance-capacitance (RC) networks, or physical models integrated with data-driven algorithms are easier to build than white-box models because simplified equations describing physical effects are used, and they can provide interpretable results; however, development tools need to be improved and methods standardized.

White-box models, which remain the most widely used, are useful tools for architectural design, energy performance evaluation and prediction, and system component and management optimization [11]. Software programs typically used for white-box model development include TRNSYS, EnergyPlus, and IDA Indoor Climate and Energy (IDA-ICE). One problem that can arise during model development is the divergence between simulated and measured results due to uncertainties in the design variables [9]. To reduce or eliminate this so-called performance gap and increase reliability, calibration is required, which is defined as the process of adjusting the input parameters in the model with the goal of improving agreement with real data [12,13]. The building envelope properties, infiltration rates, internal heat gain density, and some specific HVAC system variables are often used as adjustable parameters in this process, which can be very computationally intensive, especially when each parameter needs to be fine-tuned. To reduce the complexity, the process can focus on the most influential parameters, which are determined by sensitivity analysis [14]. Calibration can generally be performed manually or automatically. In manual methods, improvements are often achieved through trial and error in adjusting some key parameters of the model, but these methods also require expert knowledge and evidence-based processes (building audits). Automatic methods are usually formulated as optimization problems that can be solved using optimization techniques such as genetic algorithms. The calibration process can have one or multiple objectives, with hourly indoor temperatures and monthly electricity consumption being the most common. The calibration performance of energy-consumption simulation models is typically evaluated using the normalized mean bias error ($nMBE$) and coefficient of variation of the root mean squared error (*cvRMSE*), recommended by the American Society of Heating, Refrigerating and Air Conditioning Engineers (ASHRAE) and the International Performance Maintenance and Verification Protocol (IPMVP). These same indices are also widely used to assess the calibration performance of indoor temperature prediction models. However, the mean absolute error (*MAE*) and root mean squared error (*RMSE*) are more accurate and reliable metrics for this purpose and should be used instead [15]. The calibration process remains a difficult and time-consuming task due to the lack of clear guidelines and best practices [13].

### 1.1. Related Work

In recent years, there have been several studies that have investigated the use of different models and calibration methods for temperature prediction in buildings. In their study, Arendt et al. [16] compared white-box, grey-box, and black-box models and found that the black-box models provided the most accurate temperature predictions in most cases, with an average mean absolute error of the best such model of 0.4 °C, compared to 1.0 °C for the grey-box and 0.7 °C for the white-box model. The white-box model was advantageous for working conditions not included in the original data set. O'Donovan et al. [17] created a white-box model of a nearly zero-energy building (nZEB) to predict indoor temperatures at the room level, focusing on three specific rooms and accounting for occupancy schedules and window-opening strategies. After manually calibrating the model created in TRNSYS, the accuracy of the temperature prediction, as measured by the *RMSE* metric, ranged from 0.27 °C to 1.5 °C, with an average *cvRMSE* of up to 20%. Guyot et al. [18] performed a manual calibration of a white-box model of an office building with 132 thermal zones and achieved a temperature prediction accuracy of about ±2 °C compared to the measured values. The calibration was performed over one year of simulation with a time step of five minutes. Pachano and Bandera [15] created a white-box model of a school complex with 27 thermal zones for hourly predictions of heating and cooling, electricity consumption, and indoor air temperatures. They achieved an *RMSE* of 0.29 °C by dividing the calibration process of the model into several steps and automated it using a genetic algorithm. Baba et al. [19] developed an automated method to calibrate the building simulation model using the hourly indoor temperatures in multiple rooms simultaneously. By incorporating a variance-based sensitivity analysis of the most influential building parameters, a multi-objective genetic algorithm for calibration, and some newly developed evaluation criteria, they achieved a temperature prediction accuracy of 0.3 °C, measured by the *RMSE* of a school building modeled in EnergyPlus. Figueiredo et al. [20] proposed a multi-stage hybrid evolutionary algorithm for the calibration of building simulation models, which they applied to a college building. Their building model consisted of 24 thermal zones. The proposed optimization procedure was validated by comparing hourly simulated and measured data for one year in two test rooms. They obtained *RMSE* values for the rooms of 1.17 °C and 1.33 °C, respectively, and *cvRMSE* values of 4.5% and 5.4%, respectively. Royapoor and Roskilly [21] investigated the accuracy of an EnergyPlus model for a five-story office building. They emphasized the importance of using local weather data in the calibration process to minimize errors. The calibrated model predicted annual hourly indoor air temperatures with an error of ±1.5 °C and a *cvRMSE* value of less than 2%. Rosato et al. [22] developed a detailed simulation model of a laboratory test room and its HVAC system in TRNSYS and achieved an *RMSE* of 0.39 °C when comparing with measured data. They used the model to set up fault detection for the HVAC system by analyzing the variations in sensory measurements. In [23], Roberti et al. pointed out the importance of calibrating building models with more than a single parameter, especially for old houses and incomplete information about their construction. They proposed a calibration method based on a particle swarm optimization algorithm. The model created in EnergyPlus was validated with hourly indoor air and surface temperatures, with an *RMSE* between 0.4 °C and 0.8 °C. Kosak and Stadler [24] pointed out the importance of including storage mass in thermal energy modeling when predicting temperatures in thermal zones. Inappropriate settings and methods for modeling thermal mass can lead to a deviation in model prediction results of up to 26% [13]. Martínez-Ibernón et al. [25] compared two modeling approaches for predicting indoor air temperatures and relative humidity in buildings, the first considering the common modeling assumption of a completely dry building envelope and the second considering the presence of moisture in the building elements. By simulating an unheated building as a single thermal zone in TRNSYS, they obtained more accurate results in the second case (an *RMSE* of 1.35 °C compared to 1.65 °C in the first case), indicating the influence of humidity on heat balance and air temperature prediction. Martínez-Mariño et al. [26] created a combined multi-zone airflow model for buildings using TRNSYS and TRNFLOW to predict indoor temperatures and relative humidity considering occupancy and moisture gains, window and door opening, and mechanical ventilation. Using a multi-objective calibration with a genetic algorithm, they obtained average *RMSE* values for indoor temperatures of 0.51 °C and 0.48 °C (*cvRMSE* 2.27% and 2.15%) and for relative humidity of 3.58% and 2.21% (*cvRMSE* 13.8% and 6.5%), respectively, for two case study houses. They emphasized the need to model moisture buffering in walls to accurately predict indoor relative humidity. Murphy et al. [27] developed a grey-box model for nZEB using an automatic calibration algorithm and compared it to a white-box model. Although the white-box model produced slightly more accurate results, with a 1.5% difference in the *RMSE* metric, the grey-box model significantly reduced development time and effort. Berthou et al. [28] created a grey-box model using multiple RC networks. Determining the number of resistances and capacitances is the most important step in building the model. They found that the R6C2 model gave the best results for temperature prediction, in terms of both accuracy and complexity. Belić et al. [29] presented an inverse modeling approach for indoor temperature prediction in multi-zone buildings using a RC model network based on building data. They compared the results of their model with measured temperatures and arrived at an *RMSE* value of 0.33 °C for the optimized model and 0.64 °C when applying the same model to another data set with similar working conditions. Cui et al. [30] created a hybrid model by combining a RC network and an artificial neural network (ANN). The RC model was used to predict the mean indoor air temperature of a two-story single-family house, while the ANN model was used to predict the temperature difference between two thermal zones (two floors) in the house. They compared the simulation results with measurements and obtained an average *RMSE* of the hybrid model of 0.64 °C.

When simulating building energy performance, outdoor conditions, building fabric, and occupant behavior must be considered [7,31]. Solar radiation and outdoor temperatures have a great impact on indoor thermal dynamics, as noted in the study by Aguilera et al. [32]. The main uncertainty factor in the simulations is the occupants' behavior, as they have control over windows, lights, shading, and set temperatures [33,34]. Usually, occupants' actions are modeled as fixed schedules with constant values [35]. In addition, interactions with adjacent spaces must also be considered, as heat flows along the path of least resistance from areas of higher temperature to those of lower temperature [36].

A detailed simulation model can be used to detect anomalies in an HVAC system. Discrepancies between simulated thermal responses and real field measurements may indicate errors in system operation. Anomalies can be detected by analyzing deviations from fault-free trends of residuals calculated from the differences between predicted and measured values at specific time periods [37]. In recent studies, dynamic simulation models for fan coil units have been used to simulate thermal behavior and find errors in the system [38,39]. Such white-box models can generate a labeled data set for normal and faulty operation and allow implementation of data-driven techniques in automatic detection algorithms [40].

### 1.2. Contribution and Structure

This study presents the development, calibration, and validation of a white-box model for predicting indoor temperatures in a hotel building. Temperatures were simulated under real conditions using data collected from the smart room system installed in the building several years ago. The simulations took into account occupant activity, window opening, HVAC system control logic in the rooms, and external weather conditions. The goal of the model development was not only to simulate the room temperatures but also to detect anomalies in the HVAC system operation, especially related to the fan coil units in the rooms.

The literature search revealed that there are not many studies that specifically address temperature prediction in hotel rooms. HVAC systems in hotels can be very complex, room occupancy fluctuates, and guest behavior is stochastic and depends on their expectations and comfort preferences, which affects HVAC system operation and window opening. All of this introduces many uncertainties into a simulation model that is intended to be applied to every room in the building, which calls into question its accuracy and reliability. It is also not common to have such an extensive database of room-specific data, as is the case in this work, thanks to the records of the smart-room system.

The contribution of this study can be summarized in the following points:

- TRNSYS 18 software was used to develop a room-level simulation model of a hotel building in Zagreb, Croatia, and its HVAC (fan coil) system. Each room was set up as a separate thermal zone with appropriate boundary conditions and designed using the Google SketchUp 3D 2021 tool. To simplify and speed up the simulations, the model was created for a single room and applied to all rooms with similar physical characteristics by changing the room-specific parameters. In this way, the accuracy and reliability of the model in predicting indoor temperatures was increased relatively quickly through its calibration and validation with a genetic algorithm. The sensitivity analysis identified the most influential building and HVAC operating parameters, which were later used in the calibration process. Python scripts were developed to automatically create and run new simulations and process the results. Summer and winter seasons were considered, and the performances of the original (baseline) model and the calibrated model were compared.
- The simulations were fed with five-minute samples of real data collected over several years from the smart-room system installed in the hotel and integrated with the central building automation and control system. The data included temperature control logic, room occupancy, window opening, HVAC system status, and operating mode in each hotel room. Incorporating all of this data made developing the model challenging, but its accuracy is within the acceptable limits of the VDI, ASHRAE, and IPMVP guidelines. It has been shown to be applicable to any room in the building with similar characteristics but specific input parameters.
- The developed model has been used to detect anomalies associated with the operation of fan coil units in the rooms. It has successfully revealed the main anomalies in the system. However, further tests and diagnostics need to be performed to confirm the anomalies and their real causes. As a next step, the model should help develop more advanced automatic fault detection and diagnosis routines based on machine learning techniques that can be implemented in the building's smart-room system in the future.

This article is organized as follows. Section 2 first presents the materials used to build the simulation model. This includes a description of the case study hotel building, the collection of field data, and the operation and control of the HVAC system. Then, all the steps used to create the thermal model of the building and the simulation environment are explained, as well as the model calibration and validation processes. The results are presented and discussed in Section 3. Finally, conclusions are drawn in Section 4.

## 2. Materials and Methods

The objective of this study was to create a reliable white-box or simulation model of the guest rooms of the case study hotel. The model includes the rooms, each of which is considered a separate thermal zone, the HVAC system equipment in the rooms, the HVAC system control logic, and details about real-world conditions. The main purpose of the model is to predict the room air temperatures and detect possible anomalies associated with the HVAC system.

The following sections first describe the hotel building used for model development, the components of the HVAC system in the rooms and its control system, the available multi-year field data recorded independently by the installed smart-room system for each room in the building, and the weather data. The methods used to develop and calibrate the model are then presented.

### 2.1. Building Description

The case study building in the research was a hotel in Zagreb, Croatia (45°48′ N, 16°0′ E; site elevation 112 m). The building has 16 floors, 9 of which are occupied by guest rooms. All floors with rooms have the same floor plan and room layout, except for the last two. Each floor consists of 24 rooms, 20 of which have a floor area of 26 $m^2$ and 4 of which have a floor area of 32 $m^2$. The height of a room is 2.6 m. All rooms of the same size have an identical floor plan, with a bedroom and a bathroom for two people. Figure 1 shows the floor plan of a representative floor, including the overall dimensions of the building. Below and above the floors with the guest rooms are some public areas and offices, which were not considered in detail. The hotel is surrounded by other buildings on the east, north, and west sides, with the south side facing the wide avenue. The buildings on the west side are much lower, and those on the east and north sides are about half the height, partially shading the hotel and protecting it from direct wind gusts.

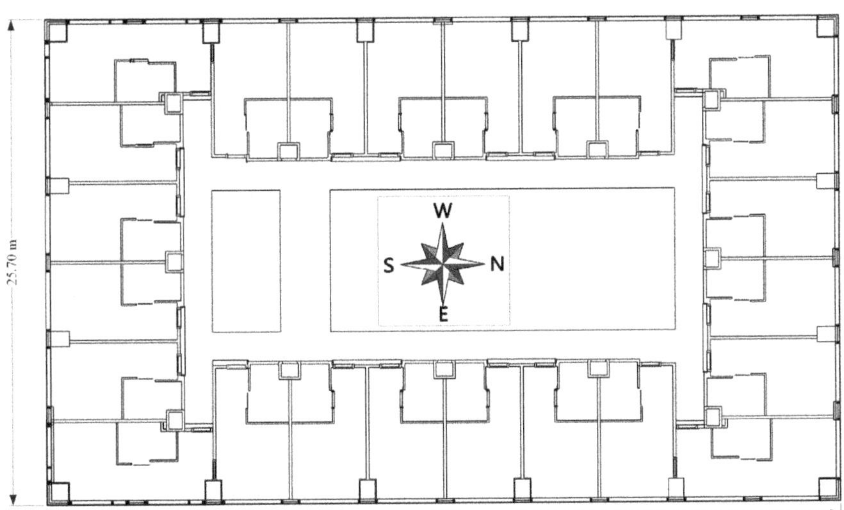


**Figure 1.** Floor plan of a single floor of the hotel.

Table 1 lists the building elements of the smaller rooms and shows the properties of each element layer, all derived from the main architectural design of the building. In Table 1, $\delta$ represents the thickness of the layer, $\lambda$ the thermal conductivity, $c$ the specific heat capacity, $\rho$ the density, and $A$ the surface area of an element. The windows consist of an aluminum frame and two layers of IZO glass, with an inert gas between them. Their overall heat transfer coefficient ($U$-value) and solar heat gain coefficient are 1.10 W/($m^2$K) and 0.33, respectively, and each window has an area of 8.65 $m^2$. Apart from the manually operated curtains, there are no shading devices on the windows.

**Table 1.** Composition of building elements and material properties in rooms.

| Elements | Material | $\delta$ [m] | $\lambda$ [W/(mK)] | $c$ [kJ/(kgK)] | $\rho$ [kg/m$^3$] | $A$ [m$^2$] |
|---|---|---|---|---|---|---|
| Ceiling /Floor | Plasterboard | 0.020 | 0.250 | 0.90 | 900 | 26.04 |
| | Air gap | 0.300 | 1.310 | 1 | 1 | |
| | Reinforced concrete | 0.300 | 2.500 | 1 | 2500 | |
| | Air gap | 0.125 | 0.570 | 1 | 1 | |
| | Wood | 0.030 | 0.180 | 1.60 | 700 | |
| Adjacent Walls | Plasterboards | 0.075 | 0.250 | 0.90 | 900 | 22.32 |
| | Thermal insulation layers | 0.150 | 0.035 | 1.03 | 100 | |
| External Wall | Plasterboard | 0.030 | 0.250 | 0.90 | 900 | 6.48 |
| | Air gap | 0.035 | 0.190 | 1 | 1 | |
| | Thermal insulation layer | 0.120 | 0.035 | 1.03 | 100 | |
| | Air gap | 0.015 | 0.090 | 1 | 1 | |
| | Glass | 0.008 | 1 | 0.75 | 2500 | |

Each room is equipped with a four-pipe fan coil unit (FCU) that can both heat and cool depending on actual temperature conditions and guest needs. The bathroom does not have an additional heating element. The FCU is connected to a central mechanical ventilation system of the hotel. All features of the HVAC system come from the designer's as-built documentation and the FCU manufacturer´s technical specifications.

A smart-room system installed in each room controls and monitors all systems in the room, such as FCU operation and lighting, and provides security through alarms. All sensory measurements and control outputs are stored by the central controller of the smart-room system for easy access and analysis. Overall management of all systems installed in the hotel is handled by the central building automation and control system.

### 2.2. HVAC System and Control Logic

The FCU in every room consists of an air filter, a fan, a hot-water heating coil, a chilled-water cooling coil, and two control valves. Each unit has 3 fan speeds and a rated capacity of 1600 W for both heating and cooling at nominal values (Table 2). It recirculates room air but is also continuously supplied with fresh air from the central mechanical ventilation system at a flow rate of 60 m$^3$/h at design temperature and humidity. The same amount of air is extracted from the bathroom. The water supply temperature in heating mode depends on the outdoor air temperature (weather compensation function under control by the central building automation and control system) and is 70 °C under nominal working conditions at −15 °C outdoors, while the water supply temperature in cooling mode is 9 °C. In the off-season, guests can switch between heating and cooling modes, while in other seasons, only one mode is available. Despite the installation of a four-pipe system, the unit can only heat during the winter season and cool during the summer season. If a guest switches to cooling mode during the winter season or to heating mode during the summer season, the unit will only ventilate the space.

**Table 2.** Operating characteristics of fan coil units.

| Parameter | Value |
|---|---|
| Air flow rate | 265/460/552 kg/h |
| Water flow rate | 68 kg/h (heating)/345 kg/h (cooling) |
| Supply water temperature | 70 °C at −15 °C (heating)/9 °C (cooling) |

Control of the FCU in this setting is based on the use of two thermostats, one for cooling mode and one for heating mode. These thermostats operate at three different levels and control the fan speeds accordingly. The activation of these speeds depends on the difference between the set temperature and the measured room temperature. The set temperature is determined by the central smart-room system control but can be changed by guests at any time. There are two modes for regulating the speed of the FCU: automatic and manual. In automatic mode, the fan speed is automatically changed according to the control logic. In manual mode, guests have the option to adjust the fan speed themselves. Guests also have the ability to turn the unit on and off at will.

The control logic of the system also takes into account special cases such as opening windows. When the window is opened, the control system automatically turns off the FCU to save energy.

### 2.3. Data Gathering

The smart-room system installed in the hotel collected data from 166 guest rooms at a sampling rate of 5 min between 2013 and 2021. While this sampling rate did not capture all possible events, the data still provide insight into the performance of the system. Table 3 lists and describes the data collected.

**Table 3.** Room-based sensory measurements from the building.

| Variable | Type | Description |
|---|---|---|
| SET TEMP. | Integer [°C] | Temperature set by guest or hotel staff |
| ROOM TEMP. | Integer [°C] | Measured room air temperature |
| HVAC SPEED | Integer [0–3] | FCU fan speed |
| HVAC STATE | Boolean | FCU status (on/off) |
| HVAC MODE | Boolean | FCU heating or cooling mode (on/off) |
| OCCUPANCY | Boolean | Presence of people in the room |
| WINDOW | Boolean | Window status (open/closed) |

Prior to developing the simulation model, the data were analyzed and processed. The data set contained missing and erroneous values that were identified and replaced with linearly interpolated values. The HVAC system also had some issues with its control logic due to system maintenance. The periods where problems occurred were excluded from the model calibration and validation processes. Preprocessing was necessary to ensure high-quality data for accurate model calibration.

In addition to the data collected at the hotel, outdoor conditions recorded by the Croatian Meteorological and Hydrological Service at a nearby weather station were also used [41]. The multi-year weather data, stored in a database as hourly averages of 10 min samples, were partially preprocessed to implement them in the model. The data used to represent the external environment are listed in Table 4.

**Table 4.** Environmental data.

| Variable | Description |
|---|---|
| OUTSIDE TEMP. | Outdoor air temperature, °C |
| IRRADIANCE | Solar irradiation on horizontal surface, W/m$^2$ |
| HUMIDITY | Relative humidity of outdoor air, % |

### 2.4. Simulation Model Development and Calibration

Based on the previous description of the hotel, its FCU system, and the available data, the building model was developed in a simulation environment. Figure 2 shows a flowchart illustrating the process of creating and calibrating the simulation model, which includes the creation of multiple simulations and model optimization. Several software solutions are used, including TRNSYS 18 [42] to develop the model and simulate the thermal behavior of the building and its FCU system, Google SketchUp 3D 2021 to define the geometry of the thermal zones, and Python to run the simulations and perform the calibration process.

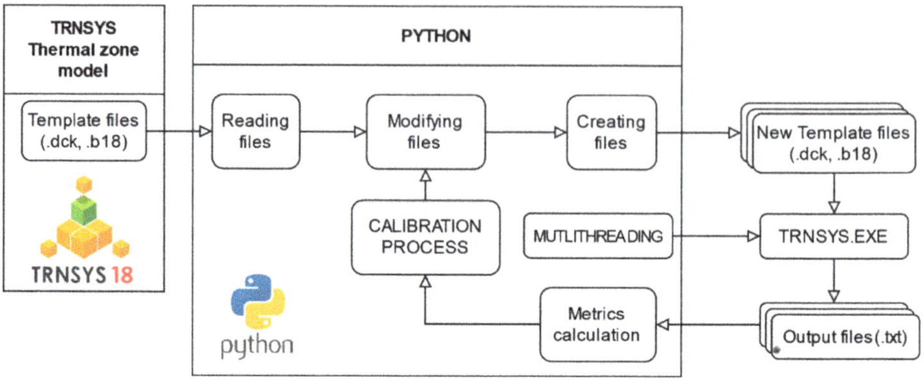


**Figure 2.** Procedure for creating simulations and performing the calibration process.

### 2.4.1. Thermal Zone Modeling

The development of a building simulation model usually begins with the division of the interior spaces into thermal zones. In this study, each room of the hotel was considered as a separate thermal zone. In addition, the bedroom and bathroom in each room were combined into a single space. This was possible because the bathrooms do not have their own heating elements and the temperatures in the bathrooms are similar to those in the bedrooms. In order to simplify and speed up the design process, while also taking into account the design of the smart-room system and the ultimate goal of the current project to upgrade this system with new features at the room level, it was decided to create a multi-zone thermal model consisting of a single room and the adjacent spaces (other rooms and public areas). Therefore, the model simulates one room at a time, taking into account the corresponding room parameters and inputs. The adjacent thermal zones are defined only to calculate the heat exchange between them and the simulated room, not to predict the temperatures in these spaces. In the simulations, the temperatures of the adjacent zones correspond to the temperatures recorded by the smart-room system (if the adjacent spaces are other rooms) and to the design temperatures (if the adjacent spaces are public areas whose actual temperatures are not included in the smart-room system records).

The shape and dimensions of the thermal zones are developed based on architectural descriptions of the building in Google SketchUp 3D 2021 using the trnsys3d plugin. The simulated room and adjacent spaces are configured in the model in terms of the room's position in the building and orientation. All non-geometric information required for model development is created in TRNBuild, including building envelope composition and material properties, internal heat gains, and infiltration and ventilation rates. The composition of the building elements is also taken from architectural descriptions. The material properties of the element layers used are listed in Table 1. All other parameters are listed in Table 5. Internal heat gains are generated only by people when they are in the room. Infiltration is defined as a constant inflow of outside air at the specified rate. When windows are open, natural ventilation occurs and is simulated by increasing the air change rate to an empirically determined value during the simulations. The room ventilation is set to draw air from the FCU (according to the manufacturer's specifications and depending on the fan speed) and fresh air from the central mechanical ventilation system.

**Table 5.** Input parameters for thermal zone modeling.

| Parameter | Value |
|---|---|
| Internal heat gain | 100 W per person [43] |
| Internal moisture gain | 35 g/h per person [43] |
| Infiltration | 0.1 air changes per hour (3 ach when windows are open) |
| Ventilation | Recirculating air + fresh air supplied and extracted at a rate of 60 $m^3/h$ |

2.4.2. Simulation Model

Figure 3 shows the simulation model for temperature prediction developed in TRNSYS 18 Simulation Studio. It consists of the multi-zone thermal model of the building described in the previous subsection, the FCU model, the control logic, the input files, and weather data processing.

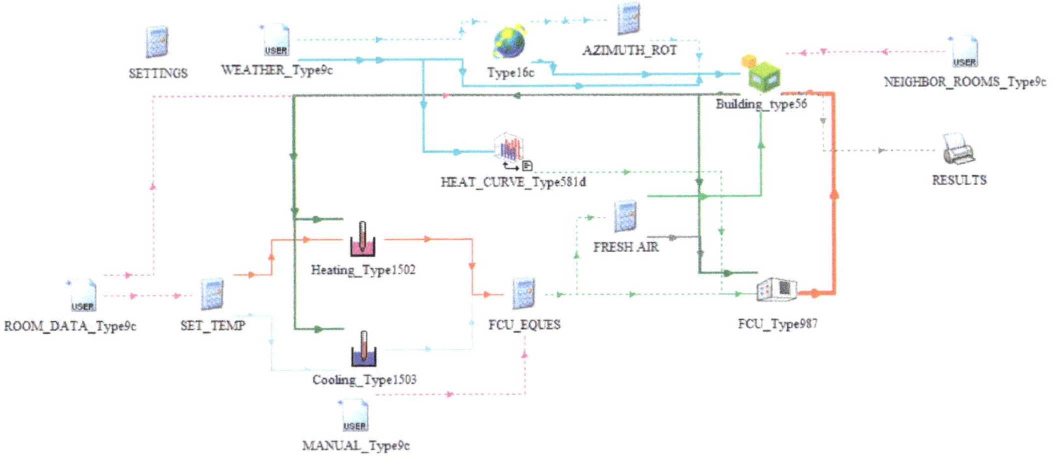


**Figure 3.** Simulation model created in TRNSYS 18 Simulation studio.

The component Type 56 BUILDING models the thermal behavior of the thermal zones—the simulated room and the adjacent spaces—taking into account their position in the building and their orientation. The component calculates the air temperatures in the rooms.

Four blocks of Type 9c components import into the model specific room and weather data for each simulation time step from the records of the smart-room system or the weather database:

- ROOM DATA contains room-specific information on setpoint temperature, occupancy, window status, HVAC status, and HVAC mode;
- MANUAL affects the HVAC system by setting the FCU fan speed desired by guests;
- NEIGHBOR ROOMS inputs the recorded temperatures of adjacent rooms and the design temperatures of public areas into the model to create boundary conditions for heat exchange with these spaces;
- WEATHER is used to create ambient conditions for the model.

Outdoor temperature and relative humidity data from the weather database are used directly in the simulation model without additional processing. Before the solar radiation data is included in the model, the solar irradiance on the zonal outdoor surfaces is calculated using the solar radiation processor Type 16c. The equation component AZIMUTH ROT defines the angle of the zonal surfaces and changes the azimuth angle of the sun depending on the position (orientation) of the currently simulated room in the building. The component Type 581d HEAT CURVE provides a weather compensation function to calculate the hot water supply temperature for the FCU.

The Type 987 component simulates the operation of the FCU based on a generated performance map with operating data provided by the manufacturer. Outputs from the component include discharge temperature, relative humidity, and airflow supplied to the room.

The room control system consists of two thermostats—one for heating mode (Type 1502) and one for cooling mode (Type 1503). The control actions are based on the outputs of two equation components. The equation component SET TEMP calculates temperature setpoints for various fan speeds and turns off the FCU based on window and HVAC status. The FCU EQUES component controls the operation of the FCU, taking into account the manual mode of the system.

The equation component FRESH AIR models the supply of fresh air to the space by the central mechanical ventilation system through the FCU. Depending on whether the FCU is active or not, the fresh air is supplied either through the FCU or directly.

Once the simulation is complete, the component RESULTS saves the simulation results, which are then processed.

### 2.4.3. Model Evaluation

The performance evaluation of the developed model includes the evaluation of the temperature prediction for each simulated room. According to [13], the mean bias error (*MBE*) and root mean squared error (*RMSE*) are common metrics for this purpose. The *MBE* reflects the extent to which the simulation model overestimates or underestimates the actual response, while the *RMSE* provides a more accurate representation of simulation performance by measuring the root of the squared error between the simulated and actual values. They are calculated using the following equations:

$$MBE = \frac{1}{n-p}\sum_{i=1}^{n}(y_{\mathbf{m,i}} - y_{\mathbf{s,i}}), \quad (1)$$

$$RMSE = \sqrt{\frac{1}{n-p}\sum_{i=1}^{n}(y_{\mathbf{m,i}} - y_{\mathbf{s,i}})^2}, \quad (2)$$

where $y_{\mathrm{m}}$ represents the measured values, $y_{\mathrm{s}}$ represents the simulated values, $n$ represents the number of samples, and $p$ represents the number of parameters to be calibrated. In this study, the value of parameter $p$ is set to zero [18]. The criterion for valid simulations is that the *RMSE* must be less than or equal to 1.5 °C [15,44].

The accuracy of energy consumption calculations is typically evaluated using normalized versions of Equations (1) and (2): normalized mean bias error (*nMBE*) and coefficient of variation in root mean squared error (*cvRMSE*). Both the *MBE* and *RMSE* are normalized using the mean of the measured values. The *nMBE* indicates the overall bias of the model but is subject to cancelation errors because it takes into account the sign of the average error vector and should not be used alone. The *cvRMSE* reflects the relative magnitude of the errors with respect to the scale of the target variable and takes into account only the distance between the values, overcoming the cancelation error. In the literature, the abbreviations *MBE* or *MBE* (%) are sometimes used when referring to *nMBE*, which may lead to confusion [17,45]. Therefore, a precise distinction between *nMBE* and *MBE* is made in this work. *nMBE* and *cvRMSE* are calculated using the following equations:

$$nMBE = \frac{MBE}{\frac{1}{n}\sum_{i=1}^{n} y_{\mathbf{m,i}}} \cdot 100(\%), \quad (3)$$

$$cvRMSE = \frac{RMSE}{\frac{1}{n}\sum_{i=1}^{n} y_{\mathbf{m,i}}} \cdot 100(\%). \quad (4)$$

Equations (3) and (4) can also be used to evaluate temperature prediction simulations and make the results comparable with other studies [15]. For the simulations to be considered valid, *cvRMSE* and *nMBE* should meet the conditions specified in the ASHRAE [46] or IPMVP [47] guidelines. The maximum allowable values depend on the time step of the calibration (monthly or hourly). The following are the conditions for the *cvRMSE* and *nMBE* metrics in the case of hourly calibration:

- ASHRAE: $cvRMSE \leq 30\%$ and $nMBE \leq \pm 10\%$;
- IPMVP: $cvRMSE \leq 20\%$ and $nMBE \leq \pm 5\%$.

2.4.4. Calibration Process

The process of calibration is an important step to improve the accuracy and reliability of the simulation model. It is part of the optimization process aimed at finding the optimal model design parameters that minimize the difference between the predicted and actual values. Genetic algorithms are often used to solve this optimization problem in order to minimize the error in temperature simulation [48].

During the development of the simulation model, case-specific files were created that contained all the information of the model. Any changes to the design parameters could be made by modifying these files. To automate the simulation and calibration process, Python was used to read and modify the files, run simulations, process the results, calculate metrics, and perform the calibration process itself (Figure 2). When modifying the files, two types of parameters need to be processed: room-specific and calibration parameters.

In order to run simulations for all observed rooms, the room orientation and Type 9c component input files (i.e., set temperatures, occupancy schedules, etc.) must be modified accordingly for each room. Once the new files were created, the simulations were run in parallel by calling on the TRNSYS 18 executable program with 100 rooms and a simulation time step of 5 min. All rooms had the same size and geometry, but they had different data inputs. The obtained results were evaluated, and the metrics were calculated.

This procedure was repeated for each change in the parameters of the calibration process made by a genetic algorithm. The parameters of the calibration process or the genes of the genetic algorithm were the parameters of the building envelope and FCU operation. The most influential of them had been previously identified by sensitivity analysis. With the calibration parameters, the input files are changed and the process requires another set of simulations for all the rooms. The hyperparameters of the genetic algorithm were adapted to this particular problem. The crossover operator was configured as a single-point crossover and the mutation operator as a single-gene mutation. Individuals were selected from a pool of solutions using tournament selection. The population was uniformly initialized.

A key component of the calibration process is a fitness function. The average value of the *RMSE* metric calculated for each simulated room was incorporated in the fitness function to provide an overall evaluation of each solution. The fitness function was defined using the following equation:

$$Fitness\ function = \frac{1}{n_{\mathbf{room}}} \sum_{i=1}^{n_{\mathbf{room}}} RMSE_{\mathbf{room,i}}, \tag{5}$$

where $RMSE_{\mathrm{room,i}}$ refers to the prediction accuracy for one room and $n_{\mathrm{room}}$ is the total number of simulated rooms. The calibration process was completed when a set of design parameters of the model was determined that resulted in the minimum value of the fitness function.

The model calibration was divided into one-month summer and winter seasons. After the calibration process was completed, the model was evaluated again for another period to validate the obtained solution. The baseline model was also included in the validation period to investigate the biases. Table 6 shows the calibration and validation periods for the winter and summer seasons.

**Table 6.** Periods of calibration and validation for winter and summer seasons.

| Process | Season | Period |
|---|---|---|
| Calibration | Winter<br>Summer | 1 January 2018–31 January 2018<br>1 July 2018–31 July 2018 |
| Validation | Winter<br>Summer | 1 December 2018–31 December 2018<br>1 August 2018–31 August 2018 |

### 2.5. Fault Detection in an HVAC System

HVAC systems in buildings can encounter problems during operation that may reduce thermal comfort and increase energy consumption. Using the case study building and the HVAC system in its rooms as an example, these problems can include a stuck FCU control valve, reduced airflow rate due to a dirty filter or fan problem, faulty FCU control due to room temperature sensor bias, window sensor failure, etc. All of these issues can lead to inadequate performance of the HVAC system and failure to achieve the desired thermal environment. To create an automated control function to detect anomalies or faults, a simulation model such as the one described in the previous subsections can be applied to identify major differences between measured temperatures and simulated temperatures that persist over an extended period of time (e.g., 3 or more Kelvin averaged over several hours or days, depending on the fault), taking into account fan speed, HVAC mode, HVAC state, window opening, and room occupancy. To this end, a set of if–then rules and thresholds must be established. Additional verification of the fault detection results can be conducted by comparing the evaluated performances in the room in question with the performances in a similar room.

## 3. Results and Discussion

In this section, the results of the calibration and validation of the simulation model for both the winter and summer seasons are presented and discussed. The results for the baseline model and the calibrated model show the distribution of temperature simulation accuracy across all 100 simulated rooms. In addition, line plots of temperatures for representative rooms are included for both models. Finally, an example of an HVAC system malfunction in a guest room that was detected by the created model is presented.

Figure 4 depicts the distribution of *RMSE* values for all simulated rooms. The results are shown for each season for both the baseline model and the calibrated model. In the winter-season period chosen for the calibration process (Table 5), the baseline model produced errors between 0.91 °C and 1.16 °C with a median value of 1 °C, while the errors of the calibrated model in the same period ranged between 0.67 °C and 0.83 °C with a median value of 0.74 °C. This corresponds to an improvement of 0.25 °C on average. The errors varied in values between rooms due to the different input settings and conditions for each room. For the validation process, the baseline model gave slightly better results than for the calibration process, averaging 0.1 °C for all rooms. The errors of the calibrated model did not change during the validation process, and the results remained the same as during the calibration process, with an improvement of 0.23 °C over the baseline model. During the summer-season calibration, the baseline model had errors between 0.87 °C and 0.98 °C with a median of 0.93 °C, while the errors of the calibrated model were between 0.73 °C and 0.82 °C with a median of 0.76 °C. The results of the calibrated model are improved by 0.17 °C on average. The results of the validation process also confirmed that the calibrated model is not overfitted and can be used in other seasons.

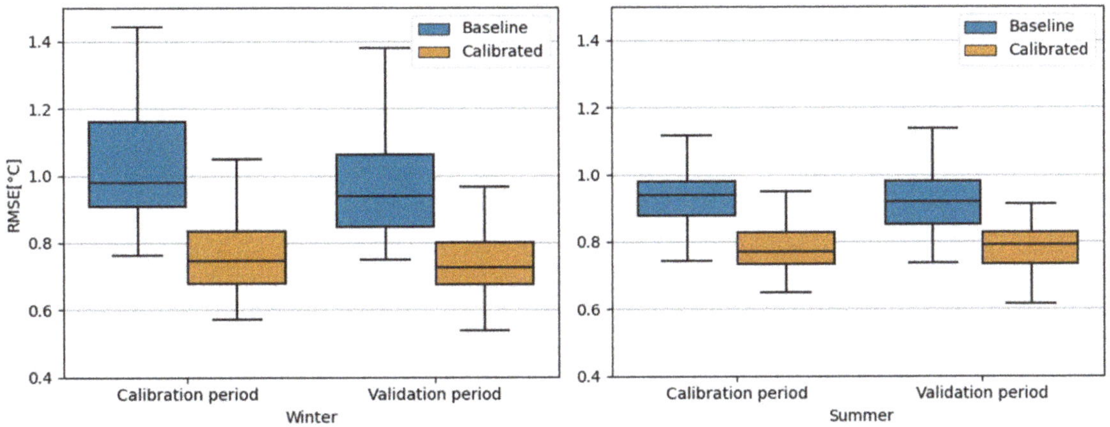


**Figure 4.** *RMSE* metrics for baseline and calibrated models in the winter and summer seasons.

Figure 5 illustrates the distribution of *MBE* values in the winter and summer seasons selected for the calibration and validation processes for both the baseline model and the calibrated model. For the winter-season calibration process, the baseline model produced results between −0.59 °C and −0.11 °C, with a median value of −0.36 °C, while the errors of the calibrated model ranged between −0.25 °C and 0.34 °C, with a median value of 0.04 °C. These results indicate that the calibrated model improved the *MBE* metrics and neither underestimated nor overestimated the measured temperatures in most rooms. On the other hand, the baseline model consistently underestimated the measured temperatures. The results for both models for the validation process differed from those for the calibration process only in distribution, but the median value remained in the same range. The *MBE* results for the validation process were more compactly distributed and had a smaller spread of up to 0.1 °C compared to the results for the calibration process.

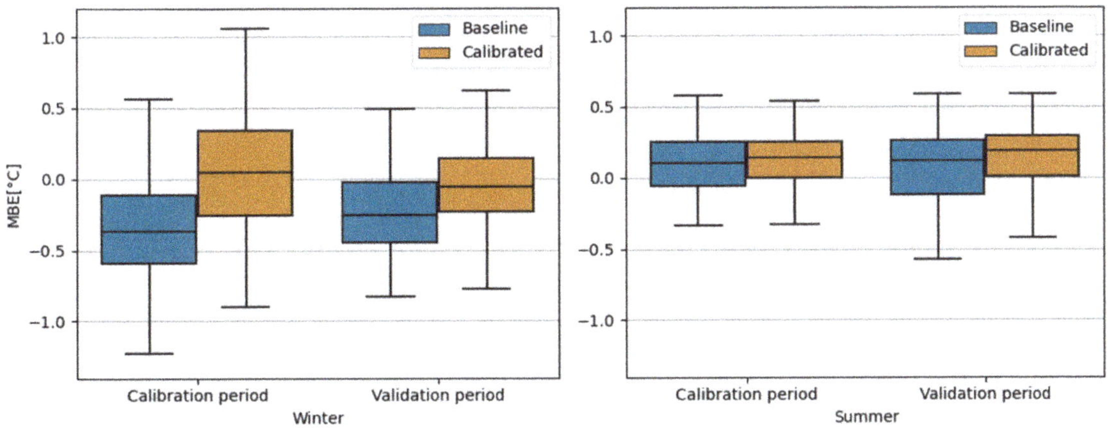


**Figure 5.** MBE metrics for baseline and calibrated models in the winter and summer seasons.

For the summer-season calibration process, the *MBE* results for the baseline model and the calibrated model ranged from −0.05 °C to 0.25 °C, with a median value of about 0.1 °C. This shows that the baseline model was already providing good solutions and the calibrated model was not improved in this respect. Both models overestimated the measured temperatures by 0.1 °C on average. The two models differed in the divergence of results for the summer-season validation period, but the median values remained in the same range. During the summer season, both models overestimated temperatures, with no improvement from the calibrated model.

Table 7 shows the mean values of all metrics used to evaluate the performance of the baseline model and the calibrated model for the winter and summer seasons. Based on the results for the calibration process, the *cvRMSE* of the calibrated model was 22% better than that of the baseline model. During the validation process, the *cvRMSE* metric was improved by 19%. The same percentage improvements were obtained for the *RMSE* metric. For the *nMBE* metric, the values of the calibrated model for the calibration and validation processes outperformed the baseline model by 42% and 59%, respectively. Although the performance of the calibrated model deteriorated during the summer season for the *nMBE* metric, the overall performance improved. The calibration process achieved better results for temperature prediction.

**Table 7.** Averages of metrics for baseline and calibrated models. For clarity, all acronyms are provided in the Abbreviations list.

| | Calibration | | | Validation | | |
|---|---|---|---|---|---|---|
| | **Baseline m.** | **Calibrated m.** | **Improvement** | **Baseline m.** | **Calibrated m.** | **Improvement** |
| *cvRMSE* | 4.60% | 3.59% | +22% | 4.40% | 3.56% | +19% |
| *RMSE* | 1.01 °C | 0.79 °C | +22% | 0.98 °C | 0.79 °C | +19% |
| *nMBE* | −0.66% | 0.38% | +42% | −0.37% | 0.15% | +59% |
| *MBE* | −0.14 °C | 0.08 °C | +43% | −0.09 °C | 0.03 °C | +67% |

The calibrated model had to be evaluated in periods other than those used for the calibration process to confirm its validity and stability, because during calibration, it is possible to overfit the parameters of the model to the measured data from that period. Therefore, the validation process is crucial to confirming the calibration results. The baseline model also needed to be tested during the validation period to detect any deviations in the corresponding data set, such as those caused by interruptions in system operations or temporary hotel closures that could result in different temperature behavior than during the calibration period. The calibration and validation results were similar for both the baseline model and the calibrated model. However, differences were found between rooms due to various factors such as orientation, guest comfort preferences, and window openings. In addition, hotel room occupancy during the summer and winter months can affect the accuracy of the model, with higher occupancy resulting in higher system utilization.

The performance of both the baseline model and the calibrated model met the requirements of the German standard VDI 6020, which specifies a maximum temperature prediction error of 1.5 °C. The models also met ASHRAE and IPMVP guidelines regarding hourly calibration. However, it is important to note that these guidelines are primarily for validation of energy consumption models and not for indoor temperature prediction models. The ASHRAE guideline limits the values of the *cvRMSE* and *nMBE* metrics to 30% and 10%, respectively, while the IPMVP sets these values to 20% and 5%, respectively. The calibrated model achieved an average temperature prediction error of $0.79 \pm 0.14$ °C measured with the *RMSE* metric and $3.58 \pm 0.7\%$ measured with the *cvRMSE* metric, which is well below the allowable limits. Both the *RMSE* and *cvRMSE* metrics were improved by an average of 20% compared to the baseline model results.

The average errors obtained are generally of the same order of magnitude as in the literature reviewed, but a direct comparison of the results is only possible when modeling the same building. Timing and fixing the values of some input parameters, modeling a smaller number of thermal zones, and simpler systems make it easier to obtain better prediction results. In this study, each of the 100 simulated rooms is a special case with its own occupancy, window-opening schedule, and FCU operating conditions. Therefore, the most important indicators of the model's performance are compliance with existing guidelines and case-specific tolerances of errors.

Although the presented values of the metrics indicate that the calibration was successful, the developed model could not accurately predict the temperature changes in the rooms at each simulation time step and in each situation. Figures 6 and 7 compare the measured and simulated temperatures of a selected room for the calibrated model and the baseline model, respectively. Figure 6 presents the winter-season performance in January 2018, while Figure 7 shows the summer-season performance in July 2018.

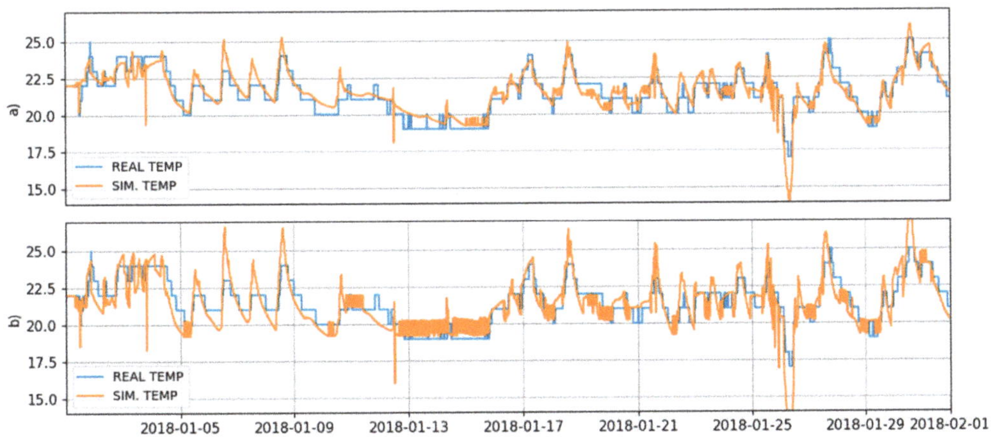


**Figure 6.** Comparison of measured (REAL) and simulated (SIM.) temperatures for the winter-season calibration period: (**a**) calibrated model, (**b**) baseline model.

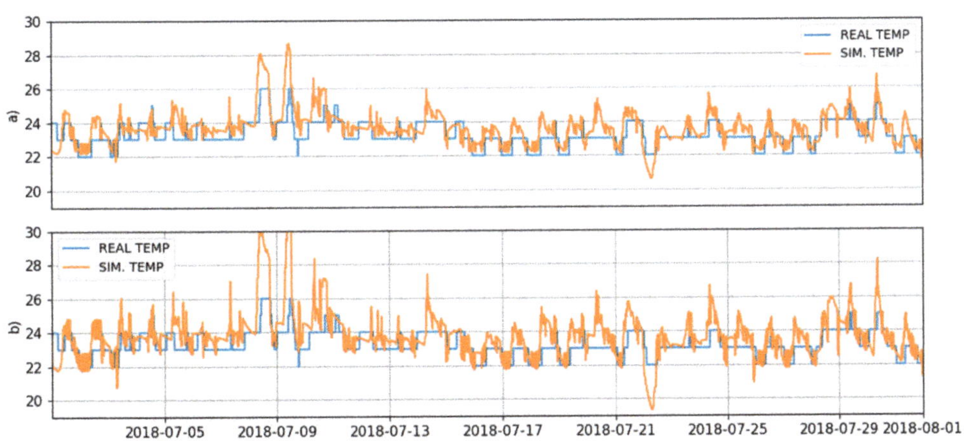


**Figure 7.** Comparison of measured (REAL) and simulated (SIM.) temperatures for the summer-season calibration period: (**a**) calibrated model, (**b**) baseline model.

Throughout the year, the temperature in the rooms was typically between 20 °C and 23 °C, as determined by the hotel management. Any deviation from this range can be attributed to certain external and internal influences. The greater increase in temperature on sunny, warm days and the decrease in temperature when windows are open on colder days and nights can be clearly seen in the figures. When windows are closed, temperatures generally return to default values. The calibrated model performed better than the baseline model in tracking these temperature variations, with simulated temperatures closer to measured temperatures. The difference between the models is more pronounced during periods when the FCU is not operating and passive temperature regulation is taking place.

The calibration process has helped reduce temperature deviation and improve the overall accuracy of the model. Despite its improved performance, the calibrated model still incorrectly predicts the dynamics of room temperatures at certain times. There are several possible reasons for this problem. These include the simplifications used in creating the model, uncertainties associated with the building fabric and its characteristics, and the fact that the measured temperatures were recorded in integer values by the central control of the smart-room system, resulting in a lack of information needed to precisely determine the control logic of the FCU system. The rounding of the values of the measured temperatures affected the uncertainty analysis, as the actual measured values remained unknown. In addition, the model calculates the room temperatures while assuming that the air in the room is well mixed. This may not be the case in the real world, and the measured temperatures may only reflect local conditions, depending on the position of the sensor in the room.

The obtained results indicate the need for more accurate modeling of the thermal response of the rooms during periods of strong solar radiation, as well as during periods of natural ventilation, taking into account the wind speed and direction. The opening of the windows was simulated by increasing the number of air changes per hour. The moment the system registers that the windows are open, the number of air changes in the model increases to a fixed, predefined value. The number of air changes per hour of 3 $h^{-1}$ resulted in the smallest average error considering all simulated rooms. This value was applied to all rooms, regardless of their orientation. In reality, not all facades are exposed to the wind at the same time. In addition, the wind does not always blow at a constant speed. Some spaces will therefore have a stronger wind influence than others, and this influence varies over time. Under these circumstances, the model overestimates the temperature changes in rooms on the downwind side of the building and may underestimate them in the rooms directly on the wind path. Further investigation should be undertaken and further model improvements should be made to address these issues.

The developed simulation model was used in the next step to identify anomalies or faults in the HVAC system of individual rooms, particularly in the context of FCUs. Figure 8 shows an example where the model successfully identified an anomaly in a particular room during the summer season by finding large discrepancies between simulated and measured air temperatures. The air temperature in the room was higher than it should be, resulting in uncomfortable conditions, even though the FCU was operating intensively (the actual fan speed was often at level 3, as indicated in the graph of REAL FAN SPEED). During the same periods, simulated temperatures were consistently lower, as were simulated fan speeds (the speed was at level 1 most of the time, as shown in the graph of SIM FAN SPEED). The windows in the room were mostly closed (the graph of WINDOW). Additional testing was performed to confirm the anomaly. This included an evaluation of the operation of the FCU and a comparison of the temperature of the room with the temperatures of other rooms with similar characteristics. The results show that the average temperature in the room with the anomaly was above 26 °C, while average temperatures in the comparison rooms were between 22 °C and 25 °C. An analysis of the FCU´s operation showed that the unit ran longer and more intensely in the room with the anomaly than in the other rooms, without providing comfortable conditions, suggesting a malfunction of the system. It was also found that the specified conditions were met in the compared rooms. The same principle was applied to detect faults in other rooms. Although the model can detect anomalies with the implemented detection algorithm, further tests and diagnostics should be performed to confirm faults and find their real causes.

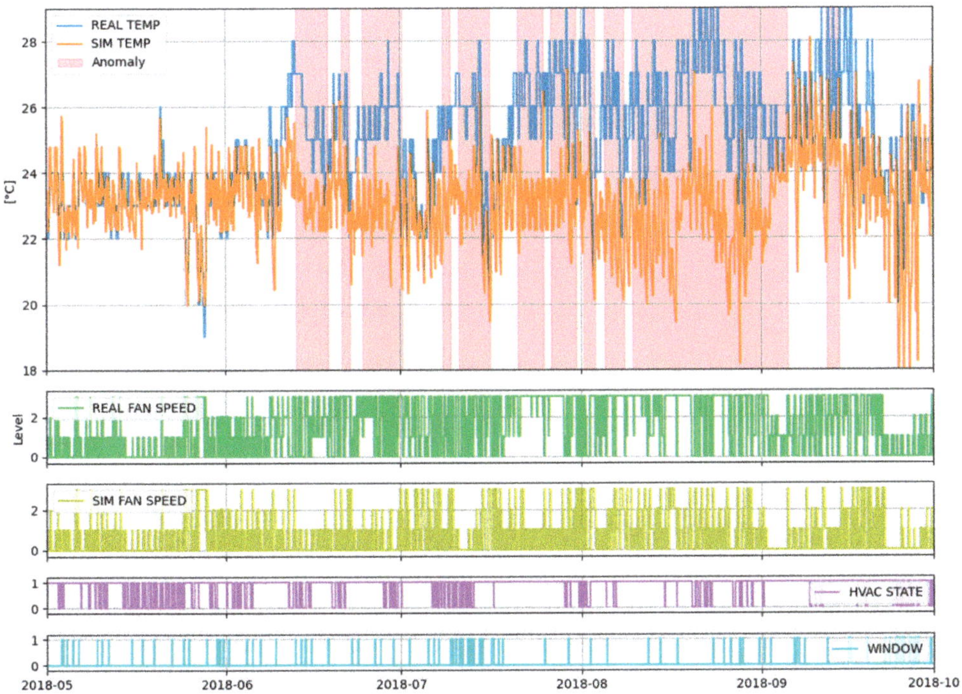


**Figure 8.** Measured (REAL) and simulated (SIM) temperatures with marked periods of anomalies in the HVAC system. Other signals affecting the operation of the system are also shown.

It should be noted that faults are not always detectable because temperatures may remain at the desired level due to favorable internal and external conditions, as is the case on cooler summer days. Another case where a fault may not be detected is when the HVAC system is not operating or is operating at a partial load. Faults are more likely to be detected under extreme conditions when the HVAC system must operate at nominal capacity. To improve the fault detection capabilities of the system, it is important to consider all signal conditions that can greatly affect the performance of the system.

## 4. Conclusions

Significant improvements in the energy efficiency of HVAC systems can be achieved through the use of smart technologies and control algorithms that optimize their performance and avoid unnecessary energy waste. One of the methods of improving control functions is the application of automatic fault detection and diagnosis routines. Their development is based on reliable simulation models of the system.

This study focused on the development and calibration of a white-box model of the guest rooms and associated HVAC systems (fan coils) in the case study hotel in Zagreb, Croatia, to predict the indoor temperatures under real conditions. The simulations took into account occupant behavior, window openings, and temperature control logic. All of this room-specific data came from an extensive database created over the course of several years by measurements from a smart-room system installed in the building. Real outdoor weather conditions recorded over the same period provided the environmental conditions for the model, which was calibrated with a genetic algorithm based on a five-minute time step to improve temperature predictions. The simulations yielded an average accuracy of $0.79 \pm 0.14$ °C as measured by *RMSE*, with a 20% improvement for *RMSE* and *cvRMSE* compared to the baseline model. For metrics accounting for positive and negative values, there was a 59% improvement for *nMBE* and 67% improvement for

*MBE*. Occupant behavior was the largest source of error in the simulations, as they have control over windows, set temperatures, and HVAC system operation in the case of manual mode selection. The actions of the occupants, particularly the opening of windows, pose a challenge to the model's predictions of the dynamics of temperature changes, which could be improved by more detailed modeling of the thermal responses of the rooms. Another potential source of error in the prediction is the fact that the temperatures measured in the field were recorded in integer values, so some of the information needed to determine the control logic was lost. Nevertheless, the developed model met the requirements of the current guidelines and can be used for further analysis of the HVAC system and optimization of energy consumption in the case study building.

The model is also capable of detecting major anomalies in the analyzed HVAC system. Anomaly detection is based on finding major discrepancies between simulated and measured temperatures. To confirm the anomalies found, additional tests were performed analyzing the temperatures and operation of the HVAC system in different rooms.

Future work consists of developing more advanced methods for automatic fault detection and diagnosis by integrating machine learning techniques and implementing model predictive control features, which is a promising approach for energy savings in buildings.

**Author Contributions:** Conceptualization, I.Š. and I.W.; methodology, D.P., I.Š., S.L. and I.W.; software, D.P.; validation, D.P., I.Š., S.L. and I.W.; formal analysis, D.P., I.Š., S.L. and I.W.; investigation, D.P., I.Š., S.L. and I.W.; resources, D.P., I.Š., S.L. and I.W.; data curation, D.P., I.Š., S.L. and I.W.; writing—original draft preparation, D.P.; writing—review and editing, D.P., I.Š., S.L. and I.W.; visualization, D.P. and I.W.; supervision, I.Š. and I.W.; project administration, I.Š., S.L. and I.W.; funding acquisition, I.Š. All authors have read and agreed to the published version of the manuscript.

**Funding:** This work was supported in part by the European Regional Development Fund (ERDF) under grant agreement number KK.01.2.1.02.0303, project Adria Smart Room. TRNSYS software was purchased as part of the HEXENER project, funded by the Croatian Science Foundation (IP-2016-06-4095).

**Data Availability Statement:** Not applicable.

**Conflicts of Interest:** The authors declare no conflict of interest. The funders had no role in the design of the study; in the collection, analyses, or interpretation of data; in the writing of the manuscript; or in the decision to publish the results.

## Abbreviations

The following abbreviations have been used in this manuscript:

| | |
|---|---|
| ASHRAE | American Society of Heating, Refrigerating and Air Conditioning Engineers |
| FCU | Fan Coil Unit |
| GHG | Greenhouse Gas |
| HVAC | Heating, Ventilation, and Air Conditioning |
| IDA ICE | IDA Indoor Climate and Energy |
| IPMVP | International Performance Maintenance and Verification Protocol |
| nZEB | Nearly Zero Energy Building |
| TRNSYS | Transient System Simulation Tool |
| *cvRMSE* | Coefficient of Variation of the Root Mean Squared Error |
| *MAE* | Mean Absolute Error |
| *MBE* | Mean Bias Error |
| *nMBE* | Normalized Mean Bias Error |
| *RMSE* | Root Mean Squared Error |

## References

1. International Energy Agency (IEA). *Buildings*; IEA: Paris, France, 2022. Available online: https://www.iea.org/reports/buildings (accessed on 13 April 2023).
2. International Energy Agency (IEA). *Promoting Energy Efficiency Investments: Case Studies in the Residential Sector*; OECD/IEA and AFD: Paris, France, 2008.

3. Mariano-Hernández, D.; Hernández-Callejo, L.; Zorita-Lamadrid, A.L.; Duque-Perez, O.; García, F.S. A Review of Strategies for Building Energy Management System: Model Predictive Control, Demand Side Management, Optimization, and Fault Detect & Diagnosis. *J. Build. Eng.* **2021**, *33*, 101692. [CrossRef]
4. Drgoňa, J.; Arroyo, J.; Figueroa, I.C.; Blum, D.H.; Arendt, K.; Kim, D.; Ollé, E.P.; Oravec, J.; Wetter, M.; Vrabie, D.; et al. All You Need to Know about Model Predictive Control for Buildings. *Annu. Rev. Control* **2020**, *50*, 190–232. [CrossRef]
5. Zakula, T.; Armstrong, P.R.; Norford, L.K.; Norford, L.K. Modeling Environment for Model Predictive Control of Buildings. *Energy Build.* **2014**, *85*, 549–559. [CrossRef]
6. Yao, Y.; Shekhar, D.K. State of the Art Review on Model Predictive Control (MPC) in Heating Ventilation and Air-Conditioning (HVAC) Field. *Build Environ.* **2021**, *200*, 107952. [CrossRef]
7. Matetić, I.; Štajduhar, I.; Wolf, I.; Ljubic, S. A Review of Data-Driven Approaches and Techniques for Fault Detection and Diagnosis in HVAC Systems. *Sensors* **2023**, *1*, 1. [CrossRef] [PubMed]
8. Cacabelos, A.; Eguía, P.; Febrero, L.; Granada, E. Development of a new multi-stage building energy model calibration methodology and validation in a public library. *Energy Build.* **2017**, *146*, 182–199. [CrossRef]
9. Yu, J.; Chang, W.-S.; Dong, Y. Building Energy Prediction Models and Related Uncertainties: A Review. *Buildings* **2022**, *12*, 1284. [CrossRef]
10. Guyot, D.; Giraud, F.; Simon, F.; Corgier, D.; Marvillet, C.; Tremeac, B. Overview of the Use of Artificial Neural Networks for Energy-related Applications in the Building Sector. *Int. J. Energy Res.* **2019**, *43*, 6680–6720. [CrossRef]
11. Delač, B.; Pavković, B.; Lenić, K.; Mađerić, D. Integrated Optimization of the Building Envelope and the HVAC System in NZEB Refurbishment. *Appl. Therm. Eng.* **2022**, *211*, 118442. [CrossRef]
12. Fabrizio, E.; Monetti, V. Methodologies and Advancements in the Calibration of Building Energy Models. *Energies* **2015**, *8*, 2548–2574. [CrossRef]
13. Chong, A.; Gu, Y.; Jia, H. Calibrating Building Energy Simulation Models: A Review of the Basics to Guide Future Work. *Energy Build.* **2021**, *253*, 111533. [CrossRef]
14. Chen, R.; Tsay, Y.-S. An Integrated Sensitivity Analysis Method for Energy and Comfort Performance of an Office Building along the Chinese Coastline. *Buildings* **2021**, *11*, 371. [CrossRef]
15. Pachano, J.E.; Bandera, C.F. Multi-Step Building Energy Model Calibration Process Based on Measured Data. *Energy Build.* **2021**, *252*, 111380. [CrossRef]
16. Arendt, K.; Jradi, M.; Shaker, H.R.; Veje, C. Comparative Analysis of White-, Gray- and Black-Box Models for Thermal Simulation of Indoor Environment: Teaching Building Case Study. In Proceedings of the 2018 Building Performance Modeling Conference and SimBuild co-organized by ASHRAE and IBPSA-USA, Chicago, IL, USA, 26–28 September 2018.
17. O´Donovan, A.; O´Sullivan, P.D.; Murphy, M.D. Predicting Air Temperatures in a Naturally Ventilated Nearly Zero Energy Building: Calibration, Validation, Analysis and Approaches. *Appl. Energy* **2019**, *250*, 991–1010. [CrossRef]
18. Guyot, D.; Giraud, F.; Simon, F.; Corgier, D.; Marvillet, C.; Tremeac, B. Building Energy Model Calibration: A Detailed Case Study Using Sub-Hourly Measured Data. *Energy Build.* **2020**, *223*, 110189. [CrossRef]
19. Baba, F.M.; Ge, H.; Zmeureanu, R.; Wang, L. Calibration of Building Model Based on Indoor Temperature for Overheating Assessment Using Genetic Algorithm: Methodology, Evaluation Criteria, and Case Study. *Build. Environ.* **2022**, *207*, 108518. [CrossRef]
20. Figueiredo, A.; Kämpf, J.; Vicente, R.; Oliveira, R.; Silva, T. Comparison between Monitored and Simulated Data Using Evolutionary Algorithms: Reducing the Performance Gap in Dynamic Building Simulation. *J. Build. Eng.* **2018**, *17*, 96–106. [CrossRef]
21. Royapoor, M.; Roskilly, T. Building Model Calibration Using Energy and Environmental Data. *Energy Build.* **2015**, *94*, 109–120. [CrossRef]
22. Rosato, A.; Guarino, F.; Guarino, F.; Filomena, V.; Sibilio, S.; Maffei, L. Experimental Calibration and Validation of a Simulation Model for Fault Detection of HVAC Systems and Application to a Case Study. *Energies* **2020**, *13*, 3948. [CrossRef]
23. Roberti, F.; Oberegger, U.F.; Gasparella, A. Calibrating Historic Building Energy Models to Hourly Indoor Air and Surface Temperatures: Methodology and Case Study. *Energy Build.* **2015**, *108*, 236–243. [CrossRef]
24. Kossak, B.; Stadler, M. Adaptive Thermal Zone Modeling Including the Storage Mass of the Building Zone. *Energy Build.* **2015**, *109*, 407–417. [CrossRef]
25. Martínez-Ibernón, A.; Aparicio-Fernández, C.; Royo-Pastor, R.; José-Luis Vivancos, J.L. Temperature and humidity transient simulation and validation in a measured house without a HVAC system. *Energy Build.* **2016**, *131*, 54–62. [CrossRef]
26. Martínez-Mariño, S.; Eguía-Oller, P.; Granada-Álvarez, E.; Erkoreka-González, A. Simulation and Validation of Indoor Temperatures and Relative Humidity in Multi-Zone Buildings under Occupancy Conditions Using Multi-Objective Calibration. *Build. Environ.* **2021**, *200*, 107973. [CrossRef]
27. Murphy, M.D.; O'Sullivan, P.D.; da Graça, G.C.; O'Donovan, A. Development, Calibration and Validation of an Internal Air Temperature Model for a Naturally Ventilated Nearly Zero Energy Building: Comparison of Model Types and Calibration Methods. *Energies* **2021**, *14*, 871. [CrossRef]
28. Berthou, T.; Stabat, P.; Salvazet, R.; Marchio, D. Development and Validation of a Gray Box Model to Predict Thermal Behavior of Occupied Office Buildings. *Energy Build.* **2014**, *74*, 91–100. [CrossRef]

29. Belić, F.; Slišković, D.; Hocenski, Ž. Detailed Thermodynamic Modeling of Multi-Zone Buildings with Resistive-Capacitive Method. *Energies* **2021**, *14*, 7051. [CrossRef]
30. Cui, B.; Fan, C.; Munk, J.D.; Mao, N.; Xiao, F.; Dong, J.; Kuruganti, T. A Hybrid Building Thermal Modeling Approach for Predicting Temperatures in Typical, Detached, Two-Story Houses. *Appl. Energy* **2019**, *236*, 101–116. [CrossRef]
31. Chen, Y.; Guo, M.; Chen, Z.; Chen, Z.; Ji, Y. Physical energy and data-driven models in building energy prediction: A review. *Energy Rep.* **2022**, *8*, 2656–2671. [CrossRef]
32. Aguilera, J.J.; Andersen, R.K.; Toftum, J. Prediction of Indoor Air Temperature Using Weather Data and Simple Building Descriptors. *Int. J. Environ. Res. Public Health* **2019**, *16*, 4349. [CrossRef]
33. Delzendeh, E.; Wu, S.; Lee, A.; Zhou, Y. The Impact of Occupants' Behaviours on Building Energy Analysis: A Research Review. *Renew. Sustain. Energy Rev.* **2017**, *80*, 1061–1071. [CrossRef]
34. Calì, D.; Andersen, R.K.; Andersen, R.K.; Müller, D.; Olesen, B.W. Analysis of Occupants' Behavior Related to the Use of Windows in German Households. *Build. Environ.* **2016**, *103*, 54–69. [CrossRef]
35. Bonte, M.; Thellier, F.; Thellier, F.; Lartigue, B. Impact of Occupant's Actions on Energy Building Performance and Thermal Sensation. *Energy Build.* **2014**, *76*, 219–227. [CrossRef]
36. Belazi, W.; Ouldboukhitine, S.-E.; Chateauneuf, A.; Bouchaïr, A. Thermal Modeling of the Occupied Multi-Zone Buildings Taking into Account the Uncertainties of Occupant Behavior. *Case Stud. Therm. Eng.* **2022**, *33*, 101978. [CrossRef]
37. Parzinger, M.; Hanfstaengl, L.; Sigg, F.; Spindler, U.; Wellisch, U.; Wirnsberger, M. Residual Analysis of Predictive Modelling Data for Automated Fault Detection in Building's Heating, Ventilation and Air Conditioning Systems. *Sustainability* **2020**, *12*, 6758. [CrossRef]
38. Pourarian, S.; Wen, J.; Veronica, D.A.; Pertzborn, A.; Zhou, X.; Liu, R. A Tool for Evaluating Fault Detection and Diagnostic Methods for Fan Coil Units. *Energy Build.* **2017**, *136*, 151–160. [CrossRef]
39. Ranade, A.; Provan, G.; Mady, A.E.-D.; O'Sullivan, D.T.J. A Computationally Efficient Method for Fault Diagnosis of Fan-Coil Unit Terminals in Building Heating Ventilation and Air Conditioning Systems. *J. Build. Eng.* **2020**, *27*, 100955. [CrossRef]
40. Rosato, A.; Guarino, F.; Sibilio, S.; Entchev, E.; Masullo, M.; Maffei, L. Healthy and Faulty Experimental Performance of a Typical HVAC System under Italian Climatic Conditions: Artificial Neural Network-Based Model and Fault Impact Assessment. *Energies* **2021**, *14*, 5362. [CrossRef]
41. Croatian Meteorological and Hydrological Service. Available online: https://meteo.hr/index_en.php (accessed on 15 February 2022).
42. Klein, S.A. *TRNSYS 18: A Transient System Simulation Program*; Solar Energy Laboratory, University of Wisconsin: Madison, WI, USA, 2017.
43. American Society of Heating, Refrigerating and Air-conditioning Engineers (ASHRAE). *2009 ASHRAE Handbook-Fundamentals (SI Edition)*; ASHRAE Inc.: Atlanta, GA, USA, 2009.
44. VDI-Fachbereich Technische Gebäudeausrüstung. *Requirements to be Met by Calculation Methods for the Simulation of Thermal-Energy Efficiency of Buildings and Building Installations Standard DINGCD22*; Verlag des Vereins Deutscher Ingenieure (VDI): Düsseldorf, Germany, 2001.
45. Ruiz, G.R.; Bandera, C.F. Validation of Calibrated Energy Models: Common Errors. *Energies* **2017**, *10*, 1587. [CrossRef]
46. American Society of Heating, Refrigerating and Air-conditioning Engineers (ASHRAE). *ASHRAE Guideline 14-2014: Measurement of Energy, Demand, and Water Savings*; ASHRAE: Atlanta, GA, USA, 2014.
47. Efficiency Valuation Organization (EVO). *International Performance Measurement and Verification Protocol Volume I: Concepts and Options for Determining Energy and Water Savings*; EVO: Washington DC, USA, 2012.
48. Ruiz, G.R.; Bandera, C.F.; Temes, T.G.A.; Gutierrez, A.S.O. Genetic Algorithm for Building Envelope Calibration. *Appl. Energy* **2016**, *168*, 691–705. [CrossRef]

buildings

MDPI

*Article*

# The ANN Architecture Analysis: A Case Study on Daylight, Visual, and Outdoor Thermal Metrics of Residential Buildings in China

**Shanshan Wang [1,2], Yun Kyu Yi [3,*] and Nianxiong Liu [4]**

[1] Medical Architecture and Environment Research Unit, School of Architecture and Urban Planning, Beijing University of Civil Engineering and Architecture, Beijing 100044, China; wangshanshan@bucea.edu.cn
[2] Beijing Key Laboratory of Green Building and Energy-Efficiency Technology, Beijing 100044, China
[3] School of Architecture, University of Illinois Urbana-Champaign, Champaign, IL 61820-5711, USA
[4] School of Architecture, Tsinghua University, Beijing 100084, China; phlnx@tsinghua.edu.cn
* Correspondence: ykyi@illinois.edu

**Abstract:** Selecting an appropriate ANN model is crucial for speeding up the process of building performance simulation during the design phase of residential building layouts, particularly when evaluating three or more green performance metrics simultaneously. In this study, daylight, visual, and outdoor thermal metrics were selected as main green performance. To find the suitable ANN model, sensitivity analysis was used to obtain a set of proper parameters applied to the ANN structure. To train the ANN model with a higher predicting accuracy, this paper tested four different scenarios of ANN parameter setups to find some general guidelines about how to set up an ANN model to predict DF, sunlight hours, QuVue and UTCI. The results showed that an ANN model with a combined output variable demonstrated better average prediction accuracy than ANN models with a separated output variable. Having two times the number of training samplings compared to the number of input variables can lead to a high accuracy of prediction. The ideal number of neurons in the hidden layer was approximately 1.5 times the number of input variables. These findings of how to improve the ANN model may provide guidance for modeling an ANN for building performance.

**Keywords:** artificial neural network (ANN); sensitivity analysis; daylight factor (DF); sunlight hours; QuVue (sky view ratio); universal thermal climate index (UTCI)

## 1. Introduction

Comparing the ANN methods with other types of simulation tools in building performance, the ANN method has some strong points in terms of simplicity, calculation speed, and learning from limited data sets [1–4]. Recently, many papers have used the artificial neural network (ANN) as a prediction model in the field of building performance, for instance, in energy use, daylight, and energy demand [5–12].

Some studies have investigated several kinds of expert-knowledge-based algorithms for ANNs [13,14], including multiple linear regression (MLR), Gaussian process regression (GPR), support vector machine (SVM), boosted tree, random forest, and so on, to compare their performance in the prediction of building performance [15,16]. Each method has its own purpose and feature to accomplish a prediction assignment. Moreover, in each algorithm, there exist some principles to make a better model to achieve ideal prediction outcomes.

The research of the ANN structure has been focused on how to enhance the performance of its models' prediction. Some studies conducted different training algorithms to achieve high accuracy of the ANN model's performance. Kwok et al. revealed that some input variables could significantly improve the accuracy of the model [17]. Mustafaraj et al. found that in building and validating models for predicting dry bulb temperature and relative humidity for different time steps, when the step-ahead time scales increased, the ANN models had a lower accuracy performance [18]. Jovanović et al. investigated three ANN models (FFNN, RBFN, and ANFIS) to improve heating energy consumption prediction accuracy, and found that the ensemble, by combining the outputs of member networks, achieved better prediction results [19]. Buratti et al. compared the prediction performance of IAQ in naturally ventilated buildings, drawing the conclusion that feed-forward backpropagation neural network models provided a greater accuracy of $R^2$ and a higher-quality prediction with a lower RMSE compared with the MLR models [20]. Deng and Chen conducted five training algorithms according to MAE and $R^2$ and found that Levenberg–Marquardt predicted the thermal comfort in indoor environments in ten offices and ten apartments/houses in Indiana, USA [21]. In past decades, optimal ANN models have been researched deeply.

In the meantime, previous studies related to ANNs have also focused on the principle behind how the input and output variables or the number of neurons or layers were set up, raising questions about constructing optimal ANN models in the area of predicting building performance. The hidden neuron is one of key elements to conducting an optimal ANN model. Yokoyama et al. conducted a study on the relationship between the number of neurons in the hidden layer and the MAE between the predicted and actual thermal sensation to determine a reasonable number of hidden neurons [22]. Moreover, the number of neurons in the input layer and in the hidden layer, the number of sample cases, and the order of inputs were also mostly investigated. Huang et al. employed a forward selection method to choose individual candidate variables. They found that an oversized network with a large input order and number of hidden layers can enhance the prediction error of ANN models with high frequency noise [23]. Rocío Escandón et al. concluded that the sample size ensured reliable prediction results according to a specific building category, finding that a ratio of 2.5 between the sample size and the number of characteristic parameters (input variables) could increase the accuracy [24]. The combination of each element of an ANN model related to building performance has been discussed. Due to the complexity and diversity of ANN prediction problems, it is required to conduct more research on improving the accuracy related to the building area. The determination of the ANN architecture for building energy consumption in order to make a quick prediction was prevalent in many studies [25–28], but in the green building design stage, how to establish the proper architecture of an ANN prediction model for green performance was the focus of only few studies.

In previous studies, the normalized method was commonly conducted to construct the ANN models; however, the normalization combination of the input and output variables was seldom investigated regarding the accuracy performance of the models.

In summary, the ANN is a common machine-learning algorithm that gives a highly accurate prediction performance and is efficient as well as computationally time saving. An in-depth discussion about improving the precision capacity of the ANN model in the studies mentioned above on the ANN structure is also an important area because it is critical for building a highly accurate prediction model. However, most of these papers focused on one factor of the ANN structure, such as the number of input variables or a comparison of one pair of factors (number of hidden layers vs. hidden neurons) to improve the prediction capacity. It is rare to find one single model covering many possible factors or across different performance areas. A comprehensive comparison of the main factors in the ANN structure should be considered in order to find which factors can improve the ANN prediction from different factors.

Based on the Assessment Standard for Green Building [29] and the Assessment Standard for Healthy Building [30], daylight, visual comfort, and outdoor thermal metrics are key considerations during the building design stage. These factors often present conflicting requirements to some extent. Therefore, it is essential to simultaneously consider these metrics when predicting building performance. This study's objective was to find optimized

ANN models to predict the daylight factor (DF), sunlight hours, QuVue (sky view ratio), and universal thermal climate index (UTCI) of residential buildings.

The contribution and novelty of this paper may provide guidance for modeling the ANN to improve the accuracy of building performance prediction.

This study was holistically designed to find the proper way to construct the ANN model for building performance. Some common measures from building simulations were used to see how these different measures can be modeled in an ANN. The sensitivity analysis method was conducted to identify the most important factors to consider in building performance. The objective of the study was to explore the possibility of predicting performance in similar building layouts and create a rule for selecting ANN model parameters that is replicable and widely applicable. For example, multiple performance predictions were carried out for architectural building layout designs that had similar numbers of buildings and exceeded three prediction categories.

## 2. Materials and Methods

This study commenced by employing parameterized modeling of the building layout to identify the independent variables, variables, and constraints. The independent variable was limited to the building volume ratio or total building area. The next step involved parametric performance simulation, which included constructing lighting simulation models, site sunshine models, visual field models, and thermal environment models. This preparation was conducted to collect data for five performance analyses. In the training phase of the ANN model, three types of independent variables from the parameterized model were used as input data: the position coordinates, x-axis and y-axis coordinates of the building unit's relative field, and its height, z. The output data consisted of calculation values from the simulation model, specifically the lighting coefficient, sunshine, vision, and outdoor thermal climate index. By conducting a four-part sensitivity analysis using the ANN architecture, improvements in prediction accuracy were achieved.

Figure 1 summarizes the overall methodology of this study including the random choice of building parameters in building models, simulation of the proposed models in Rhino-Grasshopper (Version 6) software, and training of the ANN models for five building performance estimations.

### *2.1. Input Data Collection*

In the cold climate of Beijing, China, the test case involved twelve buildings situated on a site with a total area of 143,358 $m^2$ (as shown in Figure 2). Among these buildings, ten residential buildings (No. 1–6 and No. 8–11) were included in the test, while two commercial buildings (No. 7 and No. 12) were excluded from the calculations (as depicted in Figure 2). It is worth noting that the test did not take into account the urban context surrounding the site due to its location in an open field.

In this study, test data were collected from high-rise residential buildings in Beijing, China. The test buildings included 917 resident units in 10 buildings. Two of the high-rise buildings were for commercial and office use (building numbers 7 and 12). Overall, the residential buildings had 3.1 m of floor-to-floor height and around 0.3–0.4 for the window-to-wall ratio (WWR; Figure 2). More information about the setup for the building simulation models was found at reference [31].

The following section discusses how the training dataset was obtained using different computational simulation tools in detail (Figure 3). In this study, considering the rectangular shape of the buildings, the methodology scope was bounded to residential buildings. The 12 test buildings' spatial position variables (x_n, y_n, and z_n) meant that 36 variables were used (Table 1). The tests used default three-layer ANN models, which were made up of an input layer, a single hidden layer, and an output layer.

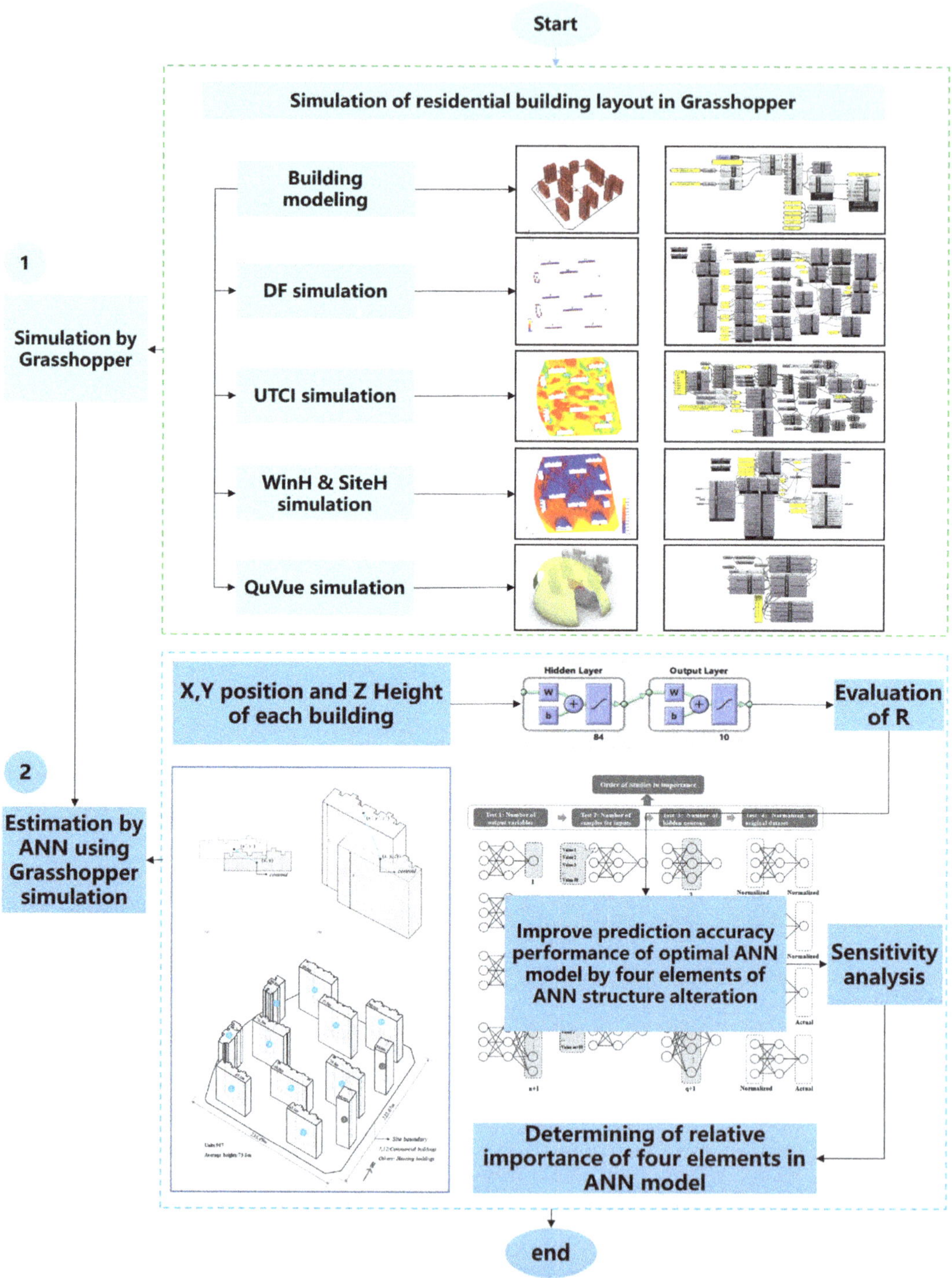


**Figure 1.** Flowchart of this study.

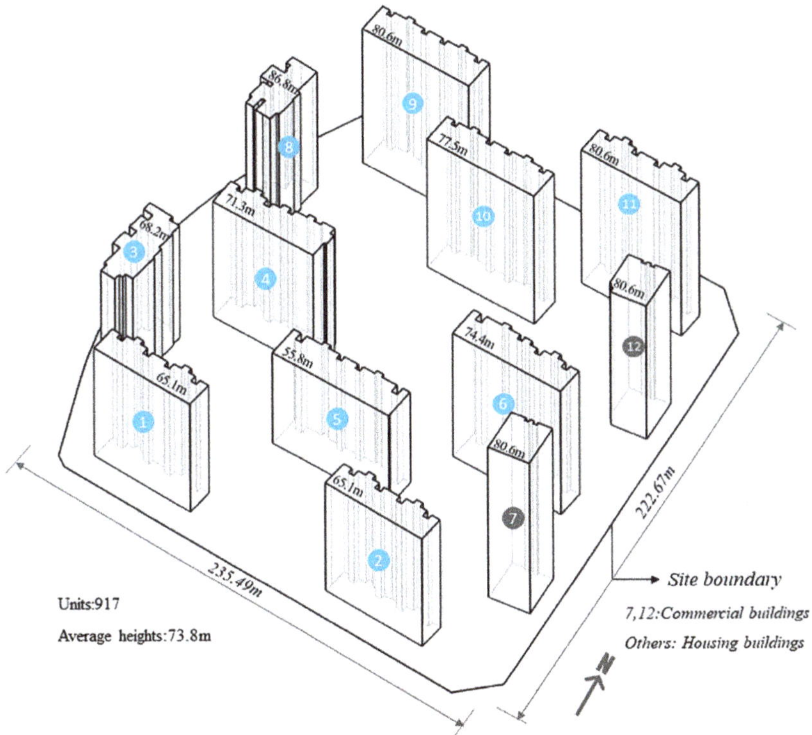


Figure 2. Test case building layout.

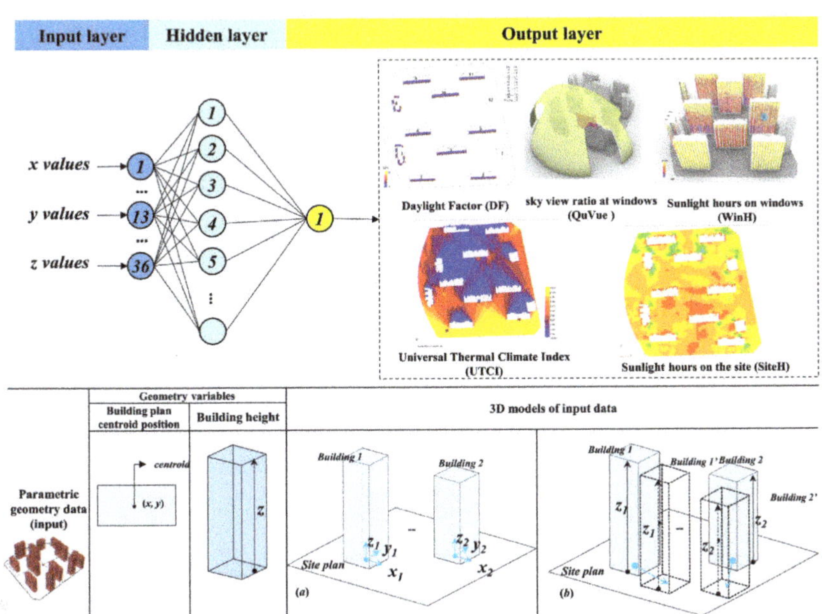


Figure 3. Input parameters for ANN model.

**Table 1.** Input and output parameters of ANN models.

| Input and Output Parameters of ANN Models | Quantity | Breakdown |
|---|---|---|
| Input Variables | 36 | x1, x2, x3, . . ., x12<br>y1, y2, y3, . . ., y12<br>z1, z2, z3, . . ., z12 |
| Output Variables | 5 | Daylight factor (DF)<br>Sky view ratio (QuVue)<br>Window sunlight hours (WinH)<br>Site sunlight hours (SiteH)<br>Universal thermal climate index (UTCI) |

The geometry-related variables were used as the input variables and different simulation results were used as the outputs for training the ANN. The controlling geometry was scripted through the non-uniform rational basis spline (NURBS) CAD (computational-aided design) tool that automated the process of updating its geographical configuration. The variables of the input parameters were the spatial position of each building (n), which were symbolized as x_n, y_n, and z_n. The objective of optimizing the design of the residential building layout was to meet the specified design requirements, such as controlling the buildings so as not to intersect with other buildings, buildings not being moved beyond the site boundary, and the total building floor area being between ±10% of the original case. The range of values was determined according to the site status and the Standard for Urban Residential Area Planning and Design [32], the test buildings were able to move along both the x- and y-axes, and the possible moveable range from −10 to 20 m was based on the initial building location. The average height of a floor was 3.1 m. The building height (z) can change from 34.1 to 99.2 m according to the High-Rise Building Code for The Fire Protection Design of Buildings [33].

The generation of random input values in these experiments was governed by the building standard to ensure control. It is important to note that these experiments were solely conducted for obtaining ANN models and not intended for actual design purposes. The input variables were randomly generated by a statistics tool according to the variable range mentioned above.

### 2.2. Output Data Collection

This research applied Rhino-Grasshopper to simulate five performance measures: DF, QuVue, WinH, SiteH, and UTCI. The details of the five stimulation tools with their main settings are shown in Table 2. This section aims to collect data from the results of Rhino-Grasshopper V6. The following section discusses the five measures used as outputs for training the dataset. The five measures were all related to the building indoor and outdoor environmental conditions, such as daylight, view, and thermal conditions for indoors and outdoors. These five measures included the daylight factor (DF), sunlight hours on the window (WinH), QuVue (sky view ratio at windows), sunlight hours on the site (SiteH), and the universal thermal climate index (UTCI). More details on each measurement are discussed below.

The daylight factor (DF) was simulated with Ladybug in Grasshopper, and measurement points were located on the first floors of all 10 test residential buildings. A 1.5 × 1.5 m grid 0.75 m above the floor level was used for calculation, which consisted of 2888 measurement points in total. The average values of the DF of all points were considered as the output values of the DF.

The sunlight hours on the window (WinH) were also calculated by Ladybug. The time duration of sunlight hours was measured from 8:00 a.m. to 16:00 p.m. on January 21st (the coldest day according to the Chinese building code). Each window on the first floor of the 10 test buildings had one average measuring value 0.8 m above the floor.

**Table 2.** Characteristic settings in the simulation.

| Metrics | Simulation Tools | Constant Items | Values |
|---|---|---|---|
| DF | Ladybug (ver. 0.061) (Radiance) | Location and weather file<br>Grid size<br>Distance from base surface | Beijing<br>1 × 1 m<br>0.75 m |
| | | Sky | Uniform CIE sky |
| | | Radiance parameters | -ps 8, -pt 0.15, -pj 0.6, -ds 0.5, -dt 0.5, -dc 0.25, -dp 64, -ab 0, -aa 0.15, -ar 32, -as 32, -lr 4, and -lw 0.05 |
| | | Window width-to-height ratio | 1.2/1 |
| WinH | Ladybug (ver. 0.061) | Date and time<br>Simulation time steps per hour<br>Grid size | Jan 21 8:00–16:00<br>1<br>3 × 3 m |
| SiteH | Ladybug (ver. 0.061) | Site grid size (SiteH)<br>Date and time<br>Simulation times step per hour | 2.5 × 2.5 m<br>Jan 21 8:00–16:00<br>1 |
| UTCI | EDDy3D (blueCFD) | Wind direction<br>Boundary type<br>Boundary inner rectangle<br>Boundary outer radius<br>Boundary height<br>Mesh size<br>Mesh type<br>CFD turbulence model<br>Pressure model | 0, 45, 90, 135, 180, 225, 270, and 315°<br>cylindrical domain<br>400 m<br>1000 m<br>250 m<br>357,568<br>OpenFOAM's blockMesh and snappyHexMesh<br>kOmegaSST<br>SIMPLE (Semi-implicit method for pressure-linked equations) |
| Sky view ratio | QuVue | Test surface<br>Measuring point | South/east side windows<br>Center of each window |

To calculate how much open sky view each residential unit had, the study used a calculator called QuVue. It can calculate the open sky view more realistically than other measures [12]. It used the same measuring point setup as for the sunlight hours on the window (WinH). The results of the QuVue calculation were the average values of all measuring points.

Sunlight hours on the site (SiteH) were the sunlight duration on the site from 8:00 a.m. to 4:00 p.m. on the coldest day of the location (Jan 21), same as the measuring condition of WinH. The site was meshed with a grid size of 2.5 × 2.5 m. The measuring height was set at 1.5 m from the ground. A total of 4454 measurement points and the average values of points were the output parameter.

The universal thermal climate index (UTCI), consisting of temperature, relative humidity, solar radiation, and wind speed, is an outdoor thermal metric that was used for the test. For the UTCI calculation, the measuring grid size was 1.5 × 1.5 m at the height of 1.5 m from the ground. A total of 7938 measured points were used, and an average value of the points was used for the test. To calculate the UTCI, the study used Eddy3D in Grasshopper, which used a CFD (computational fluid dynamics) engine to calculate the wind speed and direction for the UTCI calculation.

The average values were surveyed because, in building design standards in China, most metric values are considered averages, along with minimum and maximum values. In future studies, other solution methods for values will be given greater attention.

As discussed above, test 4 was to find how the normalizing input and output dataset could improve prediction. As shown in Table 3, the input and output values consisted of a wide range of values that might reduce the accuracy of the ANN model [34,35]. To mitigate this problem, some studies have proposed that all the input and output data could be normalized using a min–max approach, which checks whether the computation and performance of the ANN models could be enhanced by normalizing all the input and output data ranging from 0 to 1. The min–max normalization equation is shown as follows:

$$\beta\_nor = (\beta\_i - \beta\_min)/(\beta\_max - \beta\_min), \quad (1)$$

where β_max is the maximum value of the attribute and β_min is the minimum value of the attribute.

**Table 3.** Statistical measures for inputs and outputs in building performance.

| Variables | Average | Std. Dev | Minimum | Maximum |
|---|---|---|---|---|
| Inputs | | | | |
| $x_n$ (m) | 3.872 | 8.423 | −10.583 | 19.417 |
| $y_n$ (m) | 18.532 | 12.249 | −3.400 | 39.644 |
| $z_n$ (m) | 44.036 | 20.211 | 9.300 | 80.083 |
| Outputs | | | | |
| SiteH (h) | 0.488 | 0.026 | 0.432 | 0.544 |
| UTCI | 0.574 | 0.177 | 0.074 | 0.935 |
| DF (%) | 6.615 | 12.115 | 3.359 | 72.935 |
| WinH (h) | 4.789 | 0.980 | 2.667 | 6.894 |
| QuVue (%) | 32.448 | 7.981 | 11.538 | 45.527 |

*2.3. Development of ANN Models*

To find the best way to build the ANN model for different building performances, it is important to conduct a comparative study of the ANN models with different built environment measures to gain general insight into the relationship between the prediction results, the training model's setup, and the dataset. The ANN model's performance was measured by the correlation coefficient (R) between the predicted data and actual data to evaluate the performance of the established ANN models.

As shown in Figure 4, the variables of the input layers were represented as $a_n$ and the output layer variables were represented as $b_n$. Each variable contained the same number of sample sizes that had different values ($Value_n$). The hidden layer consisted of "n" numbers of neurons, represented as $c_n$.

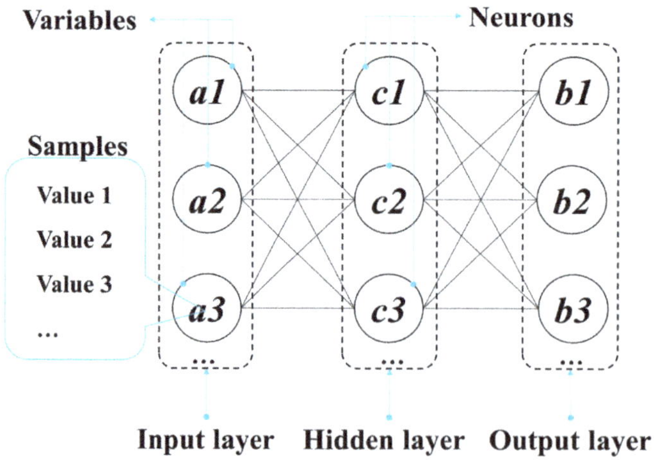


**Figure 4.** The definition of the standard structure of an ANN model.

Figure 5 demonstrates the four tests discussed above. Test 1 showed a different number of output layer variables ($b_n$). In test 2, the different numbers of sample sizes ($Value_m$) were tested to discover the difference between the number of samples and accuracy. For test 3, a q number of neurons was used for the hidden layer neurons ($c_q$). In test 4, four ANN models of different input and output variables (normalized vs. normalized, actual vs. normalized, actual vs. actual, and normalized vs. actual) were constructed.

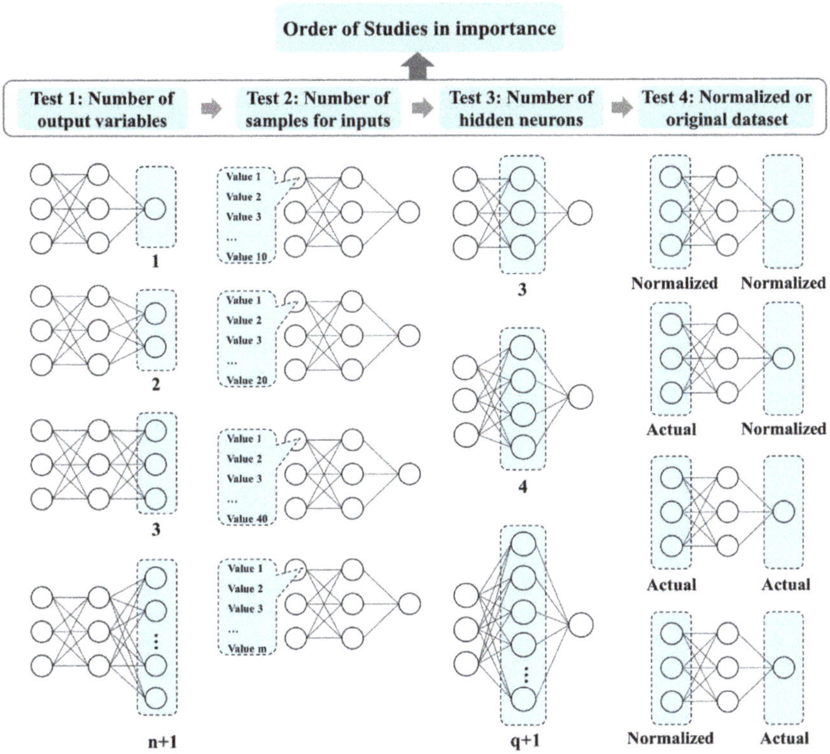


**Figure 5.** Overall view of neural network sensitivity analysis.

The procedure of using ANNs to perform prediction and the evaluation of the performance of the ANNs involved the four tests are as follows:

Test 1: number of output variables. This test was designed to question how many training output variables were reasonable with a fixed number of training input variables. It is common in building performance measures that several different measures are used to evaluate performance. For instance, in LEED v4 daylight, sDA (spatial daylight autonomy) and ASE (annual sunlight exposure) are needed, and both are related to the same input variables such as the window size and location. What is the ideal method to build an ANN model that has the same input variables? Would it be better to build two measures in one ANN model or two separate models? Test 1 was designed to understand this question of building an independent ANN model for each measure, or the possibility of combining them together. The purpose of this test was to improve the prediction efficiency if one ANN model has the capacity for higher prediction accuracy with more output values.

Test 2: number of samples for inputs. This test was designed to understand whether having more training input cases would increase the accuracy of the ANN models. This purely depends on the complexity of the problem and commonly requires more cases to improve accuracy. However, there is a question of what number of cases is good enough. The answer depends on the other elements such as the number of output and input variables. Escandón and others indicate that a ratio of 2.5 between the sample size and the number of characteristic parameters (input variables) would be a reasonable sample case to train a model [24]. The proposed question was whether this relation between the input variables and the number of training datasets would be linear. Would there be a point where the linear curve line would become a plateau? The goal of this test was to use proper quantities of samples to lead the optimal ANN models in predicting performance.

Test 3: number of hidden neurons. It is known that there is a range of ratios between the number of input variables and the number of hidden neurons that can develop accuracy. As in test 2, would a greater number of hidden neurons increase the accuracy, or would there be a point that increasing the number of hidden neurons would not further improve the accuracy? This test was intended to find some rules showing how many neurons in the hidden layers are necessary to reach high ANN model accuracy.

Test 4: normalized or original datasets. This test compared the combination of normalization and actual input and output values to decide which pair of groups achieve better accuracy. The normalization of variables was applied in a wide range of values, especially in different orders of magnitude. However, the difference in the ANN model accuracy of normalized or actual values revealed the significance of the normalization of the dataset. Upon reviewing the version of MATLAB utilized in our research, we found that the normalization step was not included in the procedure. For each test, the training process was repeated 30 times for all ANN models, and the average value was employed to facilitate comparison among different types of datasets for result validation purposes. This test aimed to discover the rules of the normalization of datasets to achieve optimal ANN models in terms of prediction accuracy ability.

This study hoped to gain certain insight into the stability and performance of ANN models through the four tests discussed above. The selection of every test led to a comparison in the end, which yielded a summary and principles to show the characteristics of how the importance of each element influenced the prediction of the ANN models.

In this research, ANN-MLP models were used to train the ANN models. For the ANN model, a three-layer feedforward network with sigmoid hidden neurons and linear output neurons that could fit multi-dimensional problems was adopted. The transfer function was a hyperbolic tangent function in the hidden neurons, and a linear function was used in the output neurons.

For training the ANN model, MATLAB's Deep Learning Toolbox was used for the test. To find the overall performance in terms of accuracy, this study conducted 30 different trainings for each test, which allowed us to reach more stable correlation coefficients R. In the ANN calculating procedure, there were four R values for testing, training, validation and all data. In order to find network performance, the average of all the R data is applied in this paper. Basic ANN model setups were employed and revised to reflect the different configurations for different tests.

A base ANN model (base case) setup was used to perform the comparison work among the four tests, and the following setups for the ANN model were used for this study (Table 4). Because the test case had 12 buildings (n = 12; Figure 3), a total of 36 variables, including from x_1 to x_12, y_1 to y_12, and z_1 to z_12, were used as the input variables. As discussed in Section 2.2, the performance metrics DF, WinH, QuVue, SiteH, and UTCI were the five output parameters. For the sample size of the ANN, we randomly selected 52 samples, which was about 1.5 times the number of input variables. The number of hidden neurons was set to 108, which was about 3 times the number of input variables. More detailed information is discussed in Table 2. The values of inputs were normalized to the interval (0, 1).

**Table 4.** Base ANN model setup.

| ANN | No. of Inputs | Number of Neurons | Layer | Training Function | Transfer Function Hidden Neurons | Transfer Function Output Neuron |
|---|---|---|---|---|---|---|
| | 36 | 108 | 3 | Levenberg–Marquardt backpropagation algorithm (trainlm) | Hyperbolic tangent function | Linear function |
| Data Division | Training: 70% of dataset | | | | | |
| | Simulation: 15% of dataset | | | | | |
| | Validation: 15% of dataset | | | | | |

### 2.3.1. Number of Output Variables

The first test investigated the accuracy with a number of output variables. It is common to have one output variable as a training dataset. However, sometimes it is beneficial to have more than one output variable as a training dataset. If the outputs are similar measures that use the same input variables, then it would be beneficial to combine two independent ANN models into one. For instance, in daylight simulation, sDA and ASE use the same input parameters to calculate the measures, and if one ANN model can predict both measures, then it would be beneficial to reduce the computational time and power. For this reason, this study developed three models to test their accuracy with a different number of outputs.

Three models (models A, B, and C) were built with the same inputs for building the geometry parameters and a different number of output variables. The comparison of models A, B, and C by different assemble approaches provides insight into how to increase the accuracy of the predictions. The ensemble approach was based on the idea that by combining similar forecasters, it would be possible to improve the overall forecasting accuracy, which was used to improve performance models in another study [36].

All three models had the following ANN setups to keep more parts of the structure of the ANN models accordant. A two-layer feedforward network with sigmoid hidden neurons and linear output neurons that could fit multi-dimensional problems was adopted. It also had the same 36 input variables and 108 neurons in one hidden layer (3 times the number of input variables). The only difference was the number of output variables.

Model A had five independent ANN models for five different outputs (DF, WinH, QuVue, SiteH, and UTCI). Figure A1 shows the ANN structure used for model A. Model B grouped the output variables into two with related measures. One group was composed of indoor measures including DF, WinH, and QuVue (Figure A2). Another group of measures included the outdoor conditions of SiteH and UTCI (Figure A3). Model C included all five output variables in one model as shown in Figure A4. It used one model to train five outputs including indoor measures of DF, WinH, QuVue, and the outdoor conditions of SiteH and UTCI.

Each model conducted 30 trainings and the R values of the forecasting models were analyzed by R distribution analysis as shown in Figures 6–8. From the R distribution histograms, the 95% confidence interval of the error distributions is located in different ranges for models A, B and C. The average values ($\mu$) and standard deviations ($\sigma$) are listed in each histogram. All of the standard deviation values are smaller than 0.12, meaning a low degree of dispersion, and the average R can be used to express the average accuracy performance of each model.

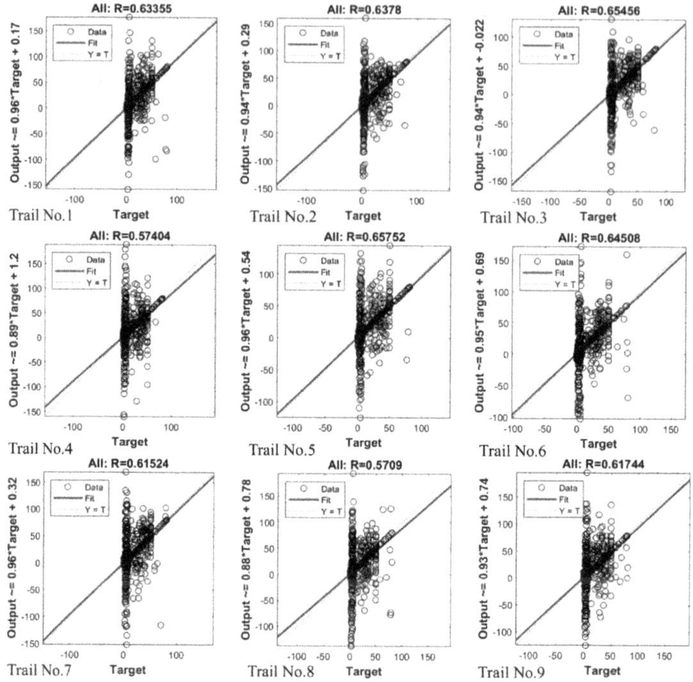


**Figure 6.** R value alteration of trials 1–9 in model A.

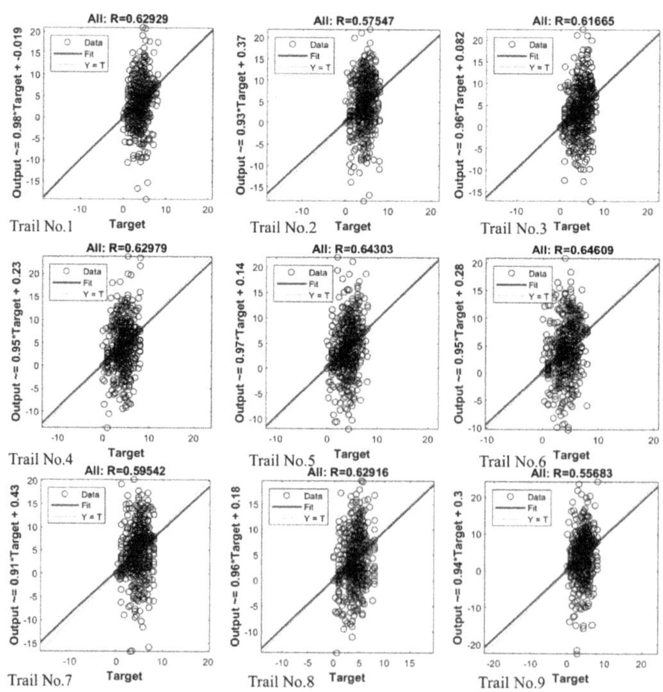


**Figure 7.** R value alteration of trials 1–9 in model B.

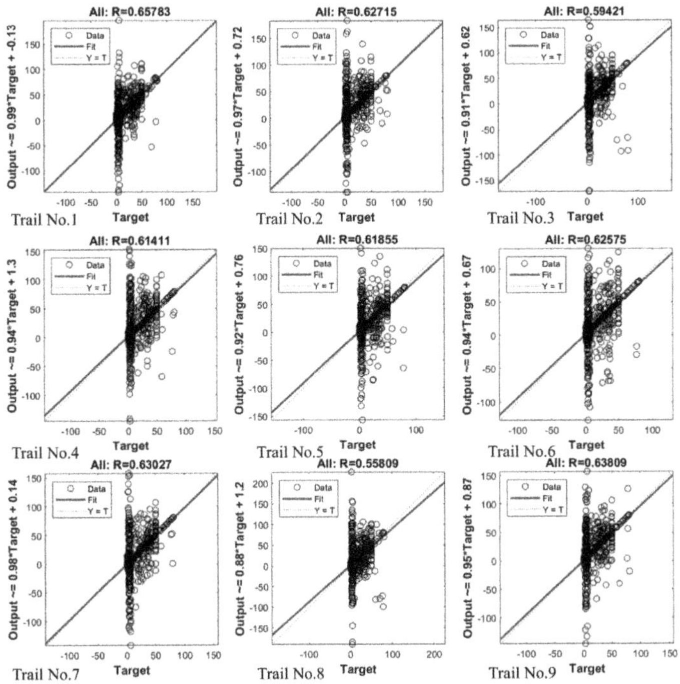

**Figure 8.** R value alteration of trials 1–9 in model C.

2.3.2. Number of Training Samples

It is commonly acknowledged that the number of samples is considered a critical factor impacting ANN performance [37]. This test was designed to understand the correlation between the number of samples and its accuracy. The test conducted three different models with output training datasets that included DF, QuVue, and WinH. To eliminate influence from other parameters, all three models used the same training functions as mentioned above with 108 hidden neurons and one hidden layer.

The number of input samples used for the test were 20, 30, 40, 52, 72, and 108. Based on the input variables (36), as a reference, a sample size 2 to 3 times the number of variables was recommended for the test, which was around 90 samples. Each test applied 30 independent trainings to find the R values. As shown in Tables A1–A3, 30 trials were conducted to reach a stable accuracy of the R value.

2.3.3. Number of Hidden Layer Neurons

The number of hidden layer neurons depends on the ANN model's complexity. A redundant number of neurons in hidden layers will lead to overfitting the ANN model and take more time for training. The numbers of hidden layer neurons used for the test were 36, 54, 72, 90, 108, and 144, which were 1, 1.5, 2, 2.5, 3, and 4 times the number of input variables (36).

2.3.4. Normalized or Original Datasets

The normalization of the dataset in ANN models has been reported in previous studies to have stability and efficiency in creating a better performing neural network. However, for some cases, the normalized dataset has little significant influence on the accuracy of the neural network models [22]. To find whether normalization would improve the prediction of building performance or not, the normalized and raw datasets were considered as both inputs and outputs.

Like in tests 2 and 3, the base ANN model with a sample size of 52, 108 neurons in the hidden layer were used. The test investigated three different models that contained different output datasets including DF, QuVue, and WinH. The input dataset for all three models used the same 36 variables.

These three models' input and output training datasets were modified to have four different training datasets. Dataset 1 normalized both the input and output datasets, dataset 2 normalized the input but not the output, dataset 3 did not normalize the input but normalized the output, and dataset 4 did not normalize datasets for either the input or output. As in the previous three tests, training was conducted 30 times for all ANN models, and the average value was used to compare the different types of datasets.

### 2.4. Performance Evaluation

The correlation coefficient R, a regression index, was used to evaluate the performance of the ANN, which measures the correlation between the predicted and actual value [38,39]. R values vary between −1 and +1; however, if an R value is closer to 1, then a more positive linear relationship and a high network performance can be obtained. Generally, the root-mean-squared error (RMSE), mean-squared error (MSE), mean absolute percentage error (MAPE), coefficient of determination ($R^2$), and MAE are commonly used performance metrics of ANN models together with the regression R [40–42]. The MSE is the average squared difference between the ANN predictions and performance simulations. The MSE is used to calculate the average squared difference between the estimated value and the simulated value, lower values of which indicate that the data fit better. The training process can automatically stop when the mean-squared error (MSE) of the validation samples is stabilized. Escandón et al. used a regression analysis with a coefficient of regression (R) and relative errors to show the reliability of the developed ANN model in predicting the energy performance and thermal comfort of a social housing stock in southern Europe [24], similar to previous studies related to a building stock [43]. Therefore, in this paper, we also used the R value to determine the performance of the ANN models [44,45].

## 3. Results

### 3.1. The Results of the Four Developed ANN Models

We explored the possibilities of the performance prediction problem of similar building layouts and provided a replicable and widely used ANN model parameter selection rule. For example, under a similar number of complex architectural layouts, more than three kinds of performance predictions are made. For different indicators, the independent variables were selected based on Tables 2 and 3.

#### 3.1.1. Number of Input and Output Variables' Results

The frequency of distribution of x_n, y_n, and z_n variables was displayed in Figure 9. The frequency of the x_n and y_n variable values were between 0.1 and 0.2, except for the value range between −15 and −10 m (Figure 9, left). As can be seen in Figure 9 (right), the z_n variables were randomly grouped into two, one group in the 0.1–0.14 frequency range and another group in a frequency between 0.04 and 0.08.

Figure 10 describes the ratio distribution achieved in the simulation results of five output metrics, which were WinH, DF, QuVue, UTCI, and SiteH. The highest frequency for the WinH simulation result ranged from 2 to 4 h. The simulation results between 0 and 2 h had the lowest frequency value, 0.10. For the DF simulation results, the maximum frequency was 0.45 when the DF value was between 0 and 4%. The lowest frequency was when the DF value was between 4 and 6%. For QuVue, the number of frequencies increased considerably from 0.05 to 0.30 when the QuVue result values were between 0 and 40%. The range of QuVue results from 40% to 50% had about the same frequency ratio as the range from 30% to 40%. The range of UTCI result values was from

0.8 to 1.0, in which the highest frequency was recorded at 0.45. Then, the frequency from 0.2 to 0.4 in UTCI ranked second, at 0.4. The frequencies of 46–49% and 49–52% in SiteH simulation outcomes were similar, exactly 0.35 for each, while in 0–46% and 52–55%, the numbers were considerably lower, averaging between 0.1 and 0.15.

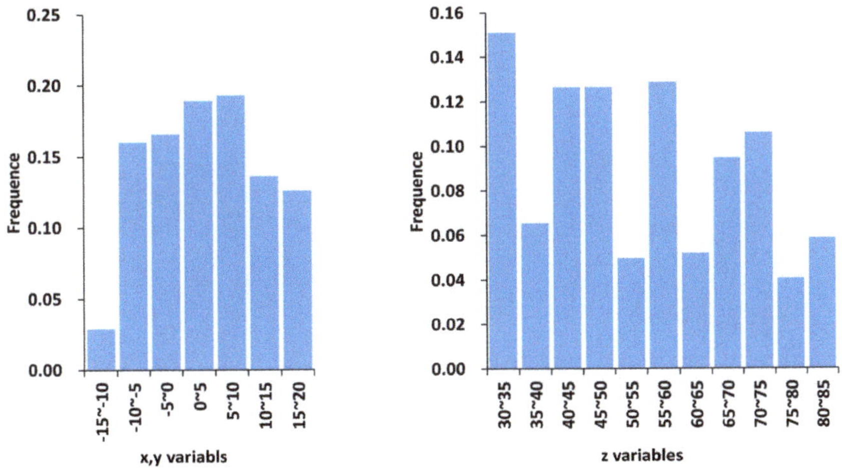


**Figure 9.** Histogram of values of input variables.

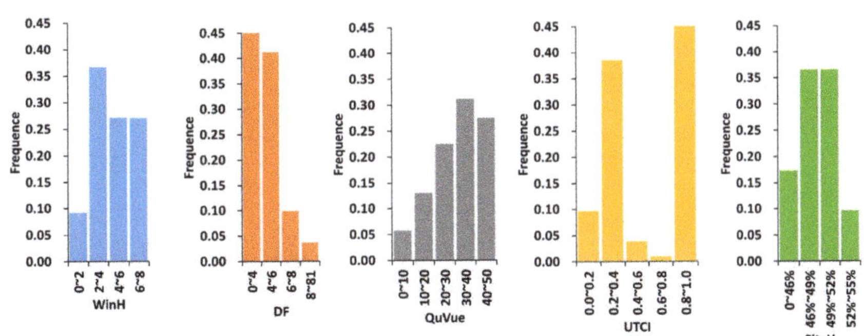


**Figure 10.** Frequency of output values from the five simulation results.

Figure 11 and Table 5 show the R value of the average of 30 trainings for all datasets, which considers the testing datasets and validation datasets. For model A, the R value for each of the five models was 0.226, 0.518, 0.623, 0.323, and 0.503. The average R value of all five ANN models was 0.438. For model B, the R value was 0.605 for group 1 and 0.507 for group 2. The average R value of the two ANN models was 0.556. Model C, which included all five output variables, had an R value of 0.621. The result was that model C made a better prediction than the other two models. Despite the fact that the absolute R value was typically low in most test scenarios, the patterns of variable transformation remained dependable. As a result, greater weight should be given to the relative R value.

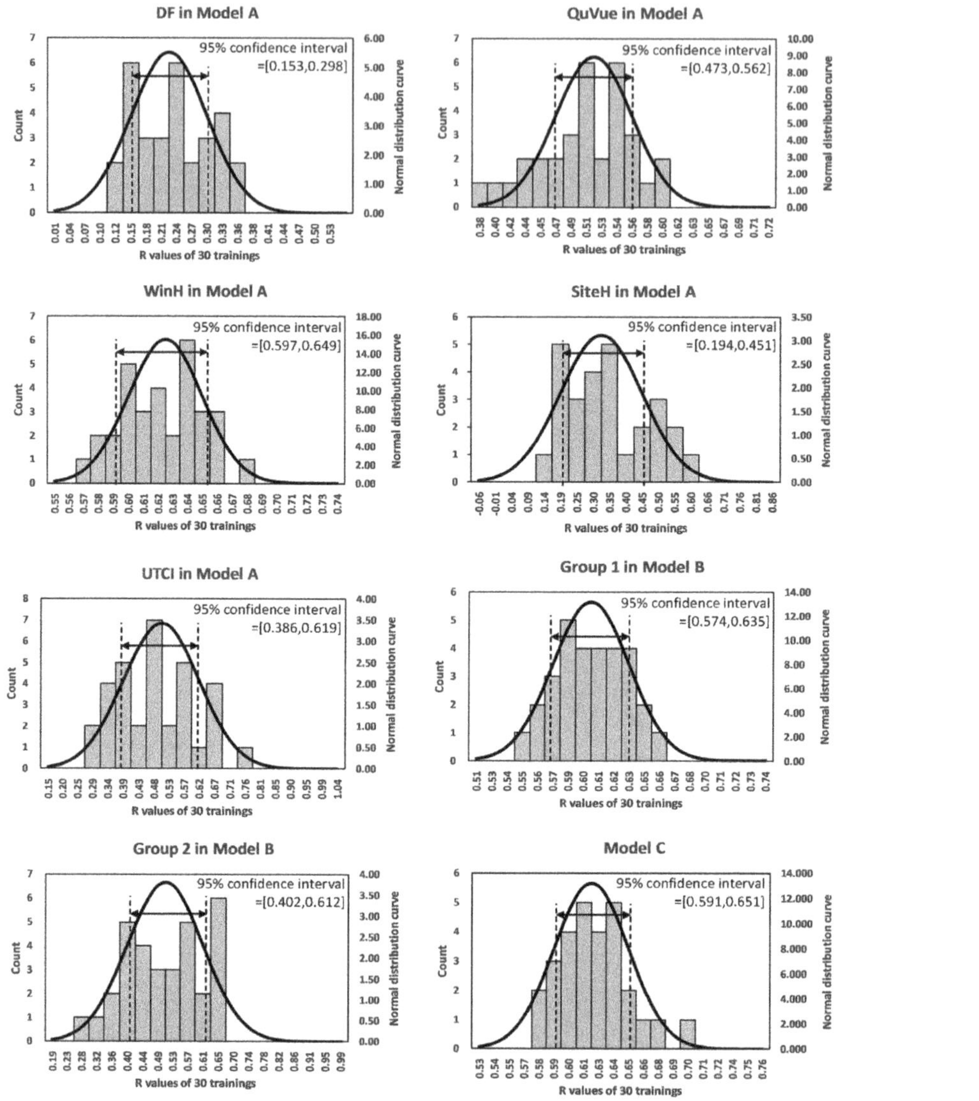


**Figure 11.** R distribution analysis for three models.

**Table 5.** R value of three models.

| | Model A | | | | | Model B | | Model C |
|---|---|---|---|---|---|---|---|---|
| | DF | QuVue | WinH | SiteH | UTCI | Group 1 (DF, QuVue, and WinH) | Group 2 (SiteH and UTCI) | |
| R | 0.226 | 0.518 | 0.623 | 0.323 | 0.503 | 0.605 | 0.507 | 0.621 |
| Avg. R | | | 0.438 | | | 0.556 | | 0.621 |

#### 3.1.2. Number of Training Samples' Results

As expected, all three tests showed an increase in the R value as more samples were used. Figure 12a shows the scatter distribution of the DF ANN model, Figure 12b represents the scatter plot of the QuVue ANN model, and Figure 12c shows the plot of the WinH ANN model. When the sample size was in the range of 20 to 52, the DF ANN model's R value was increased by 0.0038, which was almost flat. QuVue and WinH showed increases of 0.1314 and 0.157, respectively. When the sample size increased to 70, the comparative increase in the R value between the sample sizes of 52 and 70 was significant. For the DF, it increased by 0.565, which was more than 100 times the increase observed in the sample size of 20 to 55.

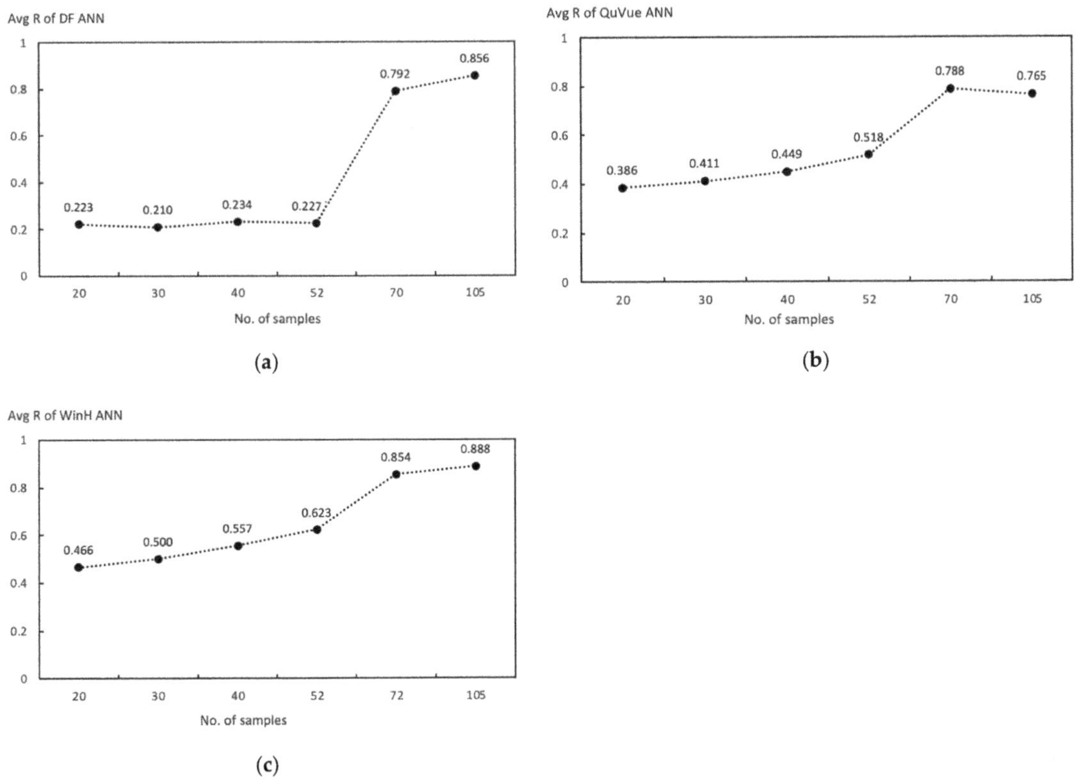


**Figure 12.** Plot of R value and sample size for (**a**) DF ANN model, (**b**) QuVue ANN model, and (**c**) WinH ANN model.

For QuVue and WinH, the increases were 0.271 and 0.231, which were 2.06 and 1.47 times that of the sample size of 20 to 52. However, when the sample size increased to 108, the increase of the R value was not as high as the increase in the case of 70 samples. The DF model was improved by 0.0637 from the sample size of 70, and for QuVue, it was decreased by 0.0233 from the sample size of 70. For WinH, it was increased by 0.0341. It was interesting to see an overserve between the sample size and the accuracy, which showed a nonlinear relation.

Generally, the DF, QuVue, and WinH ANN prediction models showed a similar trend in their R values with respect to the number of training samples. When the number of samples was below 52, the prediction accuracy was low, but it increased significantly once it reached 70, and could achieve an accuracy of 0.8 with an appropriate sample size. As the sample size continued to increase, the level of R improvement becomes limited. Therefore, the ANN prediction model can achieve efficient and effective results as long as the sample size is appropriate.

### 3.1.3. Number of Hidden Layer Neurons' Results

The test constructed three different models with an output training dataset that included DF, QuVue, and WinH. To eliminate the influence from other parameters, all three models used the same training function, number of samples, and number of hidden layers. As in test 2, each ANN model conducted 30 independent trainings to find the stable R values as shown in Tables A4–A6.

The results showed interesting outcomes. For the DF ANN model (Figure 13a), as the number of hidden neurons increased from 36 to 108, the R value of the ANN models decreased by 0.1835. Then, the R value increased by 0.0328 when the number of hidden neurons increased from 108 to 144. However, when the number of neurons was 36, the R value was 0.1507 higher than when the number of hidden neurons was 144.

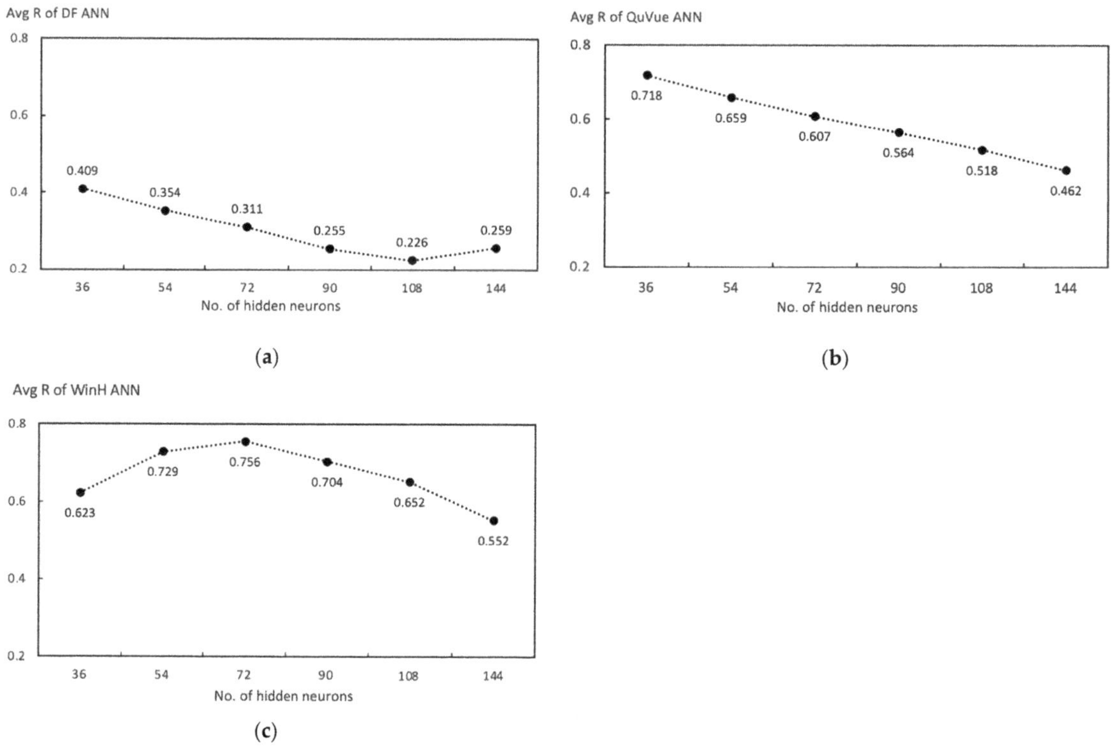


**Figure 13.** Plot of R value and number of hidden neurons for (**a**) DF ANN model, (**b**) QuVue ANN model, and (**c**) WinH ANN model.

The QuVue ANN model (Figure 13b) also had decreasing R values as the number of hidden neurons increased. When the number of hidden neurons was 36, the average QuVue ANN model's R value was 0.7175, and it decreased to 0.4622 when the number of hidden neurons was 144. The average R value in the WinH ANN model (Figure 13c) was 0.7286 when number of hidden neurons was 36, and it decreased to 0.5524 with 144 neurons.

In all three cases, with more neurons in the hidden layer, the R value decreased, except in the DF ANN model when the number of neurons was 144, where the R value was improved compared to when the number of neurons was 108. However, it was interesting to observe that when the number of neurons in the hidden layer was the same as the number of input variables, the models could have higher R values than the recommended hidden number of neurons. Figure 13a,b exhibits a distinct pattern, indicating that the model achieved its highest accuracy when the number of neurons was set to 36. This observation implies that as the number of neurons increased, there was a noticeable decline in the model's prediction accuracy.

In general, the R values of the DF, QuVue, and WinH ANN prediction models exhibited distinct trends in relation to the number of hidden layer neurons. Both the DF and WinH ANN models demonstrated inflection points in their R values. As the number of hidden layer neurons increased, the R value of the DF and QuVue ANN models exhibited a downward trend, with QuVue showing a linear decline. On the other hand, the R value of the WinH ANN model initially rose and then declined. In the range discussed in this study, only the WinH ANN model identified a more suitable number of hidden neurons, estimated to be less than 36.

#### 3.1.4. Normalized Dataset Results

The average and standard deviation of the R value of the four cases in the DF ANN model did not show a difference. Datasets 1 and 4 had a similar R value of 0.2467 and 0.2472, and datasets 2 and 3 had an R value of 0.2257 and 0.2282. Interestingly, normalizing both the input and output and not normalizing either did not show a significant difference. Also, having a combination of normalizing and not normalizing the dataset did not perform better than the other two dataset options. Overall, the standard deviations among the four different options were not significant enough to find better dataset options among the four cases (Figure 14a).

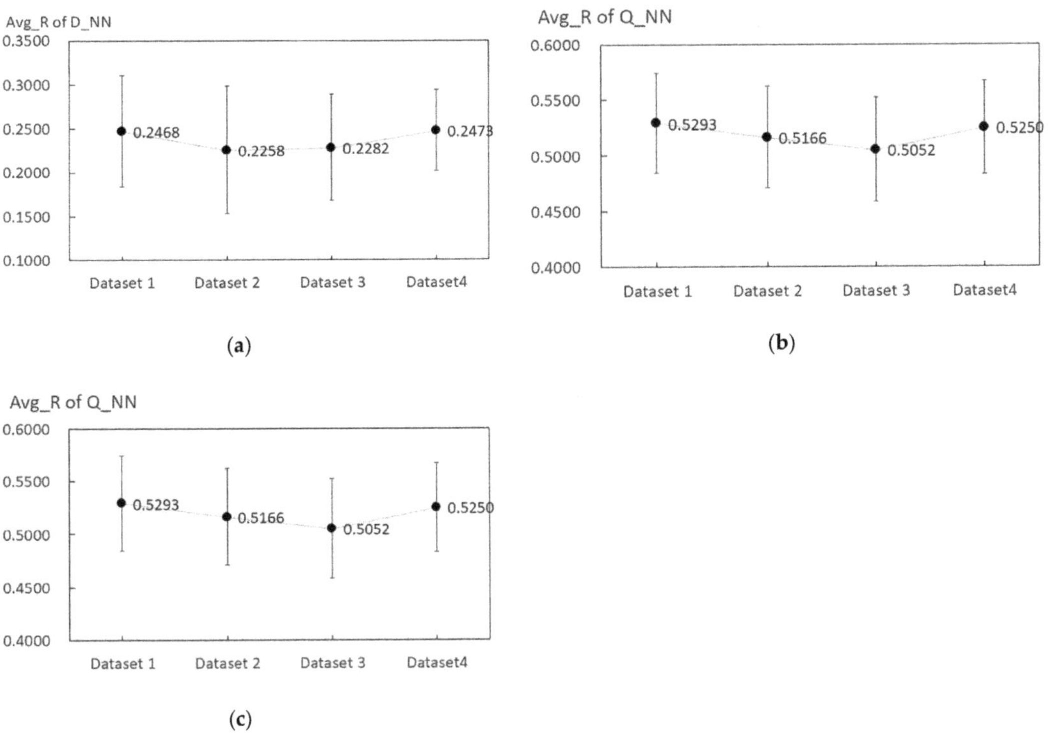


**Figure 14.** Average R value between different dataset configurations for (**a**) DF, (**b**) WinH, and (**c**) QuVue.

Figure 14b shows the average R values of the four cases within the standard deviation, which shows that it was difficult to make one case perform better than the other cases. Similar to the DF model, datasets 1 and 4 had a higher R value than datasets 2 and 3 in the QuVue ANN model. Dataset 1 was 0.529, and dataset 4 was 0.525, which indicates no significant difference between the normalized dataset and non-normalized dataset. As in Figure 14c, all the average values of the four cases were within the standard deviation, which means that not one option performed better than other dataset options. Overall, from the three models, all three results showed that there was no significant difference between the normalized dataset and non-normalized dataset.

### 3.2. Test Results Analysis

In this section, all test results were compared to find the overall influence of the performance of different ANN configurations on their prediction accuracy. For the relative comparison, a base case ANN model was used to compare the different tests, which had 52 samples for the input dataset and 108 neurons in the hidden layer, with 36 input variables being normalized and the output variables not being normalized.

Based on the base case, the percentage of the altering value of each different ANN model's R value was compared to each other (Figure 15). Figure 15a shows the summary of test 2 of the different number of samples sizes' impact on accuracy. In test 2, three ANN models were tested, which had the same input variables but different output variables. The figure shows the average R values of three models with a different sample size. It can be seen in Figure 15a that having a greater number of samples in the inputs can enhance the R value. However, the trend line was not linear after the sample size increased to more than 72, and the slope was not rising sharply. Compared to the base case (52 samples), the R value of the sample size of 72 increased significantly from 0.62 to 0.85. Based on the test, around two times the number of variables was a reasonable range to select as the sample size.

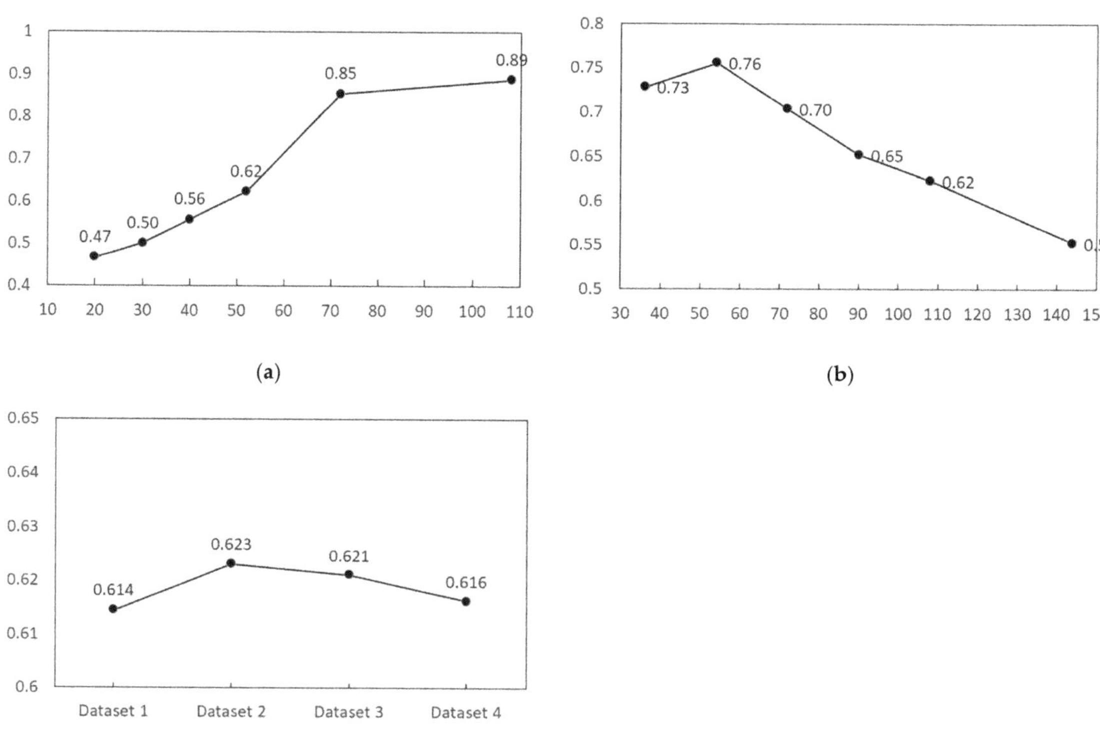


**Figure 15.** Plot of average R value with different (**a**) dataset sample sizes, (**b**) numbers of neurons in the hidden layer, and with (**c**) normalized or real datasets.

Figure 15b shows the average R value of test 3, which tested a different number of neurons in the hidden layer. As in test 2, test 3 used three different ANN models, which had the same input variables but different output variables. The figure shows the average R value of the three models with a different number of neurons in the hidden layer. The base case had 108 neurons in the hidden layer, which yielded an R value of 0.62. The figure shows that having fewer neurons than recommended by the literature review improved the R value. However, having fewer neurons also decreased the R value. The test shows that having 52 neurons achieved the highest R value of 0.76, which was about 1.5 times the number of input variables.

The summary of test 4 is shown in Figure 15c. This investigated the normalization of the dataset in the ANN models. As in the previous tests, three ANN models were built and tested with different configurations of the dataset. Four different datasets were tested: dataset 1 normalized both the input and output parameters, dataset 2 normalized the input but not the output, dataset 3 did not normalize the input data but normalized the output, and dataset 4 did not normalize either the input or output variables. As Figure 15c shows, normalizing both the input and output variables showed the lowest R value, and next lowest one did not normalize both the input and output. The base case (dataset 2), where the input variables were normalized and the output variable was not normalized, showed a higher R value (0.623) than the other three. However, the results demonstrated no significant difference among the different formats of the dataset. Compared to tests 2 and 3, test 4 did not show a significant improvement in the R value with a different dataset format.

Table 6 shows the overall R values among the different tests for the DF ANN model. The highest improvement in the R value of the DF ANN model was 0.856 when the number of input samples was 108, which was a 277.19% improvement from the base case (0.227). When the number of hidden neurons was 36, the R value of the DF ANN model reached the value of 0.409, which was an 81.26% improvement from the base value (0.227). The normalization of inputs and outputs had the highest R value compared with the other three normalization conditions, 9.3% more than that of the base case (0.227).

**Table 6.** Comparison between tests' R value with DF ANN model.

| TEST | | Average R Value | Ratio Compare to Base Case |
|---|---|---|---|
| No. of samples (test 2) | 20 | 0.223 | −1.66% |
| | 30 | 0.210 | −7.40% |
| | 40 | 0.234 | 3.02% |
| | 52 * | 0.227 * | 0.00% * |
| | 70 | 0.792 | 249.11% |
| | 108 | 0.856 | 277.19% |
| No. of hidden neurons (test 3) | 36 | 0.409 | 81.26% |
| | 54 | 0.354 | 56.68% |
| | 72 | 0.311 | 37.62% |
| | 90 | 0.255 | 13.09% |
| | 108 * | 0.227 * | 0.00% * |
| | 144 | 0.259 | 14.55% |
| Dataset format (test 4) | dataset 1 | 0.247 | 9.30% |
| | dataset 2 * | 0.227 * | 0.00% * |
| | dataset 3 | 0.228 | 1.07% |
| | dataset 4 | 0.247 | 9.53% |

* base case result.

As in Table 7, the order of increasing percentiles of the three tests for the QuVue ANN model had the same trend as the DF ANN model. For test 2, when the number of samples was 70, the highest R value ranked at a 52.36% improvement from the base case. For test 3, when the number of neurons was 36, it had the highest improvement of 38.65% from the base case and test 4 with the R value of 2.46%.

**Table 7.** Comparison between tests' R value with QuVue ANN model.

| TEST | | Average R Value | Ratio Compare to Base Case |
|---|---|---|---|
| No. of samples (test 2) | 20 | 0.386 | −25.40% |
| | 30 | 0.411 | −20.56% |
| | 40 | 0.449 | −13.25% |
| | 52 * | 0.518 * | 0.00% * |
| | 70 | 0.789 | 52.36% |
| | 108 | 0.765 | 47.86% |
| No. of hidden neurons (test 3) | 36 | 0.718 | 38.65% |
| | 54 | 0.659 | 27.24% |
| | 72 | 0.607 | 17.36% |
| | 90 | 0.564 | 8.91% |
| | 108 * | 0.518 * | 0.00% * |
| | 144 | 0.462 | −10.69% |
| Dataset format (test 4) | dataset 1 | 0.529 | 2.46% |
| | dataset 2 * | 0.517 * | 0.00% * |
| | dataset 3 | 0.505 | −2.20% |
| | dataset 4 | 0.525 | 1.64% |

* base case result.

Table 8 indicates the R values of three models of the WinH ANN model. As in the DF and QuVue ANN models, the R value improvement in test 2 was the most significant among the three tests, which was enhanced by 42.52% compared to the base case's R value of 0.623. The second most significant R value improvement was also seen in test 3, with an enhancement of 21.27% compared to the base case's R value of 0.623. In test 4, the R value of the base case ANN model performed best.

**Table 8.** Comparison between tests' R value with WinH ANN model.

| TEST | | Average R Value | Ratio Compare to Base Case |
|---|---|---|---|
| No. of samples (test 2) | 20 | 0.466 | −25.19% |
| | 30 | 0.500 | −19.68% |
| | 40 | 0.557 | −10.68% |
| | 52 * | 0.623 * | 0.00% * |
| | 70 | 0.854 | 37.04% |
| | 108 | 0.888 | 42.52% |
| No. of hidden neurons (test 3) | 36 | 0.729 | 16.93% |
| | 54 | 0.756 | 21.27% |
| | 72 | 0.704 | 12.99% |
| | 90 | 0.652 | 4.67% |
| | 108 * | 0.623 * | 0.00% * |
| | 144 | 0.552 | −11.35% |
| Dataset format (test 4) | dataset 1 | 0.614 | −1.38% |
| | dataset 2 * | 0.623 * | 0.00% * |
| | dataset 3 | 0.621 | −0.32% |
| | dataset 4 | 0.616 | −1.09% |

* base case result.

Figure 16 shows the best percentile of increments for each model in the three tests. Among the three tests of the percentage improvement in the R value, changing the number of samples in test 2 indicated the greatest improvement in all three ANN models (WinH, DF, and QuVue). Test 3 showed the next most efficient method to improve the R value in all three ANN models. Test 4's normalization demonstrated less efficiency compared to the other two tests, where the greatest improvement was a 9.5% improvement with the DF ANN model. However, compared to the improvement from test 2, the normalization impact on the improvement was limited.

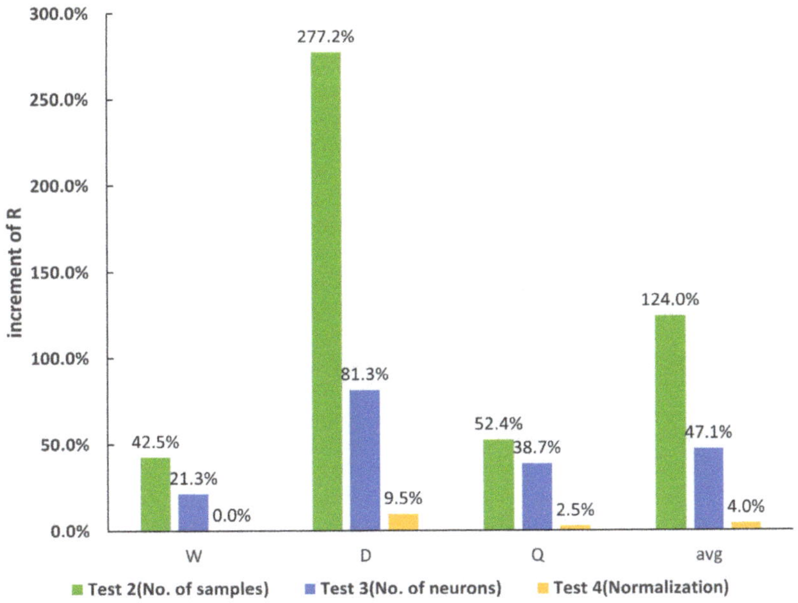


**Figure 16.** The percentile of increment for each model in three tests.

## 4. Discussion

Employing the same setups, including the training function, 52 samples, 108 hidden neurons, and 36 input parameters, test 1 found that the integration of the five performance metric prediction models (model C) had the highest accuracy performance value. The single-objective prediction model (model A) was a less recommended model because the average R value was the lowest. Combining all of the output parameters together not only improved the accuracy performance of building an ANN model, but also reduced the amount of work compared to building many models. The obtained results were not completely consistent with our initial expectations. One possible explanation for the lower performance of a single output compared to multiple outputs could be that the outcomes are more closely related to each other, which in turn may impact the final result. We see this as an area for further investigation. Given that our team's expertise does not lie in the theory of ANNs, we plan to collaborate with mathematicians or computer science experts to explore this question further. There is few research on this topic.

The quantity of samples was tested in training the ANN models in test 2. In the DF ANN model, when the number of samples increased from 52 to 108, the R value improved by 277.19%, which was the highest percentile improvement in the three models of test 2. The improvement in the R value rose from 0.5175 to 0.7885, almost 52.4%, in the QuVue ANN model, ranking second. The percentile increment of the WinH ANN was 42.5% with an R value from 0.623 to 0.888. When the number of samples (72 for QuVue and 108 for both WinH and DF) was two or three times the number of variables, the accuracy of the ANN models performed best. In Figure 15a, when the number of samples was 72, the R value was 0.85. However, increasing the number of samples will not dramatically improve the average R value. By balancing the model's complexity and training time, two times the number of input variables may be recommended. In previous research, Rocío Escandón et al. concluded that the sample size ensured reliable prediction results according to a specific building category, which means that a ratio of 2.5 between the sample size and the number of characteristic parameters (input variables) can increase the accuracy [24].

Our result showed that 2.5 may be a reasonable conclusion, but based on our test, in some cases 2.0 also can be used.

Based on test 3, the correlation coefficient (R) of the DF was improved from 0.227 to 0.409 (81.3%), thus ranking first. The average R value of QuVue improved to 0.718, 38.7% more than that of the base case, and the improvement in WinH ranked lowest. Moreover, the highest value of R was attained when the number of hidden neurons of 36 for DF and QuVue or 54 for WinH were selected. As can be seen in Figure 15b, there existed a peak value in the average R value of the three models. Based on the test, when the number of hidden neurons was about 1.5 times the number of input variables, the ANN models had a better accuracy performance than the other models. Many researchers investigated the same topic in previous studies. Sofuoglu tested eight feedforward networks with different hidden layers and numbers of neurons to search for better performance models according to R, $R^2$, and RMSE to predict the prevalence of building-related symptoms (BRSs) of office building occupants [46]. Ashtiani et al., using a varied number of hidden neurons from 1 to 100, showed that 10 hidden neurons achieved the best network performance [37]. Kerdan and Gálvez demonstrated that the number of hidden neurons is insensitive to the performance of the ANN model [47]. Moon and Jung took three steps to find the optimal artificial neural network model in terms of $R^2$ to predict the setback temperature of a building [48]. The optimal values for the number of hidden layers, number of hidden neurons, learning rate, and moment were found to be 4, 9, 0.6, and 0.9, respectively. Wang et al. discussed the effect of the time intervals of inputs and the number of hidden neurons on model accuracy using the global sensitivity analysis method for air heat pump operation systems [49]. Only a limited number of studies have investigated the link between the number of hidden neurons and the number of input variables, but test 3 did confirm a pattern concerning the accuracy of ANN model predictions.

The test of the normalization combination of variables (test 4) demonstrated little influence on the improvement in accuracy of the three models. Furthermore, the highest percentile increments were derived from the different combinations of the input and output variables in different models. From the test result, it was difficult to generalize a certain trend; the highest percentile improvement was 9.53% when the original inputs and outputs of the variables were selected in the DF ANN. In the QuVue ANN, both the normalized input and output parameters showed the highest improvement in the R value of 2.5%. Additionally, the base case (normalization input with the original variables) in the WinH ANN had a higher R value than the other three conditions. Therefore, the normalization of the input and output variables was insensitive to the accuracy performance of the ANN models [50]. In previous studies, the normalized method was commonly conducted to construct the ANN models; however, the normalization combination of the input and output variables was seldom investigated regarding the accuracy performance of the models.

From Figure 16, we can observe the summary of the percentile improvements in the three tests, which were the number of input samples, the number of hidden neurons, and the normalization combination of the variables. Among the three tests, changing the number of samples in the ANN models yielded the highest R value compared to base case.

## 5. Conclusions

This study investigated a strategy to construct a more reliable ANN model to predict a building's indoor and outdoor performance measures. A sensitivity analysis was utilized to test different strategies to build the ANN model, namely the number of output variables, the number of samples, the number of hidden neurons, and the normalization combination of the variables, that would impact the prediction of the accuracy performance. The correlation coefficients (R) of each ANN model were examined. A summary of the findings is as follows:

(1) The ideal method to build an ANN model that has the same input variables was to see if combining the performance metrics as the output variables demonstrated better prediction accuracy than modeling the ANN separately with each output variable. However, the performance indices depended on the statistical properties of the data due to the research limitations.
(2) The number of samples of input variables was sensitive to the accuracy performance of the ANN models. This relationship between the number of input variables and the number of training datasets was not linear. There existed a point where the linear curve line plateaued. The study found that two times the number of input variables in the quantity of training datasets can lead to a high accuracy of prediction.
(3) Increasing the number of hidden neurons usually led to the decreasing accuracy performance of the ANN models. However, too many hidden neurons did not further improve accuracy and even reduced it. The ideal number of neurons in the hidden layer was approximately 1.5 times the number of input variables based on the training models of R.
(4) The normalization of the input and output variables did not show a significant improvement in accuracy from the test.
(5) From Figure 16, test 2 showed the best R value. Therefore, it is possible to give an order of priority in building an ANN model. Firstly, it is possible to increase the number of dataset samples. Secondly, it is advised to increase the number of hidden neurons, and normalization is the last step to improving accuracy.

It is important to address the limitations of this work and that a careful interpretation of the findings be utilized. This study tried to capture various scenarios in building environmental conditions for generalized application. However, we acknowledge the limited scenarios of building models that were tested. We also understand the problem of overfitting in establishing ANN models. This paper did not fully investigate this matter and requires work with ridge regression and LASSO (least absolute shrinkage and selection operator) to investigate further. Previous research has indicated that decreasing the number of neurons below the input quantity can enhance the prediction accuracy of artificial neural network (ANN) models, as evidenced by higher R values. While the findings in Tables 2 and 3 of this paper support this trend, demonstrating the highest accuracy at 36 neurons, this suggests that increasing the number of neurons may diminish the model's predictive capability. To gain a more comprehensive understanding of this trend, future experiments should consider a broader range of neuron quantities for analysis. In this study, the variation of the number of hidden neurons in the ANN was explored within a range of 1–3 times the independent variable parameters. For investigations extending beyond this multiplicative relationship, further discussions and studies are warranted.

For every test, although the training procedure involved conducting 30 trials for all ANN models, the average value was used to compare among the different types of datasets to consider result validation. We will proceed with further analysis using cross-validation, specifically employing k-fold methods.

This study's tests only considered specific elements of the ANN models and further work is necessary to include other factors of constructing ANN models, such as the number of hidden layers, training algorithms, transferring algorithms, learning rate, and the moment, to enhance the performance of ANN models. The geometry variables were only considered as the input variables for this test, and further research is needed to include other input variables such as material properties. Also, the tests were based on one test case, and it would be better to include different test cases to generalize the findings.

**Author Contributions:** Conceptualization, S.W. and Y.K.Y.; methodology, S.W. and Y.K.Y.; software, S.W. and Y.K.Y.; validation, Y.K.Y.; formal analysis, S.W.; investigation, Y.K.Y.; resources, N.L.; data curation, S.W. and Y.K.Y.; writing—original draft preparation, S.W.; writing—review and editing, Y.K.Y.; visualization, S.W.; supervision, Y.K.Y. and N.L.; project administration, Y.K.Y. and N.L.; funding acquisition, N.L. All authors have read and agreed to the published version of the manuscript.

**Funding:** This research was funded by National Key R&D Program of China (2022YFC3803801 in 2022YFC3803800); Beijing Postdoctoral Research Foundation (2023-zz-143); Beijing Key Laboratory of Green Building and Energy-Efficiency Technology Open Foundation.

**Data Availability Statement:** https://pan.baidu.com/s/19vEQ6uog44ekCwu425FdFQ?pwd=1t32, accessed on 28 October 2023.

**Conflicts of Interest:** The authors declare no conflict of interest.

## Appendix A

**Table A1.** The correlation coefficients of the predicted results of D_NN for different numbers of samples.

| Trials | 20 | 30 | 40 | 52 | 70 | 108 |
|---|---|---|---|---|---|---|
| 1 | 0.0188 | 0.1855 | 0.1972 | 0.1470 | 0.8788 | 0.9026 |
| 2 | 0.1395 | 0.3126 | 0.3501 | 0.1614 | 0.8060 | 0.8550 |
| 3 | 0.3096 | 0.2885 | 0.2271 | 0.2711 | 0.6647 | 0.8821 |
| 4 | 0.0504 | 0.2291 | 0.3155 | 0.2328 | 0.7604 | 0.8856 |
| 5 | 0.2988 | 0.1632 | 0.2287 | 0.2238 | 0.7900 | 0.8736 |
| 6 | 0.1329 | 0.2202 | 0.1982 | 0.2461 | 0.8295 | 0.8544 |
| 7 | 0.2111 | 0.2573 | 0.2573 | 0.1762 | 0.7708 | 0.8631 |
| 8 | 0.0899 | 0.2218 | 0.3230 | 0.1848 | 0.8042 | 0.8236 |
| 9 | 0.3336 | 0.0183 | 0.1749 | 0.3535 | 0.8357 | 0.7970 |
| 10 | 0.1870 | 0.2392 | 0.2530 | 0.3126 | 0.8932 | 0.8800 |
| 11 | 0.3035 | 0.2583 | 0.2398 | 0.1075 | 0.7762 | 0.8373 |
| 12 | 0.2975 | 0.1596 | 0.2800 | 0.2128 | 0.7623 | 0.8351 |
| 13 | 0.0466 | 0.2958 | 0.1407 | 0.3411 | 0.8230 | 0.9303 |
| 14 | 0.3366 | 0.1636 | 0.2354 | 0.1093 | 0.8211 | 0.8775 |
| 15 | 0.2978 | 0.1069 | 0.2487 | 0.3191 | 0.7416 | 0.9318 |
| 16 | 0.2223 | 0.1867 | 0.2056 | 0.2129 | 0.8206 | 0.8824 |
| 17 | 0.3950 | 0.1952 | 0.3609 | 0.2893 | 0.7800 | 0.9153 |
| 18 | 0.3583 | 0.3000 | 0.2720 | 0.3134 | 0.6747 | 0.8164 |
| 19 | 0.4499 | 0.2178 | 0.1180 | 0.1595 | 0.9013 | 0.9083 |
| 20 | 0.2205 | 0.1789 | 0.2236 | 0.2884 | 0.6889 | 0.8185 |
| 21 | 0.1894 | 0.2580 | 0.2945 | 0.1288 | 0.7355 | 0.7946 |
| 22 | 0.2422 | 0.1877 | 0.3235 | 0.2373 | 0.7078 | 0.8788 |
| 23 | 0.1129 | 0.1660 | 0.1250 | 0.2627 | 0.7393 | 0.9054 |
| 24 | 0.2602 | 0.1059 | 0.2336 | 0.2262 | 0.8353 | 0.6396 |
| 25 | 0.2425 | 0.3335 | 0.2706 | 0.1976 | 0.8357 | 0.8043 |
| 26 | 0.4117 | 0.1720 | 0.2597 | 0.3191 | 0.8560 | 0.8957 |
| 27 | 0.2458 | 0.2519 | 0.1160 | 0.1396 | 0.7926 | 0.8464 |
| 28 | 0.1131 | 0.2542 | 0.1359 | 0.2765 | 0.7989 | 0.8921 |
| 29 | 0.2161 | 0.1640 | 0.1690 | 0.1284 | 0.8427 | 0.7878 |
| 30 | 0.2138 | 0.3077 | 0.1737 | 0.1947 | 0.7900 | 0.9262 |
| Avg | 0.2231 | 0.2101 | 0.2337 | 0.2269 | 0.7920 | 0.8557 |

**Table A2.** The correlation coefficients of the predicted results of Q_NN for different numbers of samples.

| Trials | 20 | 30 | 40 | 52 | 70 | 108 |
|---|---|---|---|---|---|---|
| 1 | 0.4219 | 0.3951 | 0.4614 | 0.5178 | 0.8168 | 0.7242 |
| 2 | 0.4056 | 0.3727 | 0.4309 | 0.5081 | 0.7738 | 0.8423 |
| 3 | 0.3731 | 0.3685 | 0.3934 | 0.5177 | 0.8037 | 0.7921 |
| 4 | 0.4845 | 0.5412 | 0.4426 | 0.5327 | 0.7843 | 0.7808 |
| 5 | 0.3736 | 0.4413 | 0.4347 | 0.5751 | 0.7794 | 0.7690 |
| 6 | 0.4378 | 0.4723 | 0.4369 | 0.5342 | 0.8806 | 0.7667 |
| 7 | 0.3951 | 0.3269 | 0.4419 | 0.4949 | 0.8055 | 0.8032 |

**Table A2.** *Cont.*

| Trials | 20 | 30 | 40 | 52 | 70 | 108 |
|---|---|---|---|---|---|---|
| 8 | 0.4363 | 0.5038 | 0.4639 | 0.4915 | 0.6429 | 0.8173 |
| 9 | 0.4178 | 0.3715 | 0.4672 | 0.5124 | 0.7195 | 0.7900 |
| 10 | 0.5499 | 0.3883 | 0.4843 | 0.5571 | 0.7774 | 0.5913 |
| 11 | 0.3938 | 0.4564 | 0.4720 | 0.3734 | 0.8290 | 0.7203 |
| 12 | 0.5078 | 0.4399 | 0.4112 | 0.4914 | 0.6024 | 0.8688 |
| 13 | 0.2033 | 0.5138 | 0.4205 | 0.5279 | 0.8485 | 0.8387 |
| 14 | 0.4194 | 0.4746 | 0.3403 | 0.4819 | 0.8588 | 0.7497 |
| 15 | 0.3526 | 0.4044 | 0.5141 | 0.5254 | 0.8030 | 0.6779 |
| 16 | 0.4173 | 0.4742 | 0.4832 | 0.5841 | 0.7287 | 0.7045 |
| 17 | 0.0435 | 0.3582 | 0.3802 | 0.4996 | 0.8337 | 0.8388 |
| 18 | 0.4505 | 0.4089 | 0.5396 | 0.4942 | 0.9035 | 0.8438 |
| 19 | 0.4690 | 0.3479 | 0.4590 | 0.4898 | 0.6988 | 0.7519 |
| 20 | 0.2469 | 0.4397 | 0.3885 | 0.5797 | 0.8411 | 0.6340 |
| 21 | 0.3673 | 0.4159 | 0.4245 | 0.5355 | 0.8180 | 0.8026 |
| 22 | 0.3026 | 0.3706 | 0.4517 | 0.5610 | 0.7827 | 0.7578 |
| 23 | 0.3901 | 0.4811 | 0.4718 | 0.5297 | 0.7716 | 0.8409 |
| 24 | 0.3645 | 0.3303 | 0.4143 | 0.5384 | 0.8054 | 0.7319 |
| 25 | 0.3618 | 0.4387 | 0.5004 | 0.4865 | 0.7997 | 0.8194 |
| 26 | 0.4432 | 0.3225 | 0.5099 | 0.5354 | 0.8465 | 0.5977 |
| 27 | 0.4954 | 0.3786 | 0.4901 | 0.4215 | 0.7762 | 0.8180 |
| 28 | 0.3415 | 0.4011 | 0.4600 | 0.5000 | 0.7455 | 0.7523 |
| 29 | 0.3862 | 0.3456 | 0.4900 | 0.5831 | 0.6845 | 0.5786 |
| 30 | 0.3300 | 0.3500 | 0.3900 | 0.5452 | 0.8554 | 0.7891 |
| Avg | 0.3861 | 0.4111 | 0.4489 | 0.5175 | 0.7885 | 0.7652 |

**Table A3.** The correlation coefficients of the predicted results of W_NN for different numbers of samples.

| Trials | 20 | 30 | 40 | 52 | 70 | 108 |
|---|---|---|---|---|---|---|
| 1 | 0.4755 | 0.5120 | 0.4793 | 0.6293 | 0.9064 | 0.8744 |
| 2 | 0.4890 | 0.4836 | 0.5928 | 0.5755 | 0.8433 | 0.8789 |
| 3 | 0.4566 | 0.5513 | 0.5163 | 0.6167 | 0.8734 | 0.8440 |
| 4 | 0.4350 | 0.5416 | 0.5844 | 0.6298 | 0.8571 | 0.8900 |
| 5 | 0.5001 | 0.4687 | 0.5214 | 0.6430 | 0.9039 | 0.9134 |
| 6 | 0.3520 | 0.4925 | 0.5805 | 0.6461 | 0.8389 | 0.8991 |
| 7 | 0.4722 | 0.4404 | 0.6001 | 0.5954 | 0.8208 | 0.8754 |
| 8 | 0.4037 | 0.5347 | 0.4707 | 0.6292 | 0.8486 | 0.8875 |
| 9 | 0.4323 | 0.5003 | 0.5723 | 0.5568 | 0.8548 | 0.9149 |
| 10 | 0.4305 | 0.4690 | 0.5692 | 0.6548 | 0.8606 | 0.9181 |
| 11 | 0.4429 | 0.4893 | 0.6102 | 0.6149 | 0.8939 | 0.8947 |
| 12 | 0.5389 | 0.5064 | 0.5628 | 0.6283 | 0.8002 | 0.8751 |
| 13 | 0.4161 | 0.4635 | 0.6139 | 0.6104 | 0.8003 | 0.9000 |
| 14 | 0.5540 | 0.5458 | 0.5070 | 0.6341 | 0.8603 | 0.8965 |
| 15 | 0.5140 | 0.5130 | 0.5859 | 0.6211 | 0.8495 | 0.8612 |
| 16 | 0.4570 | 0.4443 | 0.5870 | 0.6022 | 0.8827 | 0.8997 |
| 17 | 0.5665 | 0.5193 | 0.4885 | 0.6420 | 0.8361 | 0.8898 |
| 18 | 0.4908 | 0.4647 | 0.5343 | 0.6350 | 0.8673 | 0.9154 |
| 19 | 0.3855 | 0.4356 | 0.5163 | 0.6171 | 0.8908 | 0.8814 |
| 20 | 0.4789 | 0.5009 | 0.5561 | 0.6545 | 0.8526 | 0.8728 |
| 21 | 0.5081 | 0.5160 | 0.5910 | 0.6492 | 0.8298 | 0.8766 |
| 22 | 0.4559 | 0.5785 | 0.5638 | 0.6192 | 0.8032 | 0.9085 |
| 23 | 0.4142 | 0.5241 | 0.5667 | 0.5926 | 0.8325 | 0.9317 |
| 24 | 0.4533 | 0.4956 | 0.5829 | 0.6346 | 0.8715 | 0.8853 |
| 25 | 0.4782 | 0.5704 | 0.5633 | 0.5997 | 0.8227 | 0.8539 |
| 26 | 0.5211 | 0.5348 | 0.5790 | 0.5974 | 0.8372 | 0.8725 |
| 27 | 0.5110 | 0.4500 | 0.5413 | 0.6024 | 0.8593 | 0.8666 |

**Table A3.** *Cont.*

| Trials | 20 | 30 | 40 | 52 | 70 | 108 |
|---|---|---|---|---|---|---|
| 28 | 0.3814 | 0.4858 | 0.5462 | 0.6436 | 0.9181 | 0.8680 |
| 29 | 0.5231 | 0.4711 | 0.5579 | 0.6793 | 0.8460 | 0.9044 |
| 30 | 0.4467 | 0.5096 | 0.5542 | 0.6378 | 0.8674 | 0.8902 |
| Avg | 0.4661 | 0.5004 | 0.5565 | 0.6231 | 0.8539 | 0.8880 |

**Table A4.** The correlation coefficients (R) of the predicted results of D_NN for different numbers of hidden neurons.

| Trials | 36 | 54 | 72 | 90 | 108 | 144 |
|---|---|---|---|---|---|---|
| 1 | 0.4660 | 0.3207 | 0.3681 | 0.2689 | 0.1470 | 0.1692 |
| 2 | 0.4153 | 0.4513 | 0.3673 | 0.1800 | 0.1614 | 0.2436 |
| 3 | 0.4182 | 0.3186 | 0.3221 | 0.2213 | 0.2711 | 0.1005 |
| 4 | 0.4389 | 0.3856 | 0.3353 | 0.2470 | 0.2328 | 0.2664 |
| 5 | 0.4064 | 0.1944 | 0.2849 | 0.2488 | 0.2238 | 0.3255 |
| 6 | 0.4251 | 0.3873 | 0.3623 | 0.3336 | 0.2461 | 0.2925 |
| 7 | 0.4029 | 0.3085 | 0.3644 | 0.3187 | 0.1762 | 0.3704 |
| 8 | 0.4969 | 0.3823 | 0.3225 | 0.2691 | 0.1848 | 0.2535 |
| 9 | 0.2224 | 0.2535 | 0.3677 | 0.2403 | 0.3535 | 0.2692 |
| 10 | 0.4076 | 0.2655 | 0.3574 | 0.2614 | 0.3126 | 0.2716 |
| 11 | 0.3072 | 0.3824 | 0.3291 | 0.1806 | 0.1075 | 0.2995 |
| 12 | 0.4436 | 0.5089 | 0.1687 | 0.1571 | 0.2128 | 0.3308 |
| 13 | 0.4030 | 0.2725 | 0.2577 | 0.2349 | 0.3411 | 0.3387 |
| 14 | 0.4745 | 0.3053 | 0.3325 | 0.2423 | 0.1093 | 0.2002 |
| 15 | 0.4322 | 0.4010 | 0.2322 | 0.2710 | 0.3191 | 0.2089 |
| 16 | 0.3562 | 0.4450 | 0.3426 | 0.3727 | 0.2129 | 0.2270 |
| 17 | 0.3780 | 0.3800 | 0.3757 | 0.0901 | 0.2893 | 0.2866 |
| 18 | 0.4266 | 0.3930 | 0.2298 | 0.3055 | 0.3134 | 0.2121 |
| 19 | 0.4716 | 0.3427 | 0.2259 | 0.2607 | 0.1595 | 0.3176 |
| 20 | 0.4444 | 0.4076 | 0.2944 | 0.2604 | 0.2884 | 0.2253 |
| 21 | 0.3587 | 0.4088 | 0.3346 | 0.3300 | 0.1288 | 0.3124 |
| 22 | 0.4450 | 0.3225 | 0.3080 | 0.2913 | 0.2373 | 0.2175 |
| 23 | 0.3015 | 0.4189 | 0.3436 | 0.2960 | 0.2627 | 0.2246 |
| 24 | 0.5069 | 0.3557 | 0.2334 | 0.2213 | 0.2262 | 0.2381 |
| 25 | 0.4425 | 0.2728 | 0.2705 | 0.2758 | 0.1976 | 0.2670 |
| 26 | 0.4732 | 0.4288 | 0.3502 | 0.2399 | 0.3191 | 0.1298 |
| 27 | 0.3735 | 0.2223 | 0.3195 | 0.2430 | 0.1396 | 0.2456 |
| 28 | 0.3005 | 0.3079 | 0.3001 | 0.2677 | 0.2765 | 0.3566 |
| 29 | 0.4296 | 0.4152 | 0.2944 | 0.2706 | 0.1284 | 0.2991 |
| 30 | 0.4329 | 0.3685 | 0.8698 | 0.2605 | 0.1947 | 0.2591 |
| Avg | 0.4093 | 0.3538 | 0.3107 | 0.2553 | 0.2258 | 0.2586 |

**Table A5.** The correlation coefficients (R) of the predicted results of Q_NN for different numbers of hidden neurons.

| Trials | 36 | 54 | 72 | 90 | 108 | 144 |
|---|---|---|---|---|---|---|
| 1 | 0.7527 | 0.7276 | 0.5908 | 0.5733 | 0.5178 | 0.4628 |
| 2 | 0.7024 | 0.5618 | 0.6045 | 0.6409 | 0.5081 | 0.4541 |
| 3 | 0.6667 | 0.6455 | 0.5968 | 0.5973 | 0.5177 | 0.4511 |
| 4 | 0.6254 | 0.6695 | 0.5865 | 0.4684 | 0.5327 | 0.3933 |
| 5 | 0.7287 | 0.6849 | 0.6308 | 0.5621 | 0.5751 | 0.5361 |
| 6 | 0.6190 | 0.6235 | 0.6076 | 0.5174 | 0.5342 | 0.4677 |
| 7 | 0.7408 | 0.6642 | 0.6103 | 0.5971 | 0.4949 | 0.5404 |
| 8 | 0.7323 | 0.5689 | 0.5734 | 0.6133 | 0.4915 | 0.5287 |
| 9 | 0.7964 | 0.6905 | 0.6201 | 0.5803 | 0.5124 | 0.4713 |
| 10 | 0.6083 | 0.6884 | 0.6275 | 0.5982 | 0.5571 | 0.3499 |

**Table A5.** *Cont.*

| Trials | 36 | 54 | 72 | 90 | 108 | 144 |
|---|---|---|---|---|---|---|
| 11 | 0.7103 | 0.6760 | 0.5564 | 0.5144 | 0.3734 | 0.3986 |
| 12 | 0.6586 | 0.6526 | 0.5500 | 0.5645 | 0.4914 | 0.5046 |
| 13 | 0.7169 | 0.6835 | 0.5529 | 0.5500 | 0.5279 | 0.4789 |
| 14 | 0.7460 | 0.7313 | 0.5963 | 0.5304 | 0.4819 | 0.4776 |
| 15 | 0.7451 | 0.6831 | 0.6464 | 0.5729 | 0.5254 | 0.4924 |
| 16 | 0.7484 | 0.6410 | 0.6741 | 0.5998 | 0.5841 | 0.4482 |
| 17 | 0.7434 | 0.6258 | 0.6300 | 0.4867 | 0.4996 | 0.4654 |
| 18 | 0.7255 | 0.6384 | 0.5047 | 0.5658 | 0.4942 | 0.4902 |
| 19 | 0.6989 | 0.6312 | 0.5353 | 0.5461 | 0.4898 | 0.5080 |
| 20 | 0.7342 | 0.6478 | 0.6538 | 0.5276 | 0.5797 | 0.4931 |
| 21 | 0.7208 | 0.6063 | 0.6129 | 0.5375 | 0.5355 | 0.3973 |
| 22 | 0.7558 | 0.7317 | 0.6072 | 0.5653 | 0.5610 | 0.3710 |
| 23 | 0.7324 | 0.6627 | 0.6248 | 0.5762 | 0.5297 | 0.4733 |
| 24 | 0.7990 | 0.6443 | 0.5788 | 0.5696 | 0.5384 | 0.5477 |
| 25 | 0.6501 | 0.6232 | 0.6539 | 0.5399 | 0.4865 | 0.3436 |
| 26 | 0.740 8 | 0.7129 | 0.6839 | 0.6473 | 0.5354 | 0.5031 |
| 27 | 0.7854 | 0.6125 | 0.6159 | 0.5757 | 0.4215 | 0.4274 |
| 28 | 0.7328 | 0.6472 | 0.5886 | 0.4934 | 0.5000 | 0.4736 |
| 29 | 0.6917 | 0.7033 | 0.6990 | 0.5355 | 0.5831 | 0.4549 |
| 30 | 0.7156 | 0.6755 | 0.5887 | 0.5834 | 0.5452 | 0.4803 |
| Avg | 0.7175 | 0.6585 | 0.6073 | 0.5636 | 0.5175 | 0.4622 |

**Table A6.** The correlation coefficients of the predicted results of W_NN for different numbers of hidden neurons.

| Trials | 36 | 54 | 72 | 90 | 108 | 144 |
|---|---|---|---|---|---|---|
| 1 | 0.7670 | 0.7690 | 0.7290 | 0.6327 | 0.6293 | 0.5265 |
| 2 | 0.7017 | 0.7633 | 0.7353 | 0.6569 | 0.5755 | 0.5470 |
| 3 | 0.7274 | 0.7245 | 0.6967 | 0.6320 | 0.6167 | 0.5986 |
| 4 | 0.7358 | 0.7513 | 0.6593 | 0.6382 | 0.6298 | 0.5421 |
| 5 | 0.8041 | 0.7478 | 0.6855 | 0.5963 | 0.6430 | 0.5871 |
| 6 | 0.7565 | 0.7530 | 0.7260 | 0.6642 | 0.6461 | 0.5780 |
| 7 | 0.7407 | 0.7534 | 0.6795 | 0.6589 | 0.5954 | 0.5613 |
| 8 | 0.7808 | 0.7364 | 0.6706 | 0.6928 | 0.6292 | 0.5489 |
| 9 | 0.7932 | 0.7825 | 0.7287 | 0.6210 | 0.5568 | 0.5149 |
| 10 | 0.6418 | 0.7661 | 0.7202 | 0.6928 | 0.6548 | 0.5944 |
| 11 | 0.7736 | 0.7751 | 0.7083 | 0.6409 | 0.6149 | 0.5195 |
| 12 | 0.7686 | 0.7710 | 0.7131 | 0.6457 | 0.6283 | 0.5926 |
| 13 | 0.7227 | 0.7836 | 0.7186 | 0.6887 | 0.6104 | 0.5148 |
| 14 | 0.7536 | 0.7759 | 0.7089 | 0.6588 | 0.6341 | 0.5176 |
| 15 | 0.6542 | 0.7107 | 0.6705 | 0.6977 | 0.6211 | 0.5131 |
| 16 | 0.7913 | 0.7403 | 0.6842 | 0.6618 | 0.6022 | 0.5546 |
| 17 | 0.7840 | 0.7566 | 0.6919 | 0.6321 | 0.6420 | 0.5773 |
| 18 | 0.6381 | 0.7504 | 0.6900 | 0.6905 | 0.6350 | 0.5062 |
| 19 | 0.6574 | 0.7643 | 0.7025 | 0.6079 | 0.6171 | 0.5461 |
| 20 | 0.6402 | 0.7659 | 0.7342 | 0.6882 | 0.6545 | 0.5202 |
| 21 | 0.7780 | 0.7694 | 0.6938 | 0.5584 | 0.6492 | 0.5667 |
| 22 | 0.7887 | 0.7773 | 0.6782 | 0.6764 | 0.6192 | 0.6235 |
| 23 | 0.6744 | 0.7609 | 0.7252 | 0.6789 | 0.5926 | 0.5231 |
| 24 | 0.7875 | 0.6325 | 0.7342 | 0.6364 | 0.6346 | 0.5537 |
| 25 | 0.6422 | 0.7772 | 0.6976 | 0.6395 | 0.5997 | 0.5554 |
| 26 | 0.6304 | 0.7685 | 0.6859 | 0.6600 | 0.5974 | 0.5402 |
| 27 | 0.7283 | 0.7476 | 0.7528 | 0.6274 | 0.6024 | 0.5522 |
| 28 | 0.7586 | 0.7565 | 0.7328 | 0.6619 | 0.6436 | 0.5210 |
| 29 | 0.6799 | 0.7742 | 0.6937 | 0.6575 | 0.6793 | 0.6023 |
| 30 | 0.7564 | 0.7633 | 0.6725 | 0.6705 | 0.6378 | 0.5723 |
| Avg | 0.7286 | 0.7556 | 0.7040 | 0.6522 | 0.6231 | 0.5524 |

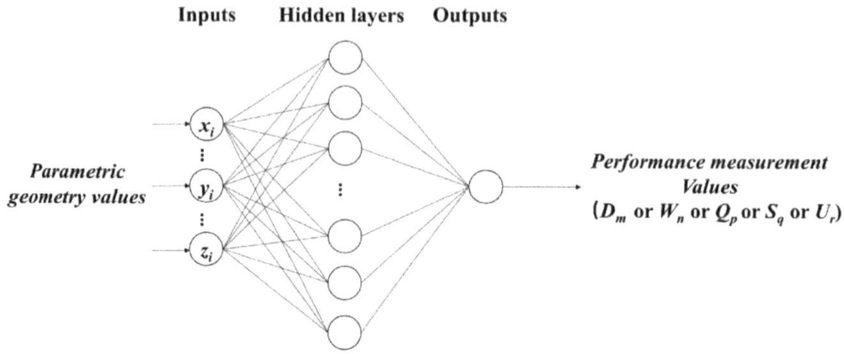


**Figure A1.** Model A: ANN model structure.

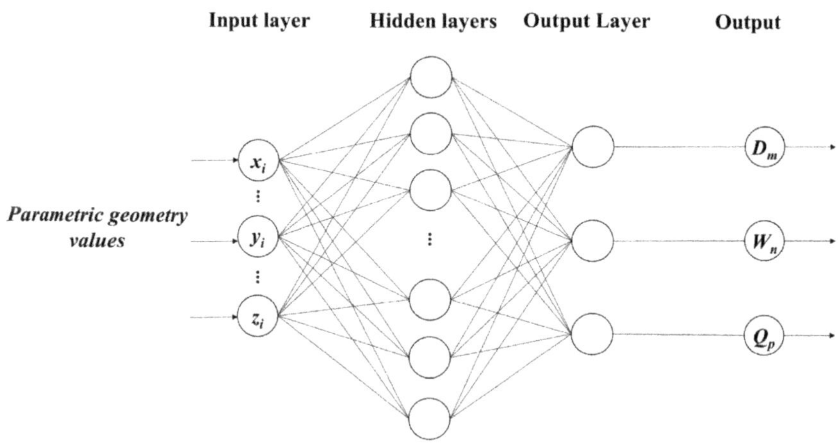


**Figure A2.** Model B group 1: ANN model for indoor measurements.

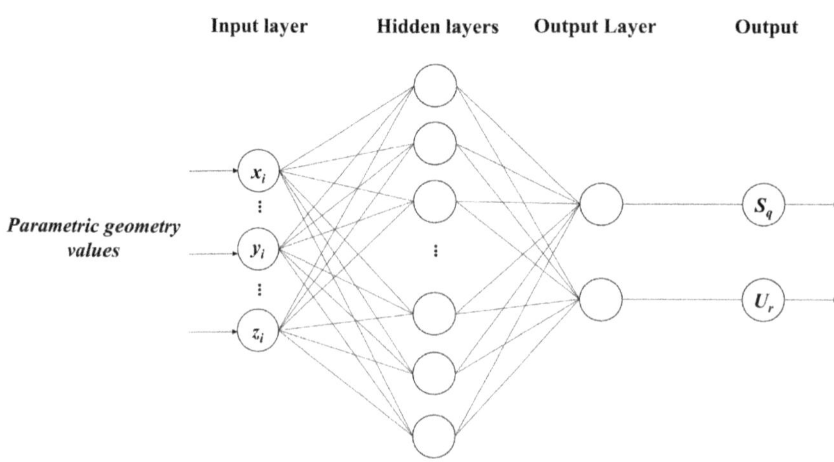


**Figure A3.** Model B group 2: ANN model with 2 outdoor measurements.

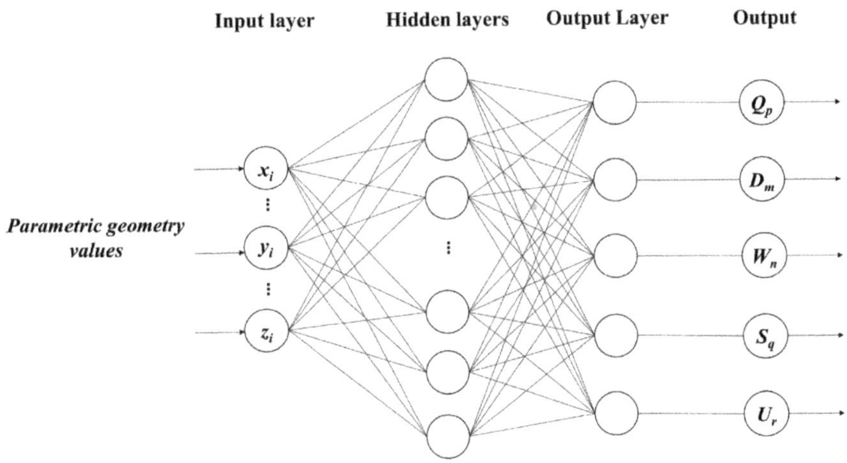


**Figure A4.** Model C: ANN model structure.

## References


1. Turhan, C.; Kazanasmaz, T.; Uygun, I.E.; Ekmen, K.E.; Akkurt, G.G. Comparative study of a building energy performance software (KEP-IYTE-ESS) and ANN-based building heat load estimation. *Energy Build.* **2014**, *85*, 115–125. [CrossRef]
2. Chokwitthaya, C.; Zhu, Y.; Mukhopadhyay, S.; Collier, E. Augmenting building performance predictions during design using generative adversarial networks and immersive virtual environments. *Autom. Constr.* **2020**, *119*, 103350. [CrossRef]
3. Ibrahim, S.; Choong, C.E.; El-Shafie, A. Sensitivity analysis of artificial neural networks for just-suspension speed prediction in solid-liquid mixing systems: Performance comparison of MLPNN and RBFNN. *Adv. Eng. Inform.* **2019**, *39*, 278–291. [CrossRef]
4. Roman, N.D.; Bre, F.; Fachinotti, V.D.; Lamberts, R. Application and characterization of metamodels based on artificial neural networks for building performance simulation: A systematic review. *Energy Build.* **2020**, *217*, 109972. [CrossRef]
5. Zhao, J.; Liu, X. A hybrid method of dynamic cooling and heating load forecasting for office buildings based on artificial intelligence and regression analysis. *Energy Build.* **2018**, *174*, 293–308. [CrossRef]
6. Moon, J.W. Performance of ANN-based predictive and adaptive thermal-control methods for disturbances in and around residential buildings. *Build. Environ.* **2012**, *48*, 15–26. [CrossRef]
7. Bui, D.K.; Nguyen, T.N.; Ngo, T.D.; Nguyen-Xuan, H. An artificial neural network (ANN) expert system enhanced with the electromagnetism-based firefly algorithm (EFA) for predicting the energy consumption in buildings. *Energy* **2020**, *190*, 116370. [CrossRef]
8. Li, Y.; Nord, N.; Zhang, N.; Zhou, C. An ANN-based optimization approach of building energy systems: Case study of swimming pool. *J. Clean. Prod.* **2020**, *277*, 124029. [CrossRef]
9. Sharif, S.A.; Hammad, A. Developing surrogate ANN for selecting near-optimal building energy renovation methods considering energy consumption, LCC and LCA. *J. Build. Eng.* **2019**, *25*, 100790. [CrossRef]
10. Luo, X.J.; Oyedele, L.O.; Ajayi, A.O.; Akinade, O.O. Comparative study of machine learning-based multi-objective prediction framework for multiple building energy loads. *Sustain. Cities Soc.* **2020**, *61*, 102283. [CrossRef]
11. Elbayoumi, M.; Ramli, N.A.; Yusof, N.F.F.M. Development and comparison of regression models and feedforward backpropagation neural network models to predict seasonal indoor PM2.5–10 and PM2.5 concentrations in naturally ventilated schools. *Atmos. Pollut. Res.* **2015**, *6*, 1013–1023. [CrossRef]
12. Kim, H.; Yi, Y.K. QuVue implementation for decisions related to high-rise residential building layouts. *Build. Environ.* **2019**, *148*, 116–127. [CrossRef]
13. Wang, L.; Kubichek, R.; Zhou, X. Adaptive learning based data-driven models for predicting hourly building energy use. *Energy Build.* **2018**, *159*, 454–461. [CrossRef]
14. Wong, S.L.; Wan, K.K.W.; Lam, T.N.T. Artificial neural networks for energy analysis of office buildings with daylighting. *Appl. Energy* **2010**, *87*, 551–557. [CrossRef]
15. Ye, Z.; Kim, M.K. Predicting electricity consumption in a building using an optimized back-propagation and Levenberg–Marquardt back-propagation neural network: Case study of a shopping mall in China. *Sustain. Cities Soc.* **2018**, *42*, 176–183. [CrossRef]
16. Walker, S.; Khan, W.; Katic, K.; Maassen, W.; Zeiler, W. Accuracy of different machine learning algorithms and added-value of predicting aggregated-level energy performance of commercial buildings. *Energy Build.* **2020**, *209*, 109705. [CrossRef]
17. Kwok, S.S.K.; Yuen, R.K.K.; Lee, E.W.M. An intelligent approach to assessing the effect of building occupancy on building cooling load prediction. *Build. Environ.* **2011**, *46*, 1681–1690. [CrossRef]

18. Mustafaraj, G.; Lowry, G.; Chen, J. Prediction of room temperature and relative humidity by autoregressive linear and nonlinear neural network models for an open office. *Energy Build.* **2011**, *43*, 1452–1460. [CrossRef]
19. Jovanović, R.Ž.; Sretenović, A.A.; Živković, B.D. Ensemble of various neural networks for prediction of heating energy consumption. *Energy Build.* **2015**, *94*, 189–199. [CrossRef]
20. Buratti, C.; Vergoni, M.; Palladino, D. Thermal Comfort Evaluation within Non-residential Environments: Development of Artificial Neural Network by Using the Adaptive Approach Data. *Energy Procedia* **2015**, *78*, 2875–2880. [CrossRef]
21. Deng, Z.; Chen, Q. Artificial neural network models using thermal sensations and occupants' behavior for predicting thermal comfort. *Energy Build.* **2018**, *174*, 587–602. [CrossRef]
22. Yokoyama, R.; Wakui, T.; Satake, R. Prediction of energy demands using neural network with model identification by global optimization. *Energy Convers. Manag.* **2009**, *50*, 319–327. [CrossRef]
23. Huang, H.; Chen, L.; Hu, E. A neural network-based multi-zone modelling approach for predictive control system design in commercial buildings. *Energy Build.* **2015**, *97*, 86–97. [CrossRef]
24. Escandón, R.; Ascione, F.; Bianco, N.; Mauro, G.M.; Suárez, R.; Sendra, J.J. Thermal comfort prediction in a building category: Artificial neural network generation from calibrated models for a social housing stock in southern Europe. *Appl. Therm. Eng.* **2019**, *150*, 492–505. [CrossRef]
25. Talib, A.; Park, S.; Im, P.; Joe, J. Grey-box and ANN-based building models for multistep-ahead prediction of indoor temperature to implement model predictive control. *Eng. Appl. Artif. Intell.* **2023**, *126*, 107115. [CrossRef]
26. Kumar, S.; Dutta, S.C.; Goswami, K.; Mandal, P. Vulnerability assessment of building structures due to underground blasts using ANN and non-linear dynamic analysis. *J. Build. Eng.* **2021**, *44*, 102674. [CrossRef]
27. Kim, J.; Kwak, Y.; Mun, S.H.; Huh, J.H. Electric energy consumption predictions for residential buildings: Impact of data-driven model and temporal resolution on prediction accuracy. *J. Build. Eng.* **2022**, *62*, 105361. [CrossRef]
28. Nutakki, M.; Mandava, S. Review on optimization techniques and role of Artificial Intelligence in home energy management systems. *Eng. Appl. Artif. Intell.* **2023**, *119*, 105721. [CrossRef]
29. *GB/T50378-2019*; Assessment Standard for Green Building. China Architecture & Building Press: Beijing, China, 2019. (In Chinese)
30. *T/ASC 02-2021*; Assessment Standard for Healthy Building. Architectural Society of China: Beijing, China, 2021. (In Chinese)
31. Wang, S.; Yi, Y.K.; Liu, N. Multi-objective optimization (MOO) for high-rise residential buildings' layout centered on daylight, visual, and outdoor thermal metrics in China. *Build. Environ.* **2021**, *205*, 108263. [CrossRef]
32. *GB50180-2018*; Standard for Urban Residential Area Planning and Design. Ministry of Housing and Urban-Rural Development: Beijing, China, 2018. (In Chinese)
33. *GB50016-2014*; High-Rise Building Code for The Fire Protection Design of Buildings (2018 Version). Ministry of Housing and Urban-Rural Development: Beijing, China, 2018. (In Chinese)
34. Zhang, T.; You, X. Improvement of the Training and Normalization Method of Artificial Neural Network in the Prediction of Indoor Environment. *Procedia Eng.* **2015**, *121*, 1245–1251. [CrossRef]
35. Choong, C.E.; Ibrahim, S.; El-Shafie, A. Artificial Neural Network (ANN) model development for predicting just suspension speed in solid-liquid mixing system. *Flow Meas. Instrum.* **2020**, *71*, 101689. [CrossRef]
36. Esonye, C.; Onukwuli, O.D.; Ofoefule, A.U.; Ogah, E.O. Multi-input multi-output (MIMO) ANN and Nelder-Mead's simplex based modeling of engine performance and combustion emission characteristics of biodiesel-diesel blend in CI diesel engine. *Appl. Therm. Eng.* **2019**, *151*, 100–114. [CrossRef]
37. Ashtiani, A.; Mirzaei, P.A.; Haghighat, F. Indoor thermal condition in urban heat island: Comparison of the artificial neural network and regression methods prediction. *Energy Build.* **2014**, *76*, 597–604. [CrossRef]
38. Hagan, M.T.; Demuth, H.B.; Beale, M. *Neural Network Design*; PWS Publishing Company: Boston, MA, USA, 1995.
39. Beale, M.H.; Hagan, M.T.; Demuth, H.B. *Neural Network Toolbox™ Getting Started Guide*; The MathWorks Inc.: Natick, MA, USA, 2018.
40. Ilbeigi, M.; Ghomeishi, M.; Dehghanbanadaki, A. Prediction and optimization of energy consumption in an office building using artificial neural network and a genetic algorithm. *Sustain. Cities Soc.* **2020**, *61*, 102325. [CrossRef]
41. Wang, R.; Lu, S.; Feng, W. A three-stage optimization methodology for envelope design of passive building considering energy demand, thermal comfort and cost. *Energy* **2019**, *192*, 116723. [CrossRef]
42. Østergård, T.; Jensen, R.L.; Maagaard, S.E. A comparison of six metamodeling techniques applied to building performance simulations. *Appl. Energy* **2018**, *211*, 89e103. [CrossRef]
43. Ascione, F.; Bianco, N.; De Stasio, C.; Mauro, G.M.; Vanoli, G.P. Artificial neural networks to predict energy performance and retrofit scenarios for any member of a building category: A novel approach. *Energy* **2017**, *118*, 999–1017. [CrossRef]
44. Bre, F.; Roman, N.; Fachinotti, V.D. An efficient metamodel-based method to carry out multi-objective building performance optimizations. *Energy Build.* **2020**, *206*, 109576. [CrossRef]
45. Khoshroo, M.; Javid, A.A.; Katebi, A. Effects of micro-nano bubble water and binary mineral admixtures on the mechanical and durability properties of concrete. *Constr. Build. Mater.* **2018**, *164*, 371–385. [CrossRef]
46. Sofuoglu, S.C. Application of artificial neural networks to predict prevalence of building-related symptoms in office buildings. *Build. Environ.* **2008**, *43*, 1121–1126. [CrossRef]

47. Kerdan, I.G.; Gálvez, D.M. Artificial neural network structure optimization for accurately prediction of exergy, comfort and life cycle cost performance of a low energy building. *Appl. Energy* **2020**, *280*, 115862. [CrossRef]
48. Moon, J.W.; Jung, S.K. Algorithm for optimal application of the setback moment in the heating season using an artificial neural network model. *Energy Build.* **2016**, *127*, 859–869. [CrossRef]
49. Wang, Z.; Liu, X.; Shen, H.; Wang, Y.; Li, H. Energy performance prediction of vapor-injection air source heat pumps in residential buildings using a neural network model. *Energy Build.* **2020**, *228*, 110499. [CrossRef]
50. Leung, M.C.; Tse, N.C.F.; Lai, L.L.; Chow, T.T. The use of occupancy space electrical power demand in building cooling load prediction. *Energy Build.* **2012**, *55*, 151–163. [CrossRef]

buildings

MDPI

Article

# Influence of Balcony Thermal Bridges on Energy Efficiency of Dwellings in a Warm Semi-Arid Dry Mediterranean Climate

Carlos Pérez-Carramiñana [1,*], Aurelio de la Morena-Marqués [1], Ángel Benigno González-Avilés [1], Nuria Castilla [2] and Antonio Galiano-Garrigós [1]

[1] Departamento de Construcciones Arquitectónicas, Escuela Politécnica Superior, Universidad de Alicante, 03690 Alicante, Spain; aureliodelamorena@gmail.com (A.d.l.M.-M.); angelb@ua.es (Á.B.G.-A.); antonio.galiano@ua.es (A.G.-G.)

[2] Centro de Investigación de Tecnología de la Edificación (CITE), Universitat Politècnica de València, Camino de Vera, s/n, 46022 Valencia, Spain; ncastilla@csa.upv.es

* Correspondence: c.perez@ua.es; Tel.: +34-615388644

**Abstract:** Thermal bridges significantly influence the energy performance of buildings. However, their impact varies depending on the type of thermal bridge, climate conditions, construction methodologies, and geometric characteristics of the building. On the Spanish Mediterranean coast, buildings with large balconies are predominant. Nevertheless, the Spanish energy efficiency regulations do not adequately specify the thermal bridges at the junctions of balconies with facades, leading to a lack of consideration for their influence in the majority of architectural projects. The objective of this study is to qualitatively and quantitatively assess the impact of such thermal bridges on the energy efficiency of buildings in a dry Mediterranean climate (BShs) within a warm semi-arid climate (BSh). As a case study, the influence of this thermal bridge is analyzed in two residential buildings located on the Mediterranean coast of southeastern Spain. The study also examines the modification of various construction parameters of this thermal bridge and determines the optimal design parameters to reduce its thermal transmittance. The results demonstrate that the energy needs caused by thermal bridges account for approximately 40% of the total annual energy needs of the studied residential buildings. Balcony thermal bridges account for 25% to 40% of the energy needs caused by all thermal bridges. The lack of differentiation in Spanish standards between balcony–facade and facade–slab edge junctions causes an imprecision in calculations equivalent to 12% of the total annual energy needs of dwellings. The novelty of this research lies in highlighting that current regulations and calculation programs need improvement to better characterize balcony thermal bridges and enhance the accuracy of building energy efficiency calculations.

**Keywords:** balcony thermal bridges; energy efficiency; hygrothermal performance; indoor thermal comfort; sustainable rehabilitation; warm semi-arid dry Mediterranean climate

## 1. Introduction

The Spanish energy-saving regulations [1,2], in accordance with European regulations [3,4], primarily emphasize five aspects: reducing the thermal transmittance of opaque enclosures and thermal bridges, reducing the thermal transmittance of glazing, controlling solar gains through glass, regulating air renewal through ventilation, and minimizing air infiltrations. The improvement of construction materials and products has allowed enhancing all these parameters. However, reducing thermal bridges is the most challenging aspect to address because it involves modifying the typical construction techniques in Spain. As some research highlights, the percentage impact of thermal bridges on the building's energy needs increases with the improvement of its thermal insulation [5]. Therefore, the increase in regulatory insulation requirements for enclosures also raises the percentage of thermal bridges in the total heat loss of the house, emphasizing the importance of their control and reduction [6].

In cold European regions, regulations require the elimination of thermal bridges due to high thermal losses and the risk of condensation, mold growth, and deterioration of building elements. In temperate and warm Mediterranean climates, correcting thermal bridges is not mandatory by regulations because condensation issues typically do not occur due to thermal bridges. However, despite the milder severity of the Mediterranean climate not making the elimination of thermal bridges essential, their influence on the energy needs of buildings is still significant [7]. Numerous studies suggest that thermal bridges in residential buildings significantly increase heating and cooling loads, though results vary depending on the climate and calculation methods [8,9]. Some estimates indicate that thermal bridges can contribute up to 20% of the total energy consumption of a home, with variations depending on climate types [10]. Other studies specify that in mild Mediterranean climates, thermal bridges may cause up to 25% of heating needs but only 3.5% of cooling needs, averaging around 8.5% annually. Therefore, cost-benefit analyses from these studies argue that the savings achieved by reducing thermal bridges in warm climates are insufficient to recover the additional construction costs incurred to eliminate them [11].

According to Spanish and European regulations, thermal bridges are areas in the building enclosure where thermal resistance is significantly lower compared to the rest of the enclosure due to materials with higher thermal conductivity, changes in enclosure thickness, or differences between internal and external areas [12,13]. The regulations consider the following thermal bridges: the facade–slab front encounter, the facade–pillar encounter, the facade–roof encounter, the facade–lower slab encounter exposed to outdoor air, the facade–ground encounter in contact with the ground, the outlines of window and door openings, and the corners [12]. Thermal bridges are calculated according to European regulations [13–16] and are measured by their linear thermal transmittance. Spanish regulations and official calculation programs [17–20] include the mentioned list of thermal bridges with their specific characteristics for calculating the building's energy needs.

However, the list of thermal bridges considered by the regulations and calculation programs does not include a very common type of thermal bridge on the Spanish Mediterranean coast: the encounters of balconies with facades. Along the Mediterranean coast of Spain, since the tourist boom of the 1960s and 1970s, there are numerous tourist apartment buildings with large balconies [21], providing spacious outdoor areas to enjoy the warm and sunny climate of the Mediterranean coast [22] and protecting the windows from the intense solar radiation of this geographic area [23–25]. These beachfront buildings were traditionally used as second homes only in the summer [26], but this architectural style has become widespread and is now applied to most residential buildings in the BShs climate [27]. Therefore, this architectural typology is very common on the Mediterranean coast, increasing the prevalence of facade–balcony encounters. This type of thermal bridge has specific construction and geometric characteristics. A balcony is an extension of the structure's slab that protrudes from the facade plane, interrupting the thermal envelope, and the construction solution to avoid the thermal bridge is complex. However, regulations and calculation programs do not differentiate between facade–balcony encounters and facade–slab front encounters, considering them the same type of thermal bridge even though they have different thermal transmittances. Additionally, the thermal bridge catalog in the regulations simplifies some construction parameters that influence the linear thermal transmittance of thermal bridges. Therefore, it is advisable to study and analyze in detail the thermal behavior of these two types of thermal bridges and their influence on the overall calculation of thermal gains and losses in buildings in this climate.

Numerous studies in colder countries highlight that balcony slabs are one of the thermal bridges that have a significant negative impact on the energy efficiency of buildings [28–31]. They affect thermal comfort [32,33] and contribute to increased condensation issues [34,35]. Their geometry and surface make them heat exchangers [36]. There are various studies emphasizing the advantages of using thermal break elements in cantilevered slabs of structures to reduce thermal bridging in balconies and overhangs [37]. These construction solutions can reduce the linear thermal transmittance of thermal bridges by more than 60% [38] and decrease heating consumption by 5% to 13% and cooling consumption by 1% [39]. However, other research limits the impact of thermal break disruptions in balconies on the overall thermal and energy performance of the building, demonstrating that their effect on annual energy consumption is relatively small [40].

In the southeastern part of Spain, these construction solutions are not employed due to their high cost. There are other more economical construction alternatives, such as thermally insulating balcony slabs from below and above. However, analyses conducted by some research studies demonstrate that balcony thermal insulation is also not economically efficient in mild climates [41].

Other construction systems with many years of application in Central Europe, such as External Thermal Insulation Systems (ETICS), are now being widely implemented on the Spanish Mediterranean coast. Their widespread adoption is primarily due to their advantages in reducing thermal bridges on facades [42,43]. There are even studies demonstrating the benefits of using insulated cornices and moldings to complete all the distinctive junctions of the exterior thermal envelope of facades with ETICSs [44]. However, thermal bridges on balconies are challenging to avoid because the ETICS is interrupted by the structure. This limits its advantages, as shown by research in other climatic zones [45]. Other studies conducted in Mediterranean climates have demonstrated that insulating from the interior has a higher cost/benefit ratio than the ETICS [46]. They also caution against the lack of optimization and cost-effectiveness of using exterior insulation with excessive thickness in mild climates [47]. However, these studies do not specify the parameters influencing the energy needs of homes in the BShs climate based on the construction system employed.

Given the lack of conclusive data and specific results for the characteristics of the BShs climate and the imprecision of Spanish regulations regarding thermal bridges on balconies, studies are needed to analyze their influence on the energy needs of buildings on the Spanish Mediterranean coast.

This work comparatively analyzes the influence of different thermal bridges in a group of recently constructed residential buildings on the Mediterranean coast of southeastern Spain to determine which thermal bridges are more predominant using thermographic images. Subsequently, it conducts a detailed analysis of the thermal bridges on balconies and slab fronts in two buildings, qualitatively and quantitatively determining their influence on energy efficiency. The study also assesses whether Spanish energy-saving regulations adequately consider these thermal bridges. To achieve this, the study employs a combination of field measurements and computer simulations. Detailed thermographic photographs were taken to qualitatively assess the thermal bridges. A thermal transmittance flowmeter was used to quantify the thermal transmittances of the facades and thermal bridges. The impact of the analyzed thermal bridges on the total annual energy needs of the inspected dwellings was evaluated through computer simulations. Finally, it conducts a detailed study to determine the influence of modifying various construction and geometric parameters of these thermal bridges to identify optimal construction solutions using computer simulations. This work considers the most common construction solutions currently used in buildings on the Spanish Mediterranean coast, with conventional ceramic brick envelopes, laminated plaster beaches, and insulation such as rock wool and expanded or extruded polystyrene. Therefore, more efficient thermal insulation materials or more innovative building solutions with phase change materials [48,49] are not considered.

The objective is to qualitatively and quantitatively determine the influence of thermal bridges at the facades' encounters with balconies on the energy efficiency of buildings in a dry Mediterranean climate (BShs) within a warm semi-arid climate (BSh). The novelty of this study lies in demonstrating that current regulations and calculation programs need improvement to characterize these thermal bridges more accurately and enhance the precision of energy efficiency calculations. The work complements and refines the typological database of thermal bridges contemplated by regulations and evaluates and determines optimal design parameters to reduce the linear thermal transmittance of the most predominant thermal bridges.

## 2. Materials and Methods

### 2.1. Methodology

The methodology consisted of three phases. First, a comparative analysis was conducted on the majority of buildings constructed in Alicante in the last two years using thermal images to identify the most significant and common types of thermal bridges in this geographic area. Subsequently, two representative buildings were selected, CS1 (case study 1) and CS2 (case study 2), for a qualitative and quantitative analysis of the chosen thermal bridges through thermal imaging and on-site measurements. The influence of the selected thermal bridges on the total energy needs of the two studied buildings was quantified using computer simulations. Finally, various modifications to the construction and geometric parameters of the selected thermal bridges were studied, and their impact on the energy efficiency of the building was calculated through computer simulations. This methodology allowed for the identification of the most significant thermal bridges in residential buildings in this geographic area, quantifying their influence on the energy needs of the homes and determining optimal construction solutions to reduce thermal gains and losses, thereby improving the energy efficiency of the buildings.

Phase 1. Twenty-six buildings constructed between 2018 and 2022 were selected based on the database of the Architects' Association of Alicante. Thermal images of their facades were taken to differentiate qualitatively the zones of the building envelope with varying thermal insulation and to detect thermal bridges based on temperature differences on the surfaces of the facades [50].

The thermal bridges considered by the regulations were identified, related to the complete or partial penetration of construction elements with different thermal conductivity or changes in the thickness of the enclosure. Thermal bridges related to differences between the internal and external areas of the enclosure at corners were not considered because they are not associated with the employed construction systems and cannot be detected thermographically.

The percentage of the facade surface occupied by each type of thermal bridge was measured. Additionally, the thermal difference between the average surface temperature of each type of thermal bridge and the average temperature of the rest of the facade was measured. With these data, the Thermal Bridge Wall Ratio (THBWR) was calculated for each thermal bridge. This index indicates the proportion between the surface area multiplied by the temperature difference for each thermal bridge and the total sum of surface areas and average temperatures for all thermal bridges. This process established a classification of the most representative and common thermal bridges in the studied buildings. In this classification, the connections of balconies with facades were differentiated from the rest of the connections at slab edges due to their specific construction and geometric characteristics.

Phase 2. Two representative buildings were selected, CS1 (case study 1) and CS2 (case study 2), to qualitatively and quantitatively analyze the most significant thermal bridges. Detailed thermographic photographs were taken to qualitatively study the selected thermal bridges (Figure 1a). The purpose of conducting thermographic images is to obtain approximate information about the temperatures of the exterior surfaces of the different areas of the facade. This allows us to detect which areas of the facade experience greater energy loss from the interior to the exterior of the building in winter and, thus, which areas of the enclosure have higher thermal transmittance. On-site measurements were taken to quantify thermal transmittances of the facades and thermal bridges by using a thermal transmittance flowmeter (Figure 1b). The purpose of using the thermal transmittance flowmeter is to approximately quantify the thermal transmittance per unit area of the enclosure, thereby allowing us to approximately estimate the linear thermal transmittance in the thermal bridges. This method allowed for more precise on-site quantification of the heat flow in different areas of the enclosures of each of the two analyzed dwellings.

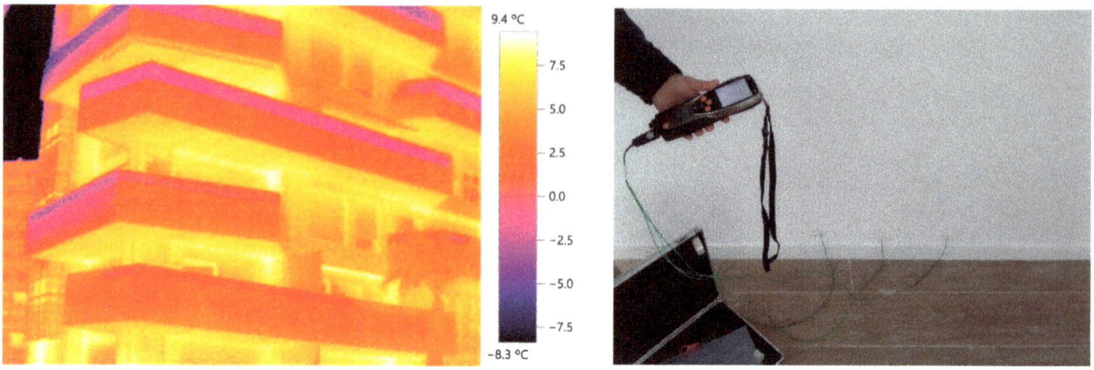


(a) (b)

**Figure 1.** On-site measurements of the studied building enclosures: (**a**) thermographic photograph; (**b**) thermal transmittance measurement.

Next, the influence of the analyzed thermal bridges on the total annual energy needs of the two examined dwellings was quantified through computer simulations. To achieve this, the energy needs of each dwelling were calculated, taking into account the actual thermal transmittance data of the building enclosures obtained from on-site measurements. The total annual energy needs caused by the linear thermal transmittance of all thermal bridges regulated by standards were independently calculated. Additionally, the total annual energy needs caused by the thermal transmittance of opaque enclosures, the thermal transmittance of glazing, solar radiation through glass, air exchanges through ventilation, and air infiltrations were also computed. Subsequently, the proportion of thermal gains and losses caused by all regulated thermal bridges that constitute the connections of balconies and the remaining floor slab edges on the facades was determined. Firstly, the total thermal gains and losses caused by all thermal bridges were calculated. Then, the gains and losses were recalculated by eliminating the thermal bridges in the floor slab edges of the facades, and the results were compared. Finally, the gains and losses were recalculated by only eliminating the thermal bridges in the connections of balconies and facades.

Phase 3. Computer simulations were used to evaluate the thermal effect of the modification of various geometrical and constructive parameters of the analyzed thermal bridges in the two selected buildings, CS1 and CS2. Before calculating the new proposals, we proceeded to adjust and validate the calculations made with the energy simulation tool. For this purpose, computer simulations were compared with actual measurements obtained from the electricity consumption records of the dwellings studied. The annual energy needs were calculated by computer simulation comparing the results obtained with the actual measurements.

Thermal bridges at balcony and slab-front junctions in the two buildings studied were specifically analyzed. It was calculated how the modification of the thermal insulation of these thermal bridges influences their linear thermal transmittance and the total energy needs of the dwellings.

Thermal insulation improvement can be achieved by increasing its thickness, reducing its thermal conductivity, or combining both solutions. In this study, it was decided to modify the insulation thickness for easier comparison of changes in construction parameters. The thermal insulation of the facades, floors, and ceilings was modified with different insulation thicknesses. This methodology allowed for quantifying the impact of the thermal insulation improvement on the reduction in the total energy needs of the dwellings and determining optimal construction solutions. Four different options were calculated for the two selected buildings, CS1 (case study 1) and CS2 (case study 2), with the current characteristics, with less insulation and with more insulation (Table 1).

**Table 1.** Summary of thermal insulation characteristics of the thermal bridges of the four calculated options in the two selected buildings: case study 1 (CS1), case study 2 (CS2).

| Isolation Situation | Option 1 Very Low Isolation | Option 2 Low Isolation | Option 3 Current Features | Option 4 More Isolated |
|---|---|---|---|---|
| | CS1/CS2 | CS1/CS2 | CS1/CS2 | CS1/CS2 |
| False ceiling indoor dwelling | -/- | 30 mm MW (0.035 W/m·K)/20 mm XPS (0.034 W/m·K) | 60 mm MW (0.035 W/m·K)/40 mm XPS (0.034 W/m·K) | 90 mm MW (0.035 W/m·K)/60 mm XPS (0.034 W/m·K) |
| Indoor floor dwellings | -/- | 20 mm EPS (0.030 W/m·K)/- | 40 mm EPS (0.030 W/m·K)/- | 60mm EPS (0.030 W/m·K)/- |
| External insulation (ETICS) slab front (alternative solution only in dwelling 2) | -/- | -/20 mm EPS (0.032 W/m·K) | -/40 mm EPS (0.032 W/m·K) | -/60mm EPS (0.032 W/m·K) |

The methodology used is shown in a flow chart below (Figure 2).

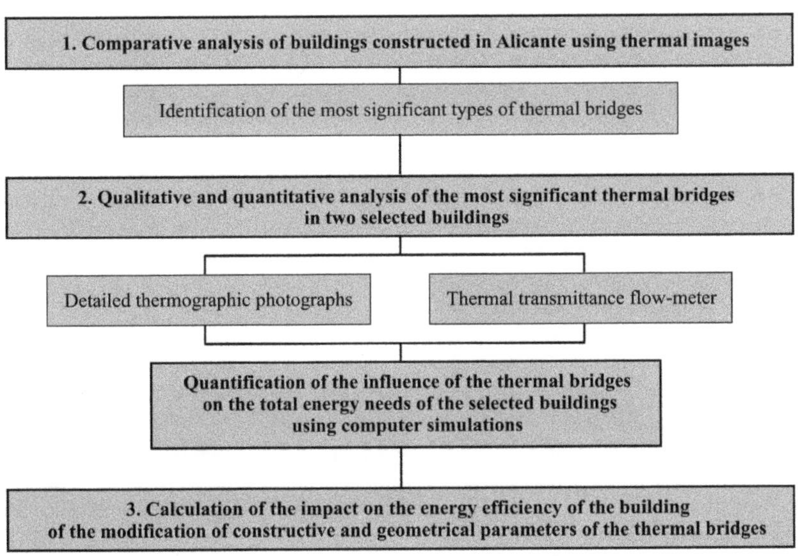


**Figure 2.** General flow chart of the method.

### 2.2. Case Studies

The buildings in the case study are located in Playa de San Juan, Alicante (Spain). The area has a dry Mediterranean climate (BShs) within the warm semi-arid climate zone (BSh) according to the Köppen-Geiger climate classification [51] (Figure 3) and a B4 climate according to Spanish legislation [52]. This climate experiences lower annual temperature fluctuations compared to most climates due to the Mediterranean Sea breeze acting as a thermoregulator. As a result, winters are mild, and summers are not excessively hot, with an average winter temperature of around 18 °C and a summer temperature of approximately 27 °C. This climate is characterized by high radiation all year round.

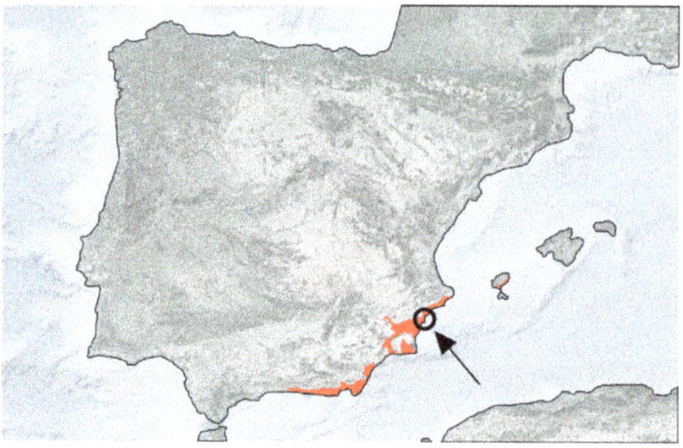

(a)

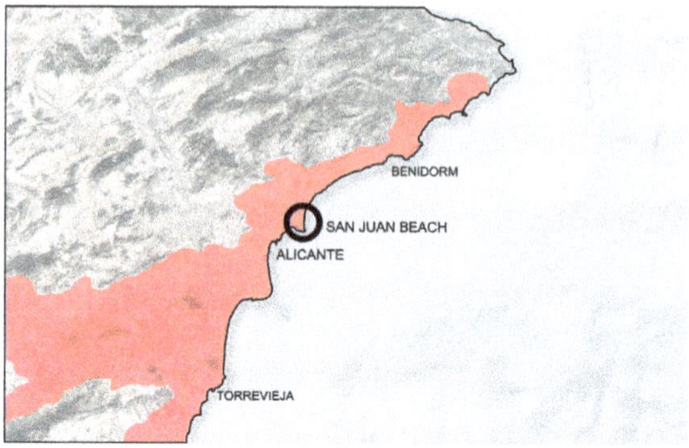


(b)

**Figure 3.** (**a**) Geographical area of BShs climate in Spain (red); (**b**) building location.

Two residential buildings were studied. The case study buildings were selected among the last residential buildings constructed in this city. Their construction finished in 2022. The architectural typology of the buildings and their dimensions, as well as the typologies of the dwellings and their dimensions, are very common in current buildings in this geographic area. The first selected building is a nine-story structure with a rectangular plan geometry measuring 60 m in length and 20 m in width. The second selected building is a four-story structure with a rectangular plan geometry measuring 55 m in length and 16 m in width. Both buildings receive ample sunlight and are surrounded by residential structures with numerous landscaped areas. The analyzed dwellings feature a living area of approximately 90 m$^2$. Each dwelling includes a living-dining room, a kitchen, three bedrooms, two bathrooms, and spacious exterior balconies on their facades (Figure 4).

(a)

(b)

**Figure 4.** (**a**) Case study buildings: (**a**) external image of building 1; (**b**) external image of building 2.

These buildings were chosen because they were recently constructed and comply with the latest energy efficiency standards. Additionally, they feature architectural and construction typologies that are very common in this geographic area. Both buildings have spacious balconies, and each is executed using one of the two most common construction techniques in Spain currently. The first one has facades made of exposed brick, thermally insulated from the inside with additional cladding. The second one has facades with a continuous External Thermal Insulation System (ETICS). This allows for the analysis of thermal bridges, considering current standards, architectural typologies, and the most common construction techniques in this area. Consequently, these results and conclusions can be extrapolated to the majority of new residential buildings in the coastal areas of Southern Spain and Europe.

### 2.3. Methodological Details and Tools

The thermographic measurements of the 26 studied buildings were conducted in winter (January and February 2023), between 23:00 h and 02:00 h. The goal was to avoid distortion of the temperature reflected from solar radiation and to maintain a thermal difference of around 15 °C between the interior of the dwellings and the exterior temperature, aiming to minimize the margin of error of the equipment. Mean outdoor temperatures were between 6 and 10 °C during the measurements. Occupied dwellings were assumed to have indoor temperatures between 21 and 25 °C at the time.

Both the thermographic measurements and the in situ thermal transmittance measurements of the two selected buildings were carried out during the night, between 23:00 h and 02:00 h, in the coldest week of the year, with heating active inside the dwellings. Mean outdoor temperatures were between 6 and 7 °C, and indoor temperatures were between 21 and 23 °C. The objective was to achieve the maximum thermal difference between the interior and exterior, ensuring the highest precision in the measurements. The thermographic measurements considered the average emissivity of the building materials of the facades, around 0.90. The humidity and outside temperatures were also measured. Measuring instruments were to be switched on five minutes before taking the measurements to allow the sensors to acclimatize to the indoor and outdoor environment.

On-site measurements with the thermal transmittance flowmeter were taken from the interior of the dwellings. Three types of thermal transmittance measurements were taken for each of the two analyzed dwellings. Measurements were conducted in the center of each facade section to determine the average thermal transmittance of the facade. Measurements were also taken in areas of the facade close to the junction with the floor slab edge, and measurements were taken in areas of the facade near the junction with the balconies. Measurements were taken at different points of the facade–slab and facade–balcony junctions to verify that windows and other thermal bridges (columns, jambs, corners, etc.) had no influence (Figures 5–7).

(a) (b)

**Figure 5.** Drawings of the two analyzed dwellings indicating the measurement points: (**a**) case study 1 (brick facade with interior insulation and cladding); (**b**) case study 2 (facade with ETICS).

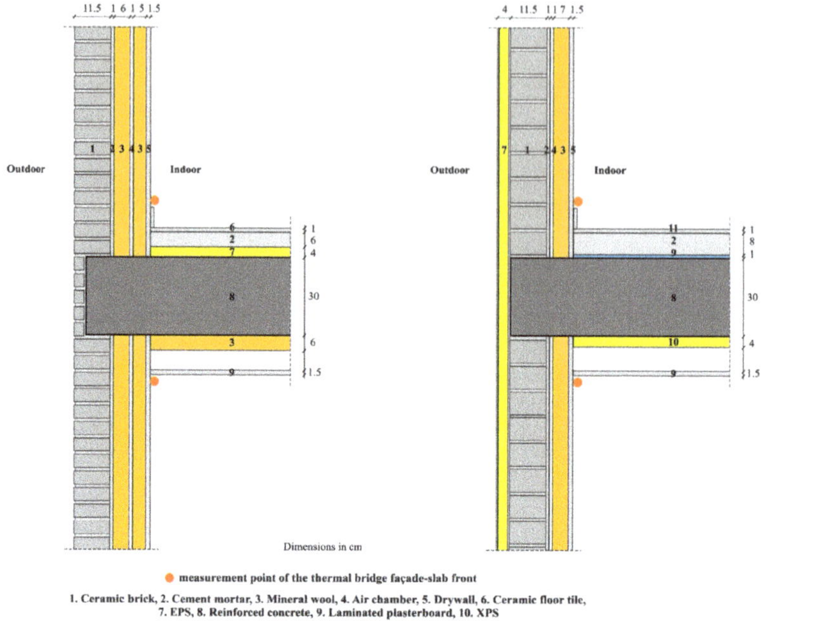


(a) (b)

**Figure 6.** Drawings of the facade–slab junction indicating the measurement points: (**a**) case study 1 (brick facade with interior insulation and cladding); (**b**) case study 2 (facade with ETICS).

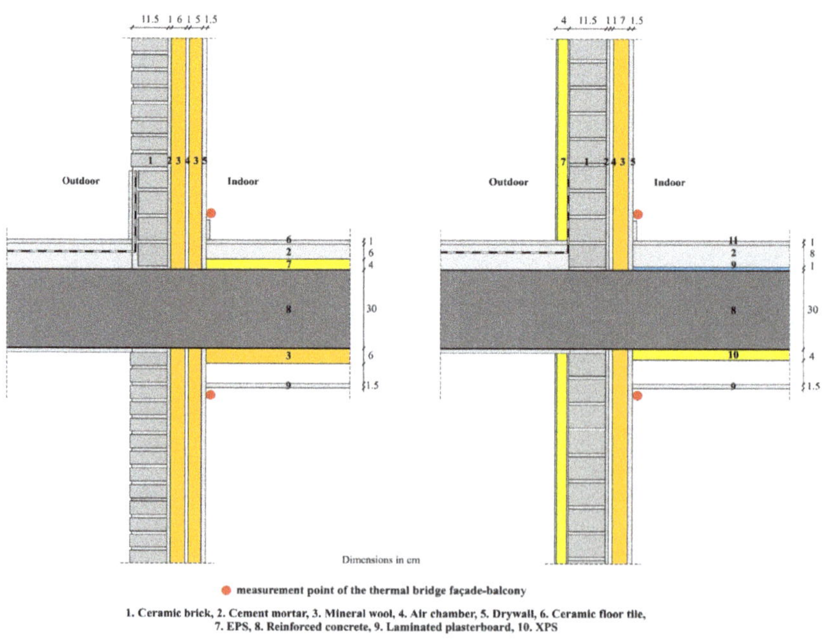


(a) (b)

**Figure 7.** Drawings of the facade–balcony junction indicating the measurement points: (**a**) case study 1 (brick facade with interior insulation and cladding); (**b**) case study 2 (facade with ETICS).

The thermographic evaluation was carried out by using a Testo 868 camera (Testo SE & Co. KGaA, Titisee, Germany). The measurements of thermal transmittance, humidity, and dry bulb temperature were carried out using a Testo 435-2 multifunctional instrument (Testo SE & Co. KGaA, Titisee, Germany) calibrated to standard settings (Figure 8 and Table 2).

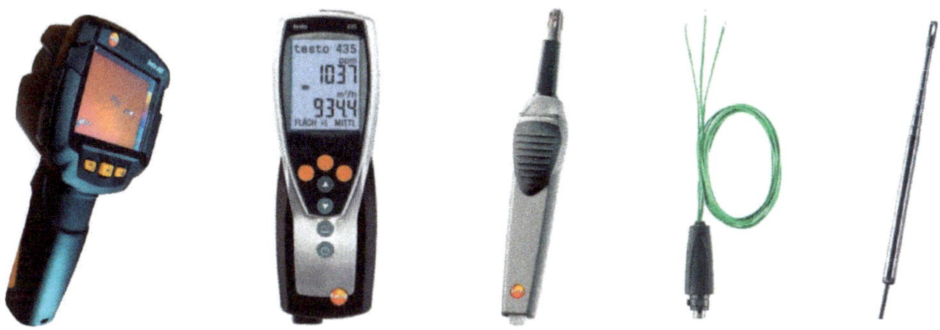


(a) (b) (c) (d) (e)

**Figure 8.** Instruments and measuring equipment used: (**a**) Testo 868 thermal imaging camera; (**b**) thermal transmittance flowmeter Testo 435-2; (**c**) humidity/temperature probe with wireless handle; (**d**) surface probe; (**e**) hot wire probe.

**Table 2.** Summary of measuring instruments and their characteristics.

| | Model | Measuring Range | Accuracy |
|---|---|---|---|
| Thermographic camera | Testo 868 | −15–+50 °C<br>+10–+95% HR | ±2 °C/ ±2%<br>±2% HR |
| Thermal transmittance flowmeter | Testo 435-2 | −20–+50 °C | - |
| Humidity/temperature probe | Testo | −20–+70 °C<br>+10–+100% HR | ±0.3 °C<br>±2% HR |
| Surface probe | Testo | −20–+70 °C | ±0.1 °C |
| Hot wire probe | Testo | −20–+70 °C | ±0.3 °C<br>±0.3 m/s |

A Líder-Calener Unified Tool [17] (version 2.0.2253.1167) was used because it is the official software of the Spanish Government for modeling and assessing the energy of buildings in Spain for the verification of compliance with the Spanish Technical Building Code (Royal Decree 732/2019) and Spanish and European energy certification standards (Royal Decree 390/2021 and Directive (EU) 2018/844). This software uses the calculation procedure included in the UNE-EN ISO 52000-1:2019 standard [53]. Although the official software was less capable of calculating and analyzing the results than other calculation programs, it allowed a better verification of the energy performance of the building in accordance with the established regulatory requirements. Moreover, this aspect is fundamental in order to be able to consider possible energy improvements that have to be made. The software program Cypetherm Bridges [54] (version v2022.e, Cype Ingenieros S.A., Alicante, Spain) was also used. This software determines the thermal transmittance in linear thermal bridges through the resolution and post-processing of a two-dimensional heat transfer analysis model using finite element methods in accordance with the European standard UNE-EN ISO 10211 [13]. Only heat exchange by conduction was considered in the analyses.

The computer calculations considered the thermal properties of the building envelope obtained from field measurements. The computer model for the calculation of the building's energy needs considered its location, dimensions, orientation, geometry, glazing position, and thermal envelope design features. The calculation method used considered the thermal transmission of the enclosure, the transmission of solar radiation through translucent surfaces using a correction factor for the solar radiation coefficient, the air transmission, and the gains and losses due to air transmission, according to current standards. The virtual simulation of the hygrothermal performance considered outdoor temperature and relative humidity data recorded on the same day as the field measurement campaign (Table 3).

**Table 3.** Summary of thermal properties of the building facades according to on-site measurements: case study 1 (CS1), case study 2 (CS2).

| Thermal Properties | Thickness (cm) | U (W/m$^2$·K) | g | Absorptivity | Air Permeability m$^3$/h·m$^2$ |
|---|---|---|---|---|---|
| | CS1/CS2 | CS1/CS2 | CS1/CS2 | CS1(CS2 | CS1/CS2 |
| Opaque facade enclosure | 25/25 | U = 0.36/0.37 | | | |
| Glass (80% of the window) | 2.0/2.2 | Ug = 2.421/2.368 | 0.58/0.60 | | |
| Frames (20% of the window) | 7.0/7.0 | Uf = 2.534/2.475 | | 0.75/0.75 | 27.00/27.00 |
| Floor | 40.0/45.0 | U = 0.85/0.90 | | | |
| Roof | 40.0/45.0 | U = 0.85/0.78 | | | |
| Air change rates by natural ventilation = 0.55 ren/h | | | | | |
| Frame air permeability = 27.00 m$^3$/h·m$^2$ | | | | | |

The calculations took into account parameters such as occupancy, air conditioning performance, usage plans and calendars, target temperatures, legally mandated supply air updates, and internal lighting loads (Table 4).

**Table 4.** Parameters considered in the computer simulation calculations: case study 1 (CS1), case study 2 (CS2).

| Parameter | Values | | | Applicable Regulation |
|---|---|---|---|---|
| | CS1/CS2 | | | |
| Occupation | People/m$^2$<br>0.05/0.05 | Metabolic rate<br>1 | Schedule<br>Activated 24/7 | DB-HE<br>Application Guide 2019 [55] |
| Cooling equipment<br>Heating equipment | Cop<br>4/4.2<br>3.5/3.6 | Months<br>6/7/8/9<br>1/2/3/4/5/10/11/12 | Schedule<br>0:00–24:00 27 °C<br>0:00–24:00 19 °C | DB-HE Annex D<br>Operational conditions and use profiles [56] |
| Mechanical ventilation<br>Natural nocturnal ventilation | Air renewals/hour<br>0.55/0.52<br>4.0 (summer) | | Schedule<br>Activated 24/7<br>0:00–08:00 100% | DB-HS3 [56] |
| Internal lightning loads | Average illumination<br>200 lux/200 lux | Power<br>2 W/m$^2$ | 0:00–07:00 10%<br>07:00–19:00 30%<br>19:00–23:00 100%<br>23:00–24:00 50% | Royal Decree 486/1997<br>Annex IV [57] |

The calculations by finite element simulations considered the thicknesses and material properties of the different layers of the facade, floors, and roofs of the two dwellings (Table 5).

**Table 5.** Parameters considered in the computer calculations by finite element simulations: case study 1 (CS1), case study 2 (CS2).

| Construction Element | Layer | Thickness (cm) | Linear Thermal Conductivity (W/m·K) | Thermal Resistance ($m^2$·K/W) |
|---|---|---|---|---|
| CS1 | | | | |
| Facade: | | | | |
| | Ceramic brick | 11.5 | $\lambda$ = 0.991 | |
| | Cement mortar | 1.0 | $\lambda$ = 0.550 | |
| | Mineral wool | 6.0 | $\lambda$ = 0.035 | |
| | Air chamber | 1.0 | | R = 0.15 |
| | Rock wool | 5.0 | $\lambda$ = 0.035 | |
| | Drywall | 1.5 | $\lambda$ = 0.250 | |
| Indoor floor: | | | | |
| | Ceramic floor tile | 1.0 | $\lambda$ = 2.300 | |
| | Cement mortar | 6.0 | $\lambda$ = 0.550 | |
| | EPS | 4.0 | $\lambda$ = 0.030 | |
| Structural slab: | Reinforced concrete | 30.0 | $\lambda$ = 1.838 | |
| Indoor false ceiling: | | | | |
| | Mineral wool | 6.0 | $\lambda$ = 0.035 | |
| | Laminated plasterboard | 1.5 | $\lambda$ = 0.250 | |
| CS2 | | | | |
| Facade: | | | | |
| | EPS | 4.0 | $\lambda$ = 0.032 | |
| | Ceramic brick | 11.5 | $\lambda$ = 0.991 | |
| | Cement mortar | 1.0 | $\lambda$ = 0.550 | |
| | Air chamber | 1.0 | | R = 0.15 |
| | Mineral wool | 7.0 | $\lambda$ = 0.035 | |
| | Drywall | 1.5 | $\lambda$ = 0.250 | |
| Indoor floor: | | | | |
| | Laminated wood flooring | 1.0 | $\lambda$ = 2.300 | |
| | Cement mortar | 8.0 | $\lambda$ = 0.550 | |
| | Acoustic foil | 1.0 | $\lambda$ = 0.050 | |
| Structural slab: | Reinforced concrete | 30.0 | $\lambda$ = 1.838 | |
| Indoor false ceiling: | | | | |
| | XPS | 4.0 | $\lambda$ = 0.034 | |
| | Laminated plasterboard | 1.5 | $\lambda$ = 0.250 | |

## 3. Results

First, this section identifies the most representative thermal bridges of the facades of buildings constructed in recent years through the thermographic study carried out. Second, the results of the qualitative and quantitative analysis of the thermal bridges in the two studied buildings through on-site measurements are shown. Subsequently, the impact of the analyzed thermal bridges on the total annual energy requirements of the dwellings was determined using computer simulations. Third, this section illustrates the effects of modifying various construction and geometric parameters of the analyzed thermal bridges on the overall energy consumption of the dwellings.

### *3.1. Thermographic Study of the Facades of Buildings Constructed in Recent Years*

The thermographic study of the analyzed facades allowed for the identification of the most representative thermal bridges. Balcony junctions with facades and slab edges on the facades constitute the type of thermal bridge with the highest proportion of surface area. These thermal bridges collectively account for an average of 44% of the surface area of all detected thermal bridges in the studied buildings. When analyzed separately, balcony junctions with facades represent 24% of the surface area with an average thermal difference of 1.4 °C. Facade slab edges represent 20% with an average thermal difference of 1.2 °C.

Integrated pillars in facades account for 18%, with an average thermal difference of 0.9 °C. Window sills represent 9% of the surface area with an average thermal difference of 0.7 °C. Roof–facade junctions make up 7%, with an average thermal difference of 1.3 °C. Cantilever junctions with facades have a 5% surface area proportion and an average thermal difference of 1.6 °C. The remaining thermal bridge types specified by regulations each represent less than 5% (Figure 9).

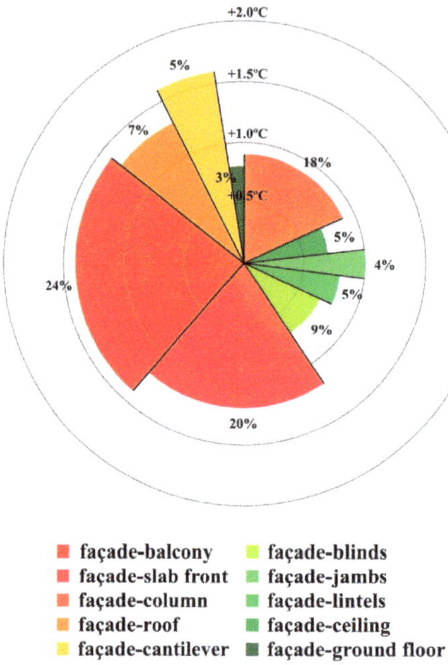


**Figure 9.** The percentage of facade surface occupied by each type of thermal bridging, and the thermal difference between the average surface temperature of the facade surface of each type of thermal bridging and the average surface temperature of the rest of the facade.

Next, the Thermal Bridge Wall Ratio (THBWR) of each thermal bridge was considered. This index indicates the proportion between the surface multiplied by the average temperature difference of each thermal bridge ($S_n \cdot \Delta T_n$) and the total sum of surfaces and average temperature differences of all thermal bridges ($\Sigma\, S_{(1,2,\ldots n,\ldots 10)} \cdot \Delta T_{(1,2,\ldots n,\ldots 10)}$). The average temperature difference of each thermal bridge was calculated by comparing the average surface temperature of each thermal bridge and the average surface temperature of the facade. The facade–balcony thermal bridge area was calculated by multiplying the thickness of the slab by the length of the balconies. The facade–floor thermal bridge area was calculated by multiplying the thickness of the slab by the length of the facade–floor junctions. The junctions of balconies with facades and the slab edges on the facades have a THBWR of 0.523, meaning they account for more than 50% of the total thermal differences caused by all thermal bridges. When analyzed separately, the junctions of balconies with facades have a THBWR of 0.305, and the slab edges have a THBWR of 0.218. Integrated pillars on the facades have a THBWR of 0.162. Roof-to-facade connections have a THBWR of 0.083. Cantilever junctions with facades have a THBWR of 0.073. Copings have a THBWR of 0.057. The remaining types of thermal bridges established by the regulations each have a THBWR of less than 0.05 (Figure 10).

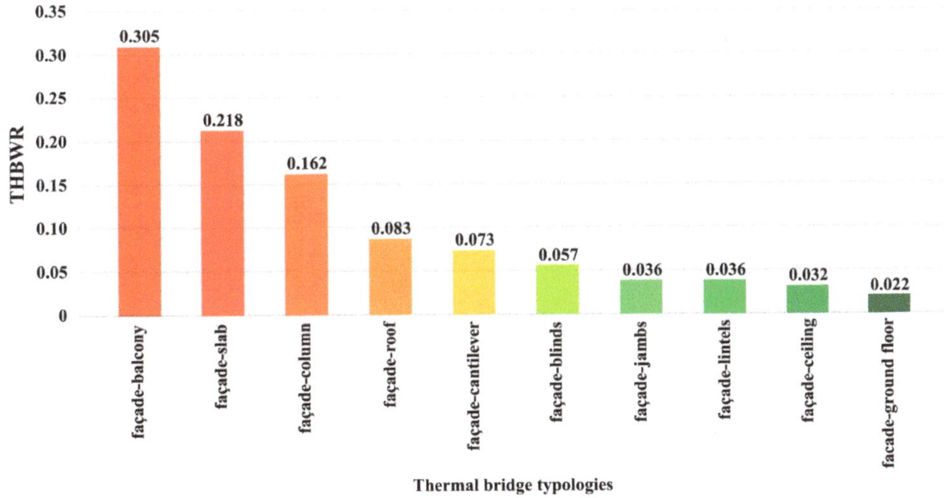


**Figure 10.** Thermal Bridge Wall Ratio Classification of the thermal bridges identified in the buildings studied depending on the Thermal Bridge Wall Ratio.

### *3.2. Influence of the Thermal Bridges Analyzed on the Total Annual Energy Needs of Studied Dwellings*

The detailed thermographic study of the buildings in the two analyzed dwellings confirmed that the slab edges and the junctions of balconies with facades are the most influential types of thermal bridges in both buildings. Differences were also observed between them based on the construction typology of the facade: a brick-faced facade with interior cladding and insulation (case study 1) and a facade with External Thermal Insulation System (ETICS) (case study 2) (Figures 6, 7 and 11).

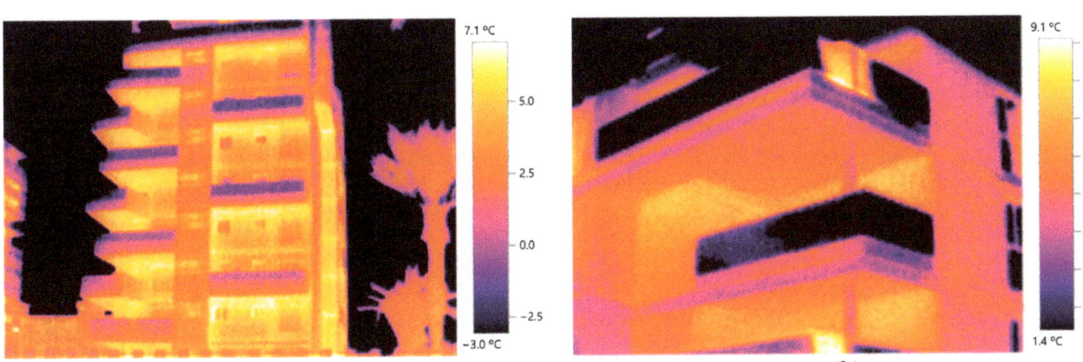


(**a**) (**b**)

**Figure 11.** A thermographic image of the facades of the two buildings analyzed in detail: (**a**) case study 1 (brick facade with interior insulation and cladding); (**b**) case study 2 (facade with ETICS).

These two types of thermal bridges constitute 45% of the total thermal bridge surface area in the building of dwelling 1 (case study 1) and 42% in the building of dwelling 2 (case study 2). Specifically, the junctions of balconies with facades account for a surface proportion of 26% and an average thermal difference of 1.3 °C in building 1 and 22% with 1.6 °C in building 2. The slab edges on the facades represent 19% with an average thermal difference of 1.2 °C in building 1 and 20% with 0.8 °C in building 2 (Figure 12).

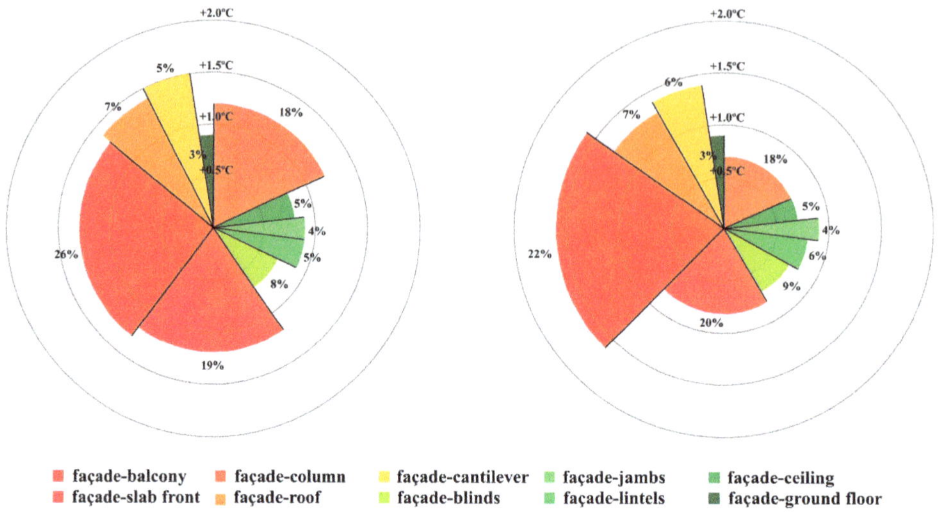


(**a**) (**b**)

**Figure 12.** The percentage of facade area occupied by each type of thermal bridging, and the thermal difference between the average facade surface temperature of each type of thermal bridging and the average surface temperature of the rest of the facade: (**a**) study case 1; (**b**) study case 2.

On-site measurements using a heat flux meter confirm significant differences between the average thermal transmittance of the facade and the thermal transmittance of the facade areas near the slab edges and the junctions with balconies (Figures 5–7). Substantial variations are also evident based on the construction typology of the facade. The average thermal transmittance on the facade of building 1 is 0.27 W/m$^2$·K, near the slab edge is 1.10 W/m$^2$·K, and near the balcony junction is 1.22 W/m$^2$·K. In building 2, the average thermal transmittance on the facade is 0.27 W/m$^2$·K, near the slab edge is 0.74 W/m$^2$·K, and near the balcony junction is 1.51 W/m$^2$·K (Table 4). From the surface thermal transmittance obtained from the on-site measurements, the linear thermal transmittance of the thermal bridge can be estimated considering the total thickness of the floor slab with the ceiling and the flooring. Subsequently, the linear thermal transmittances of the analyzed thermal bridges were calculated using finite element simulations considering the thicknesses and material properties of the different layers of the facade, floors, and ceilings of the two dwellings (Table 5). The linear thermal transmittance of the facade–slab edge junction is 0.477 W/m·K in case study 1 (Figure 13a) and 0.320 W/m·K in case study 2 (Figure 13b).

The statistical analysis for the validation of the energy model considered 504 thermal transmittance values obtained from on-site measurements inside the two analyzed dwellings, distributed over six measuring points in each dwelling. The actual measurements were compared with the results of the computer simulations for the same constructive solutions. The obtained Mean Bias Error (MBE) was 1.62. The Normalized Mean Bias Error (NMBE) was 4.95%. This value is less than the ±10% upper limit set by ASHRAE Guideline 14 in the hourly calibration criteria. The Root Mean Square Error (RMSE) was 3.86. And the Coefficient of the Variant of Root Mean Square Error (CV(RMSE)) was 8.23%. This value is lower than the 30% upper limit set by ASHRAE Guide 14. This allows validation of the calculation procedure used.

Subsequently, the annual energy needs of the two studied dwellings were calculated considering the thermal transmittance data of the enclosures obtained in situ and confirmed by computer simulations.

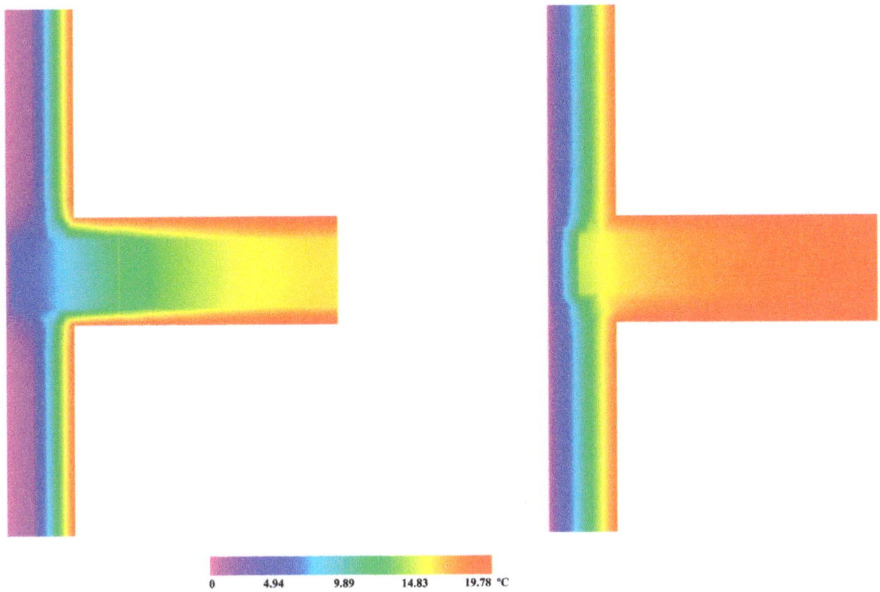


(**a**) (**b**)

**Figure 13.** Analysis of the linear thermal transmittance by computer simulation according to UNE EN-ISO 10211 of the thermal bridge facade-floor slab front: (**a**) case study 1; (**b**) case study 2.

The linear thermal transmittance of the facade–balcony junction is 0.529 W/m·K in case study 1 (Figure 14a) and 0.653 W/m·K in case study 2 (Figure 14b).

(**a**) (**b**)

**Figure 14.** Analysis of the linear thermal transmittance by computer simulation according to UNE EN-ISO 10211 of the thermal bridge facade-balcony: (**a**) case study 1; (**b**) case study 2.

Initially, energy needs were calculated without distinguishing between the thermal bridges of the facade slab edge and facade balcony, following standard procedures. Subsequently, the calculation was repeated, differentiating between these two types of thermal bridges.

The results indicate that the annual energy needs (heating + cooling) caused by the thermal bridges in dwelling 1 are equivalent to 40.8% of the total energy needs if the thermal bridges of the facade slab edge and facade balcony are not differentiated, and 44.3% if they are distinguished (Figure 15). The annual energy needs in dwelling 1 are 20.6 kWh/m$^2$·year without distinguishing the thermal bridges and 21.9 kWh/m$^2$·year when considering the thermal bridges of the facade slab edge and facade balcony separately. The difference is 1.3 kWh/m$^2$·year, which represents 6.3% of the total annual energy needs of the dwelling and entails a 15.5% difference in the calculation of the energy needs of the thermal bridges.

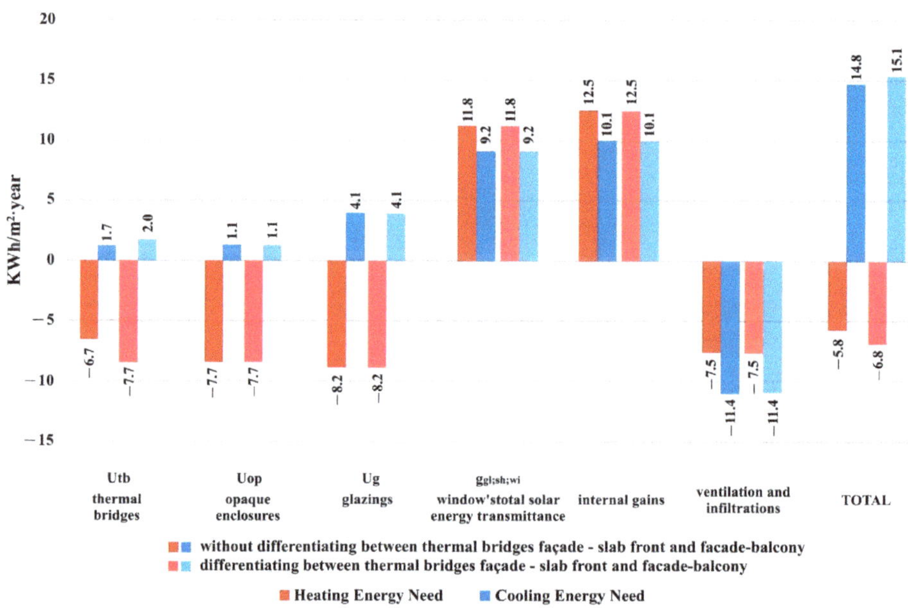


**Figure 15.** Total annual energy needs of dwelling 1 according to the type of thermal bridge.

Meanwhile, the annual energy needs (heating + cooling) caused by the thermal bridges in dwelling 2 are equivalent to 29.7% of the total energy needs if the thermal bridges are not differentiated, and 37.4% if they are distinguished (Figure 16). The annual energy needs in dwelling 2 are 15.5 kWh/m$^2$·year without distinguishing the thermal bridges and 17.4 kWh/m$^2$·year when considering the thermal bridges separately. The difference is 1.9 kWh/m$^2$·year, which represents 12.3% of the total annual energy needs of the dwelling and entails a 41.3% difference in the calculation of the energy needs of the thermal bridges.

Subsequently, the annual energy needs specifically caused by the thermal bridges of the facade slab edge and facade balcony were calculated. The results show that the combined contribution of these two types of thermal bridges accounts for 41.2% of the annual energy needs caused by all thermal bridges in case study 1 and 52.3% in case study 2. Specifically, the junctions of balconies with facades contribute to 25.8% of the energy needs in case 1 and 38.5% in case 2, while the slab edges on the facades contribute to 15.4% in case 1 and 13.8% in case 2 (Figure 17).

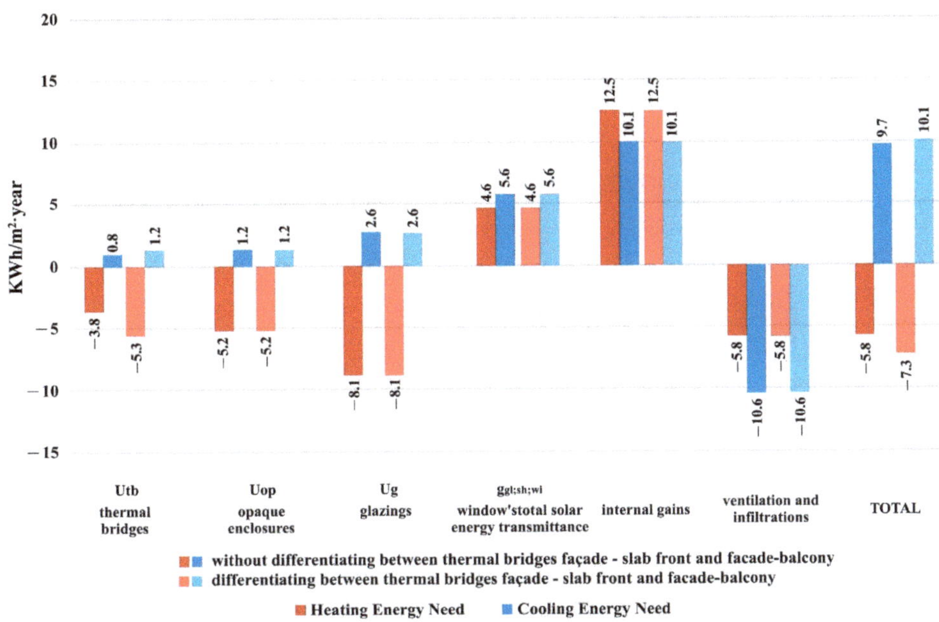


**Figure 16.** Total annual energy needs of dwelling 2 according to the type of thermal bridge.

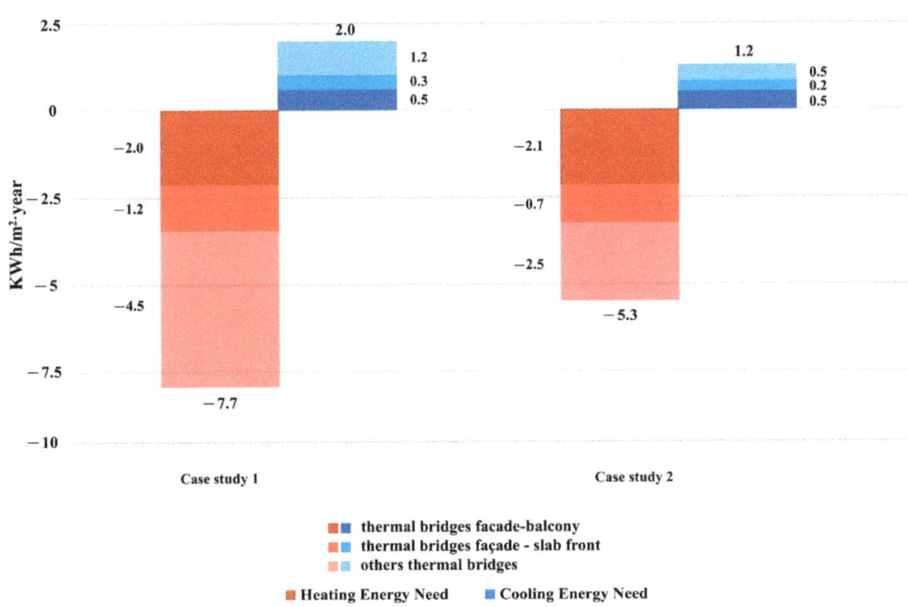


**Figure 17.** Total annual energy needs according to the type of thermal bridge in each dwelling surveyed.

### *3.3. Influence of Modification of Construction Parameters on Thermal Bridges*

The results indicate that improving the thermal insulation of thermal bridges significantly reduces their linear thermal transmittance in case 1 but only marginally in case 2. Additionally, enhancing the insulation of thermal bridges does not result in a directly proportional reduction in the linear thermal transmittance of the thermal bridge. The reduction becomes negligible for higher insulation thicknesses.

In the balcony–facade junctions, the thermal transmittance of this thermal bridge can be reduced by increasing the thermal insulation of indoor floors and ceilings near the balcony. Insulating floors and ceilings in the proximity of the balcony–facade junctions of dwelling 1 (Figure 7a) reduces the linear thermal transmittance of the thermal bridge by 34.7%; doubling the insulation reduces the transmittance by 45.3%, and tripling it reduces the transmittance by 51.4%. On the other hand, insulating ceilings near the balcony–facade junctions of dwelling 2 (Figure 7b) reduces the linear thermal transmittance by 14.8%; doubling the insulation reduces the transmittance by 20.1%, and tripling it reduces the transmittance by 23.8% (Figure 18).

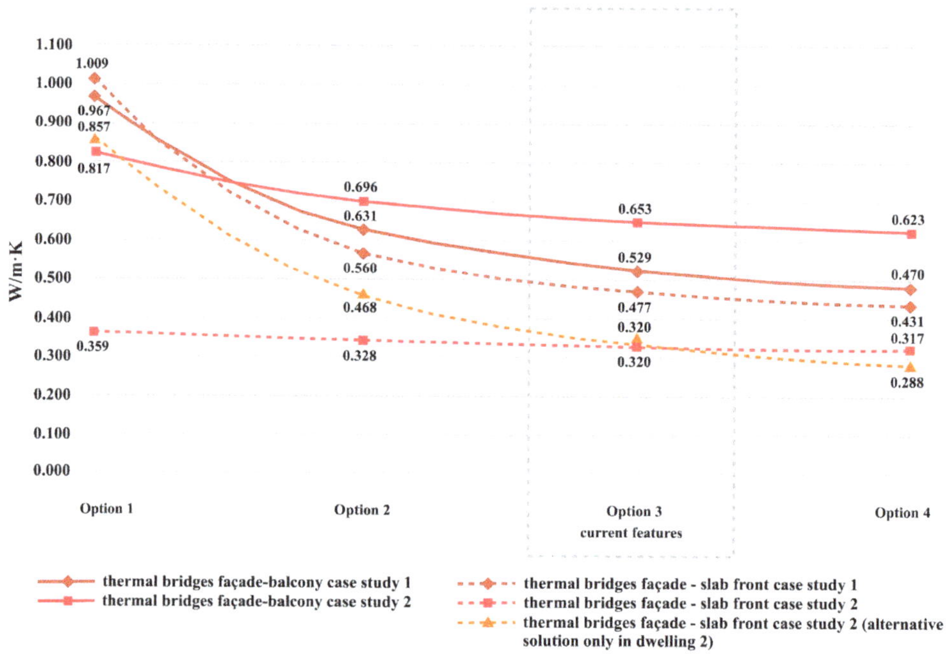


**Figure 18.** The linear thermal transmittance depending on the type of thermal bridge and the thickness of the thermal bridge insulation.

In the slab edge junctions, insulating the floors and ceilings indoors near the facade of dwelling 1 (Figure 6a) reduces the linear thermal transmittance of the thermal bridge by 44.5%; doubling the insulation reduces the transmittance by 52.7%, and tripling it reduces the transmittance by 57.3%. Conversely, insulating the ceilings near the facade of dwelling 2 (Figure 6b) reduces the linear thermal transmittance by 8.6%; doubling the insulation reduces the transmittance by 10.9%, and tripling it reduces the transmittance by 11.7%. Instead of increasing the insulation of internal ceilings, the external thermal insulation can be increased at the slab faces (alternative solution only in dwelling 2). This second option reduces the linear thermal transmittance of the thermal bridge much more. The ETICS exterior insulation of the slab edge junctions in dwelling 2 reduces the linear thermal transmittance of the thermal bridge by 45.4%; doubling the insulation reduces the transmittance by 62.7%, and tripling it reduces the transmittance by 66.4%.

The energy needs calculations for the dwellings show that insulating floors and ceilings near the junctions of the facade slab edge and facade balcony in dwelling 1 reduces the total annual energy needs by 9.6%; doubling the insulation reduces the energy needs by 12.4%, and tripling it reduces the energy needs by 15.6%. On the other hand, insulating floors and ceilings near the junctions of the facade slab edge and facade balcony in dwelling 2 reduces the total annual energy needs by 2.9%, doubling the insulation reduces the energy needs by 4.6%, and tripling it reduces the energy needs by 5.1% (Figure 19). Additionally, in facades with External Thermal Insulation Systems (ETICS), doubling the exterior thermal insulation at the junctions of the facade slab edge in dwelling 2 reduces the total annual energy needs by 5.1%, and tripling it reduces the energy needs by 6.2% (Figure 18).

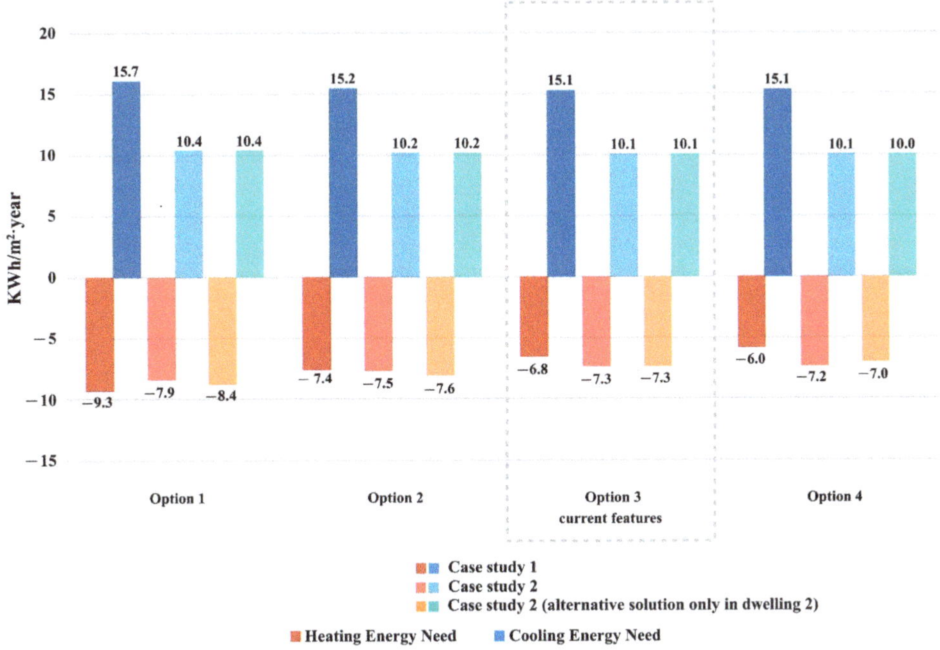


**Figure 19.** Total annual energy needs according to the thickness of the thermal bridge insulation.

## 4. Discussion

### *4.1. Constructive Study of the Current Facades*

The analysis of the obtained results highlights that the most representative thermal bridges are the slab edge junctions on the facades, constituting an average of 44% of the total thermal bridge surface area and a Thermal Bridge Wall Ratio (THBWR) of 0.523. Although Spanish regulations consider the slab edge junctions as a single type of thermal bridge, within this category, it is advisable to differentiate balcony–facade junctions from the rest of the slab edge junctions because they exhibit different thermographic measurements. The slab edge junctions on facades have higher thermal differences concerning the average surface temperature of the rest of the facade. However, the balcony–facade junctions have a larger transmission surface area. All this demonstrates that these two thermal bridges have different thermal behaviors due to significantly different construction and geometric characteristics.

In the facade–slab edge junctions, there are layers of the vertical facade cladding that cover the front of the structure. However, in the balcony–facade junction, the structural slab extends through the facade toward the exterior with layers covering the slab both below (ceiling of the overhang) and above (balcony floor). Due to these differences in construction, geometry, and thermal properties, it would be advisable to consider the balcony–facade junction as a specific type of thermal bridge. Furthermore, each of these two types of thermal bridges has a sufficiently representative THBWR. The balcony–facade junctions have a THBWR of 0.218, while the rest of the slab edge junctions have a THBWR of 0.305. Other types of thermal bridges have much lower THBWR. However, Spanish regulations do not identify the balcony–facade junctions as separate and independent thermal bridges. This reduces the precision of the calculation of the building's energy needs and complicates the specific treatment of this type of thermal bridge in building projects. This issue is particularly significant in the region of the Spanish Mediterranean coast and the BShs climate, where buildings often feature numerous balconies.

### 4.2. Influence of the Thermal Bridges Analyzed on the Total Annual Energy Needs of Studied Dwellings

The detailed thermographic study of the two buildings demonstrated that the balcony–facade junctions and facade–slab edge junctions are the most influential types of thermal bridges in this building type, constituting more than 40% of the total thermal bridge surface area. The results showed that exterior surfaces in balcony–facade junctions have a higher thermal difference from the average facade temperature compared to facade–slab edge junctions. Additionally, the facade–slab edge junctions exhibit significantly lower thermal differences when an Exterior Thermal Insulation System (ETICS) is utilized, whereas the thermal difference in balcony–facade junctions remains similar in both buildings.

These findings align with the results obtained from in situ measurements of the thermal transmittance of different areas on the facades. The analysis of measurements reveals thermal transmittance differences greater than 0.83 $W/m^2{\cdot}K$ near the facade slab edge in case study 1, compared to only 0.47 $W/m^2{\cdot}K$ in case study 2. Similarly, thermal transmittance differences are 0.95 $W/m^2{\cdot}K$ near balcony–facade junctions in case study 1 and 1.24 $W/m^2{\cdot}K$ in case study 2. These results are consistent with computer simulation outcomes, which indicate a 50% higher linear thermal transmittance of facade–slab edge junctions in case study 1 and a 23% higher transmittance of balcony–facade junctions in case study 2.

These differences stem from the lack of sufficient thermal insulation in facade–slab edge junctions in brick facades with interior insulation, whereas in the ETICS, they have an appropriate layer of external insulation. However, in balcony–facade junctions, the structural slab extends through the facade to the exterior in both studied buildings, interrupting the thermal envelope insulation in both cases. Additionally, in case study 2, the dwelling lacks thermal insulation in the floor, making the thermal bridge on the balcony worse than in case study 1, which has thermal insulation in the floor.

Due to these construction and thermal differences, it would be advisable to consider balcony–facade junctions as a specific type of thermal bridge, distinct from facade–slab edge junctions. Computer calculations demonstrated that thermal bridges account for over 40% of the total annual energy needs in dwelling 1 and around 30% in dwelling 2. Thermal losses and gains through thermal bridges are comparable to those produced by all other opaque enclosure surfaces in the dwelling. This is due to the increased requirements of current energy efficiency regulations, which have improved insulation in facades. However, as buildings enhance the insulation of their facades, the lack of insulation in unique junctions between different construction systems, such as thermal bridges, becomes more evident. The analysis of results indicates that minimizing thermal bridges would almost eliminate the heating energy needs in dwelling 1 and significantly reduce it in dwelling 2.

Furthermore, the results also demonstrated that the total energy needs of the dwellings change substantially when differentiating and correctly specifying facade–slab edge junctions and balcony–facade junctions. Distinguishing these two thermal bridges increases the energy needs related to all thermal bridges by 15.5% in dwelling 1 and the total annual energy needs by 6.3%. In the case of dwelling 2, the energy needs related to all thermal bridges increased by 41.3%, and the total annual energy needs of the dwelling increased by 12.3%. This implies that the current regulations and official calculation programs, by not considering the qualitative and quantitative differences between these two thermal bridges, result in a high degree of inaccuracy in the calculation of thermal bridges and the energy needs of buildings. This situation is particularly critical in the region of the Spanish Mediterranean coast and the BShs climate, where the surface and length of balconies are often much greater than in other Spanish and European geographical areas.

Finally, computer simulations also demonstrated that facade–slab edge junctions and balcony–facade junctions are the most decisive thermal bridges in the analyzed buildings. These two thermal bridges constitute 41.2% of the annual energy needs caused by all thermal bridges in case study 1 and 52.3% in case study 2. The comparative analysis of results also confirms that balcony–facade junctions cause much more energy needs than facade–slab edge junctions, especially in case study 2, where facade–slab edge junctions are well-insulated thanks to the ETICS. These data are consistent with the results obtained in the detailed thermographic study of the two buildings and in the in situ measurements of the thermal transmittance of different areas on the facades. These findings further highlight the severity of the regulation's failure to consider the construction and thermal specificities of balcony–facade junctions.

### 4.3. Influence of Modification of Construction Parameters on Thermal Bridges

The analysis of the results demonstrates that improving the insulation of thermal bridges does not result in a directly proportional reduction in the linear thermal transmittance of the thermal bridges. In fact, the reduction becomes negligible for very high thermal insulation thicknesses. This is because the linear thermal transmittance value of the thermal bridge is also influenced by the difference between the exterior surface of the facade and the intrados of the facade. The value cannot approach zero because there is a heat flow transmitted through the thickness of the facade slab edge or the balcony, which adds to the energy transmitted through the rest of the facade (Figure 13).

Enhancing the thermal insulation of floors and interior ceilings near facade slab edges and balconies significantly reduces the linear thermal transmittance of thermal bridges in the brick facade with interior insulation (case study 1) but has a minimal impact on the facade with an Exterior Thermal Insulation System (ETICS) (case study 2). This is because facades with ETICSs have facade slab edge thermal bridges that are well-insulated from the exterior, and consequently, their thermal transmittances are already small and challenging to improve by insulating from the interior. In facades with ETICSs, increasing the exterior insulation on facade slab edges does reduce the thermal transmittance of that thermal bridge significantly. On the other hand, in balcony–facade junctions, the thermal transmittance of the thermal bridge in the ETICS facade of case study 2 has very high values. This is due to the interruption of the ETICS, creating significant thermal bridges because the dwellings lack interior insulation for floors. In balcony–facade junctions, the only solution is to insulate the floors and interior ceilings or use costly construction systems with thermal breakers for balconies and cantilevers (thermal separation of cantilevered reinforced concrete slabs).

Consequently, improving the insulation of thermal bridges does not result in a directly proportional reduction in the energy needs of dwellings, with improvements becoming negligible for high thermal insulation thicknesses (Figure 14). This is because the energy needs also include factors such as thermal gains and losses from all opaque enclosures, internal gains, ventilation air exchanges, and air infiltrations.

Enhancing the thermal insulation of thermal bridges significantly reduces the energy needs in the brick facade with interior insulation (case study 1) but has a minimal effect on the facade with the ETICS (case study 2). This is because facades with ETICSs have facade slab edge thermal bridges that are well-insulated from the exterior, and consequently, their thermal transmittances are already small and difficult to improve. However, in balconies, the ETICS is interrupted, creating significant thermal bridges because the dwellings lack interior insulation for floors.

In conclusion, Figures 13 and 14 indicate that optimal construction solutions are those that greatly reduce the thermal transmittance of thermal bridges and the energy needs of dwellings without excessively increasing their thermal insulation.

Furthermore, in facades with ETICSs, it is better to increase the exterior thermal insulation on facade slab edges than to insulate floors and interior ceilings. However, to reduce the energy needs, it is essential to also reduce the thermal bridge in balcony–facade junctions by insulating floors and interior ceilings of dwellings in areas near balconies or by using thermal breakers in cantilevered slab structures.

Moreover, the regulations also do not consider that ETICSs are often combined with interior claddings for thermal-acoustic insulation, reducing the thickness of the ETICS as less thermal insulation is needed (Figures 6b and 7b). This results in a greater thermal bridge in facade slab edges and balconies than the regulations state because the exterior thermal insulation is continuous, but the interior insulation is interrupted.

The study has some limitations. Official regulations and calculation programs do not sufficiently include the effect of thermal radiation on balcony surfaces. In architectural typologies with deep and elongated balconies, balcony surfaces can influence thermal gains and losses as radiators. In the summer, the high solar radiation of the BShs climate can cause balcony surfaces to absorb a lot of radiation and transmit heat to the interior during peak sunlight hours. Additionally, it would be advisable to complete calculations with more combinations of construction solutions, combining ETICSs and interior thermal-acoustic insulations, using different insulation thicknesses in floors and ceilings and studying more housing typologies with varying sizes and orientations.

## 5. Conclusions

(1) This work demonstrates that the energy needs caused by thermal bridges account for approximately 40% of the total annual energy needs of the studied residential buildings, constructed under the current Spanish energy efficiency regulations. The annual energy needs (heating + cooling) caused by thermal bridges are slightly higher than 40% in buildings with a brick facade with interior cladding with insulation and slightly less than 40% in buildings with an Exterior Thermal Insulation System (ETICS).

(2) Balcony–facade and facade–slab edge junctions are the most predominant thermal bridges in the analyzed buildings. The sum of these two thermal bridges constitutes, on average, 44% of the total thermal bridge surface area and has a Thermal Bridge Wall Ratio (THBWR) of 0.523. They represent more than 40% of the annual energy needs caused by all thermal bridges in the studied building with interior insulation and over 50% in the building with the ETICS. Balcony thermal bridges account for between 25% and 40% of the energy needs produced by all thermal bridges, while facade–slab edge junctions contribute around 15%.

(3) The constructive, geometric, and thermal behavior characteristics of balcony–facade and facade–slab edge junctions are very different. The linear thermal transmittance of the balcony–facade junction is higher than that of the facade–slab edge junction. Consequently, balcony–facade junctions cause much more energy needs than slab edge junctions, especially in buildings with ETICSs. However, current Spanish energy efficiency regulations and official calculation programs do not identify the balcony–facade junction as a separate and independent thermal bridge but consider it the same as slab edge junctions.

(4) The lack of differentiation in Spanish regulations between balcony–facade and facade–slab edge junctions results in a very high lack of precision in the calculation of thermal bridges and energy needs for buildings. Correctly specifying these two thermal bridges increases the energy needs of thermal bridges by up to 41.3% and the total annual energy needs of dwellings by up to 12.3%. The lack of precision in regulations leads to inaccurate energy needs calculations, especially in regions of the Spanish Mediterranean coast and the BShs climate, where the surface and length of balconies are often much larger than in other Spanish and European regions.

(5) Improving the insulation of thermal bridges does not result in a directly proportional reduction in the linear thermal transmittance of thermal bridges or the total energy needs of the dwelling, with the reduction becoming negligible for very high thermal insulation thicknesses. Therefore, optimal construction solutions are those that greatly reduce the

thermal transmittance of thermal bridges and the energy needs of the dwelling without excessively increasing its thermal insulation.

(6) Enhancing the insulation of thermal bridges reduces the total annual energy needs of buildings with a brick facade and interior insulation by up to 15% and up to 6% in buildings with ETICSs.

(7) In facades with ETICSs, it is better to increase the exterior thermal insulation on slab edges than to insulate floors and interior ceilings. However, an ETICS significantly reduces the thermal transmittance of slab edge junctions in facades, while barely affecting the thermal transmittance of balcony–facade junctions. Therefore, to reduce the energy needs, it is necessary to also reduce the thermal bridge in balcony–facade junctions by insulating floors and interior ceilings or using thermal breakers in the structure.

(8) This study highlights the need to modify current energy efficiency regulations and official calculation programs in Spain to incorporate balcony–facade junctions as a separate and specific thermal bridge. It also underscores the need to promote new construction solutions that reduce thermal bridges in balcony–facade junctions.

**Author Contributions:** Conceptualization, C.P.-C. and A.d.l.M.-M.; methodology, C.P.-C., A.d.l.M.-M. and Á.B.G.-A.; software, C.P.-C. and A.d.l.M.-M.; validation, C.P.-C., A.d.l.M.-M., Á.B.G.-A. and A.G.-G.; formal analysis, C.P.-C. and A.d.l.M.-M.; investigation, C.P.-C., A.d.l.M.-M., Á.B.G.-A., N.C. and A.G.-G.; resources, C.P.-C. and A.d.l.M.-M.; data curation, C.P.-C. and A.d.l.M.-M.; writing—original draft preparation, C.P.-C., A.d.l.M.-M., Á.B.G.-A. and A.G.-G.; writing—review and editing, C.P.-C., A.d.l.M.-M., Á.B.G.-A., N.C. and A.G.-G.; visualisation, C.P.-C. and A.d.l.M.-M.; supervision, C.P.-C., A.d.l.M.-M., Á.B.G.-A. and A.G.-G.; project administration, C.P.-C., A.d.l.M.-M., Á.B.G.-A. and A.G.-G.; funding acquisition, C.P.-C. and Á.B.G.-A. All authors have read and agreed to the published version of the manuscript.

**Funding:** This research received no external funding.

**Data Availability Statement:** The data presented in this study are available on request from the corresponding author. The data are not publicly available due to privacy.

**Acknowledgments:** This article has been written within the framework of the collaboration agreement of the project "Research on Construction Solutions for Industrialisation and Improvement of Passive Housing Conditioning" between AEDAS HOMES and the University of Alicante.

**Conflicts of Interest:** The authors declare no conflicts of interest.

## References

1. Agencia Estatal Boletín Oficial del Estado. *Real Decreto 732/2019, de 20 de Diciembre, Por el Que se Modifica el Código Técnico de la Edificación, Aprobado Por el Real Decreto 314/2006, de 17 de Marzo*; Agencia Estatal Boletín Oficial del Estado: Madrid, Spain, 2019.
2. Agencia Estatal Boletín Oficial del Estado. *Real Decreto 390/2021, de 1 de Junio, Por el Que se Aprueba el Procedimiento Básico Para la Certificación de la Eficiencia Energética de los Edificios*; Agencia Estatal Boletín Oficial del Estado: Madrid, Spain, 2021.
3. European Union. *Directiva 2010/31/UE del Parlamento Europeo y del Consejo de 19 de Mayo de 2010 Relativa a la Eficiencia Energética de los Edificios (Refundición)*; European Union: Maastricht, The Netherlands, 2010.
4. European Union. *Directiva (UE) 2018/844 del Parlamento Europeo y del Consejo, de 30 de Mayo de 2018, Por la Que se Modifica la Directiva 2010/31/UE Relativa a la Eficiencia Energética de Los Edificios y la Directiva 2012/27/UE Relativa a la Eficiencia Energética*; European Union: Maastricht, The Netherlands, 2018.
5. Ingeli, R.; Vavrovič, B.; Čekon, M. Thermal Bridges Minimizing through Typical Details in Low Energy Designing. *Adv. Mater. Res.* **2014**, *899*, 62–65. [CrossRef]
6. Martin, K.; Erkoreka, A.; Flores, I.; Odriozola, M.; Sala, J.M. Problems in the calculation of thermal bridges in dynamic conditions. *Energy Build.* **2011**, *43*, 529–535. [CrossRef]
7. Bruno, R.; Arcuri, N.; Carpino, C. Study of innovative solutions of the building envelope for passive houses in Mediterranean areas. *Energy Procedia* **2017**, *140*, 80–92. [CrossRef]
8. Ge, H.; Baba, F. Dynamic effect of thermal bridges on the energy performance of a low-rise residential building. *Energy Build.* **2015**, *105*, 106–118. [CrossRef]
9. Ge, H.; Baba, F. Effect of dynamic modeling of thermal bridges on the energy performance of residential buildings with high thermal mass for cold climates. *Sustain. Cities Soc.* **2017**, *34*, 250–263. [CrossRef]
10. Freitas, J.; Cunha, E. Thermal bridges modeling in South Brazil climate: Three different approaches. *Energy Build.* **2018**, *169*, 271–282. [CrossRef]

11. Evola, G.; Margani, G.; Marletta, L. Energy and cost evaluation of thermal bridge correction in Mediterranean climate. *Lancet* **2011**, *43*, 2385–2393. [CrossRef]
12. Ministerio de Fomento. *DA DB-HE/3, Puentes Térmicos, Documento de Apoyo al Documento Básico, DB-HE Ahorro de Energía, Código Técnico de la Edificación;* Ministerio de Transporte, Movilidad y Agenda Urbana: Madrid, Spain, 2020.
13. *UNE-EN ISO 10211:2022;* Thermal Bridges in Building Construction. Heat Flows and Surface Temperatures. Detailed Calculations (ISO 10211:2017). AENOR, Agencia Española de Normalización y Certifiación: Madrid, Spain, 2022.
14. *UNE-EN ISO 13789:2017;* Thermal Performance of Buildings—Transmission and Ventilation Heat Transfer Coefficients—Calculation Method (ISO 13789:2017). AENOR, Agencia Española de Normalización y Certifiación: Madrid, Spain, 2017.
15. *UNE-EN ISO 14683:2017;* Thermal Bridges in Building Construction, Linear Thermal Transmittance. Simplified Methods and Default Values (ISO 14683:2017). AENOR, Agencia Española de Normalización y Certifiación: Madrid, Spain, 2017.
16. *UNE-EN ISO 6946:2017;* Building Components and Building Elements. Thermal Resistance and Thermal Transmittance. Calculation Methods (ISO 6946:2017, Corrected Version 2021-12). AENOR, Agencia Española de Normalización y Certifiación: Madrid, Spain, 2017.
17. AICIA. *Líder-Calener Unified Tool Software;* v2.0.2253.1167; Dirección General de Arquitectura, Vivienda y Suelo del Ministerio de Transportes, Movilidad y Agenda Urbana y Por el Instituto Para la Diversificación y Ahorro de la Energía (IDEA): Madrid, Spain, 2021. Available online: https://www.codigotecnico.org/Programas/HerramientaUnificadaLIDERCALENER.html (accessed on 18 December 2021).
18. *CE3X;* v2.3; Efinovatic, Centro Nacional de Energías Renovables (CENER): Madrid, Spain, 2023.
19. *CE3;* v20160906; Applus Norcontrol S.L.U.: Madrid, Spain, 2023.
20. *CERMA;* v5.11; Atecyr, Asociación Técnica española de Climatización y Refrigeración, Instituto Valenciana de Edificación: Madrid, Spain, 2022.
21. Martínez Medina, A. Arquitectura del boom turístico (1953–1979). *Canelobre* **2016**, *66*, 167–185.
22. Adams, B.; Bissio, R.; Yoke Ling, C.; Donald, K.; Martens, J.; Prato, S.; Vermuyten, S. *Spotlight. Enfoques Sobre Desarrollo Sostenible 2017;* Social Watch: Montevideo, Uruguay, 2017; pp. 110–122.
23. Grao-Gil, O. El Desafío a las Alturas Desde la Singularidad. El Apartotel Riscal de Alicante de Francisco Muñoz; pp. 1–12. Available online: https://congresopionerosarquitectos.com/api/assets/es/VI_PDF_comunicaciones_aceptadas/VI_Congreso_Pioneros_Arquitectura_moderna_-_16.pdf (accessed on 20 March 2022).
24. Navas, J. VIA 02p-120. Francisco Muñoz Llorens. ViA-Arquitectura. Available online: https://www.via-arquitectura.net/02_prem/02p-120.htm (accessed on 9 June 2022).
25. Jaén Urban, G.; Martínez Medina, A.; Oliva Meyer, J.; Oliver Ramírez, J.L.; Sempere Pascual, P.; Calduch Cervera, J. *Guía de Arquitectura de la Provincia de Alicante;* Instituto Alicantino de Cultura "Juan Gil-Albert": Alicante, Spain, 1999.
26. Martínez-Medina, A. *Mestres. Arquitectura Moderna en la Comunidad Valenciana;* Universidad de Alicante: Alicante, Spain, 2019. Available online: http://hdl.handle.net/10045/122744 (accessed on 1 January 2022).
27. Martínez Medina, A.; Calduch Cervera, J.; Agustí Lara, L.; Oliva Meyer, J. *Arquitectura del Sol en la Comunidad Valenciana;* Colegio Oficial de Arquitectos de la Comunitat Valenciana, Ed.; Universidad de Alicante y Colegio Oficial de Arquitectos de la Comunidad Valenciana: Alicante, Spain, 2001. Available online: http://hdl.handle.net/10045/102529 (accessed on 1 January 2022).
28. Alhawari, A.; Mukhopadhyaya, P. Thermal bridges in building envelopes—An overview of impacts and solutions. *Int. Rev. Appl. Sci. Eng.* **2018**, *9*, 31–40. [CrossRef]
29. Saied, A.E.; Maalouf, C.; Bejat, T.; Wurtz, E. Slab-on-grade thermal bridges: A thermal behavior and solution review. *Energy Build.* **2022**, *257*, 111770. [CrossRef]
30. Sabie, D.; Ghiaus, A. Effects of thermal bridges on the heat demand of residential buildings. *IOP Conf. Ser. Earth Environ. Sci.* **2021**, *664*, 012100. [CrossRef]
31. Vunjak, D.; Čeh, A.; Đurić, N.; Karaman, G.; Hegediš, I. The Influence of Linear Thermal Transmittance of Thermal Bridges on the Energy Performance Class of Buildings—Simplified Method. *Arch. Tech. Sci.* **2016**, *14*, 73–78. [CrossRef]
32. Rybakova, A.; Kalousek, M. Modeling of Details for Single-Layer Wall in a Passive House: The Case of Balcony Slab. *Adv. Mater. Res.* **2014**, *1041*, 171–174. [CrossRef]
33. Buday, P.; Ingeli, R.; Čekon, M. Influence of Thermal Break Element Applied in Balcony Slab on Internal Surface Temperature. *Adv. Mater. Res.* **2014**, *1057*, 79–86. [CrossRef]
34. Marincioni, V.; May, N.; Altamirano-Medina, H. Parametric Study on the Impact of Thermal Bridges on the Heat Loss of Internally Insulated Buildings. *Energy Procedia* **2015**, *78*, 889–894. [CrossRef]
35. Llomets, S.; Kalamees, T. Evaluation of the criticality of thermal bridges. *J. Build. Pathol. Rehabil.* **2016**, *1*, 11. [CrossRef]
36. Boian, I.; Tuns, I. Reducing heat losses—First step toward nZEB. *IOP Conf. Ser. Mater. Sci. Eng.* **2020**, *789*, 012005. [CrossRef]
37. Aghasizadeh, S.; Kari, B.M.; Fayaz, R. Thermal performance of balcony thermal bridge solutions in reinforced concrete and steel frame structures. *J. Build. Eng.* **2022**, *48*, 103984. [CrossRef]
38. Zhang, X.; Jung, G.; Rhee, K. Performance Evaluation of Thermal Bridge Reduction Method for Balcony in Apartment Buildings. *Buildings* **2022**, *12*, 63. [CrossRef]
39. Ge, H.; McClung, V.; Zhang, S. Impact of balcony thermal bridges on the overall thermal performance of multi-unit residential buildings: A case study. *Energy Build.* **2013**, *60*, 163–173. [CrossRef]

40. Susorova, I.; Stephens, B.; Skelton, B. The Effect of Balcony Thermal Breaks on Building Thermal and Energy Performance: Field Experiments and Energy Simulations in Chicago, IL. *Buildings* **2019**, *9*, 190. [CrossRef]
41. Sadowska, B.; Bieranowski, P. Efficiency of Different Balcony Slab Modernization Método in Retrofitted Multi-Family Buildings. *Energies* **2021**, *14*, 6666. [CrossRef]
42. Lujan, S.; Arrebola, C.; Sánchez, A.; Benito, P.; Cortina, M. Experimental comparative study of the thermal performance of the façade of a building refurbished using ETICS, and quantification of improvements. *Sustain. Cities Soc.* **2019**, *51*, 101713. [CrossRef]
43. Fernandes, C.; Brito, J.; Cruz, C. Thermal Retrofitting of Façades: Architectural Integration of ETICS. *J. Perform. Constr. Facil.* **2016**, *30*, 06015002. [CrossRef]
44. Glew, D.; Brooke-Peat, M.; Gorse, C. Modelling insulated coving's potential to reducing thermal bridging and moisture risk in solid wall dwellings retrofitted with external wall insulation. *J. Build. Eng.* **2017**, *11*, 216–223. [CrossRef]
45. Liu, X.T.; Wei, D.; Zhao, L.H. Heat transfer analysis for the energy efficiency of balcony of residential building. *J. Dalian Univ.* **2006**. Available online: https://api.semanticscholar.org/CorpusID:114656632 (accessed on 20 February 2023).
46. Santamía, J.; Girón, S.; Campano, M.A. Economic assessments of passive thermal rehabilitations of dwellingsin Mediterranean climate. *Energy Build.* **2016**, *128*, 772–784. [CrossRef]
47. Pombo, O.; Allacker, K.; Rivela, B.; Neila, J. Sustainability assessment of energy saving measures: A multi-criteriaapproach for residential buildings retrofitting—A case study of theSpanish housing stock. *Energy Build.* **2016**, *116*, 384–394. [CrossRef]
48. Mengjie, S.; Fuxin, N.; Ning, M.; Yanxin, H.; Shiming, D. Review on building energy performance improvement using phase change materials. *Energy Build.* **2018**, *158*, 776–793. [CrossRef]
49. Musial, M.; Lichołai, L.; Katunský, D. Modern Thermal Energy Storage Systems Dedicated to Autonomous Buildings. *Energies* **2023**, *16*, 4442. [CrossRef]
50. Ficco, G.; Iannetta, F.; Ianniello, E.; d'Ambrosio, F.R.; Dell'Isola, M. U-value in situ measurement for energy diagnosis of existing buildings. *Energy Build.* **2015**, *104*, 108–121. [CrossRef]
51. Chazarra, A.; Flórez, E.; Peraza, B.; Tohá, T.; Lorenzo, B.; Criado, E.; Moreno, J.V.; Romero, R.; Botey, R. *Mapas Climáticos de España (1981–2010) y ETo (1996–2016)*; Ministerio para la Transición Ecológica, Agencia Estatal de Meteorología: Madrid, Spain, 2018.
52. Ministerio de Fomento. *Documento Descriptivo Climas de Referencia*; Ministerio de Fomento: Madrid, Spain, 2017.
53. *UNE-EN ISO 52000-1:2019*; Energy Performance of Buildings. Overarching EPB Assessment. Part 1: General Framework and Procedures (ISO 52000-1:2017). AENOR, Agencia Española de Normalización y Certifiación: Madrid, Spain, 2019.
54. *Cypetherm Bridges Software*; v2023.e; Cype Ingenieros S.A.: Alicante, Spain, 2022.
55. Ministerio de Transportes; Villar Burke, R.; Sorribes Gil, M.; Jiménez González, D.; Sobaler Rodríguez, J. *DB-HE Application Guide 2019*; Ministerio de Transportes, Movilidad y Agenda Urbana; Sorribes Gil, M., Jiménez González, D., Eds.; Gobierno de España: Madrid, Spain, 2019.
56. Agencia Estatal Boletín Oficial del Estado. *Order FOM/588/2017, of 15 June, Amending Basic Document DB-HE 975 "Energy Saving" and Basic Document DB-HS "Health" of the Technical Building Code, Approved by Royal Decree 976 314/2006, of 17 March, 2017 (In Spanish: Orden FOM/588/2017, de 15 de Junio, Por la Que se Modifican el Documento Básico DB-HE 975 "Ahorro de Energía" y El Documento Básico DB-HS "Salubridad", del Código Técnico de la Edificación, Aprobado por Real Decreto 976 314/2006, de 17 de Marzo, 2017)*; Agencia Estatal Boletín Oficial del Estado: Madrid, Spain, 2017.
57. Agencia Estatal Boletín Oficial del Estado. *Real Decreto 486/1997, de 14 de Abril, Por el Que se Establecen Las Disposiciones Mínimas de Seguridad y Salud en Los Lugares de Trabajo*; Agencia Estatal Boletín Oficial del Estado: Madrid, Spain, 1997.

*buildings*

MDPI

*Article*

# Retrofit Analysis of Exterior Windows for Large Office Buildings in Different Climate Zones of China

**Sai Liu [1,2,3], Farid E. Mohamed Ghazali [1,*], Jingjing Yang [3], Zongkang Guo [3], Kejun Zeng [3] and Yixing Chen [3,4,*]**

1 School of Civil Engineering, Universiti Sains Malaysia, Nibong Tebal 14300, Malaysia
2 School of Civil and Architectural Engineering, Hunan Institute of Technology, Hengyang 421002, China
3 College of Civil Engineering, Hunan University, Changsha 410082, China
4 Key Laboratory of Building Safety and Energy Efficiency of Ministry of Education, Hunan University, Changsha 410082, China
* Correspondence: cefarid@usm.my (F.E.M.G.); yixingchen@hnu.edu.cn (Y.C.)

**Abstract:** In the energy-saving retrofit of existing buildings, investors are particularly concerned about the energy-saving performance of exterior windows and the payback period of additional costs. This study evaluates representative cities in four different climate zones in China to simulate the energy consumption of large office buildings after replacing different glass windows and conducting energy-saving and economic feasibility assessments. The research method includes the following steps: First, a baseline model of large office buildings in four cities was established using AutoBPS and OpenStudio. Then, the baseline and retrofit models of replacing glass windows were simulated using the EnergyPlus V9.3.0 to obtain multiple hourly energy consumption results. The commercial electricity and gas prices in the four cities were adjusted to calculate the total cost within 20 years after replacing different types of windows. Using the discounted payback period (DPP), net present value (NPV), and profitability index (PI) as evaluation indicators, a feasibility analysis was conducted in the four regions to evaluate the economic feasibility of replacing building windows. The simulation results show that considering economic feasibility and meeting energy-saving standards, it is more economical to choose windows with moderate U-value and SHGC value in the four regions than to choose windows with the smallest U-value and SHGC value, and that both energy savings and economic benefits are closely related to building age, with older buildings (especially those in Changsha and Shenzhen) showing greater benefits. Furthermore, the optimal window types in the four cities determined in this study can recover the investment cost within the window life, with Harbin (SC), Beijing (C), Changsha (HC), and Shenzhen (HW) with the payback period of 6.60, 15.66, 10.16, and 11.42 years, respectively. The research model established in this study provides a useful evaluation path for selecting windows for the energy-saving retrofit of large office buildings in cities in different climate zones and provides data support for the decision making of energy-saving retrofit investors.

**Keywords:** EnergyPlus; windows; building energy simulation; economic analysis

## 1. Introduction

Global warming is accelerating with the continuous expansion of human habitats and energy consumption. The net anthropogenic emissions of carbon dioxide ($CO_2$) must approach zero by mid-century (2050) to stabilize the global mean temperature at the level targeted by international efforts [1]. Decarbonizing the building sector is extremely important to mitigating climate change as the sector contributes 40% of the overall energy consumption and 36% of the total greenhouse gas emissions in the world [2]. Approximately 50% of the energy consumption in the building sector is attributed to the heating, ventilation, and air conditioning (HVAC) of 20 years in this study systems [3]. The energy consumption of HVAC systems depends on the design of the building envelope [4], and windows account for up to 60% of the heat losses in the building, making them the least energy-efficient component [5]. The energy efficiency of the same type of window varies in different climate zones, highlighting the importance of selecting appropriate windows to achieve energy-saving goals in buildings.

China's building energy efficiency work started in the 1980s. The mandatory building energy-efficiency design standards continue to improve the energy-saving rate, and the requirements for the thermal performance of the envelope structure also continue to improve. Isolating thermal conduction is an effective solution for window thermal management [6]. According to the market survey, the window types in the existing buildings are single-glazing, double-glazing, and multiple-glazing. At present, single-layer glass windows have been withdrawn from the market [7]. Double-layer or multi-layer glass windows generally reduce energy consumption by adding gas, vacuuming, coating, and other ways to isolate heat conduction [8], and the thermal insulation performance of windows has improved. With the global promotion of zero-carbon buildings, functional windows such as aerogel glass windows [9], electrochromic (EC) windows [10], phase change material (PCM) windows [11], hydrogel glass windows [12], and photovoltaic glass windows [13] have been developed. These windows possess energy-saving and other advantages by incorporating relevant functional materials into the glass. For example, high-transparency siliconized cellulose aerogel (SiCellA) has been developed from wood pulp, which can be inserted into gaps between multi-pane glasses, providing superior thermal insulation performance. The window's transparency can be adjusted by modifying the material composition, and the manufacturing cost is around USD 1 per square foot [14].

Among building energy consumption simulation software, the EnergyPlus software is the most commonly used software [15–17]. Through simulation analysis by the EnergyPlus software, many high-performance windows have better energy-saving rates than traditional windows. Vacuum gas-chromic smart windows were found to exhibit excellent thermal insulation performance across varying climate conditions [18]; PCM windows integrated with 3 mm solid-phase change materials outperform clear double-glazed windows, saving up to 17.2%, 14.0%, and 5.8% energy for HVAC systems in warm, mixed, and cold climates, respectively, resulting in the overall energy savings of 9.4%, 6.7%, and 3.2% for the entire building [19]. Another study explored the energy performance of thermochromic smart windows and radiant coolers in a medium-sized office building using EnergyPlus simulations, reporting reductions in the total energy consumption of 10.6% and 23.0%, respectively [20]. Other researchers have focused on integrating advanced materials into window systems. For instance, windows incorporating solid-solid phase change materials (PCMs) and silica aerogels were evaluated using the EnergyPlus software [21]. In Bangladesh, researchers' simulation using the EnergyPlus software revealed that semi-transparent CdTe-based building-integrated photovoltaic (BIPV) windows could reduce electricity consumption by 30–61% across various climate conditions compared to conventional windows [5]. In southern Italy, photochromic windows (PCWs) were evaluated against traditional low-emissivity (Low-E) windows and transparent double-glazed windows using the EnergyPlus software, and the study concluded that PCWs could achieve annual energy savings of up to 9.3% compared to transparent windows and 4.1% compared to Low-E windows [22]. Despite the ongoing research and promotion of these functional windows, the mainstream market for building windows in China still primarily consists of gas-filled or low-emissivity (Low-E) coated double-pane or multi-pane windows. This highlights a gap between research innovations and market adoptions and the need for further studies to bridge this divide.

Different types of windows have varying economic costs and benefits in different cities [7]. Therefore, the selection of windows for the retrofit of existing buildings is also diversified. For instance, when renovating public buildings such as government schools, cost considerations may favor using coated windows with SHGC values ranging from 0.4 to 0.6, which offer shorter payback periods [23]. In tropical climates, windows retrofitted with glass material with a lower U-value can provide better overall thermal transfer value (OTTV) for the building envelope subsystem. Reducing energy consumption translates into energy cost savings for buildings with retrofitted windows that can offset the installation costs of new glass materials [24]. In the hot summer and cold winter regions of China., the optimized variable transparency shape-stabilized phase change material (VTSS-PCM) window, the VTSS-PCM window, reduced the total annual unfavorable heat transfer (TAHT) by 30.14% and the total annual cost by 28.39% [25]. In the hot, arid climate, double-pane windows with either air or argon as the insulator gas exhibited the shortest payback period [26]. In a residential building, the operating costs of PVC windows are over 30% lower than in the case of wooden windows and almost 20% lower compared to aluminum windows [27]. In cold climates, the application of the low-emissivity window film on the outward-facing surface of the inner pane of the double-glazed windows helped to reduce heat loss through the windows in winter and unwanted heat gains in summer by almost 36% and 35%, respectively. This resulted in a 6% reduction in the building's annual energy consumption for heating purposes. However, the relatively high price of the films and the low price of district heating resulted in a rather long payback period of around 30 years [28]. The economic costs and benefits of window glass are also affected by the direction in which it is installed. In the study of 30 double-glazing configurations in the Indian composite climate zone, researchers found that based on the reduction in annual net cooling and heating costs, the optimal window orientation order is southeast, southwest, south, northeast, northwest, north, east, west [8]. In hot climates, large windows can be installed on the north side of the building, while a small window is recommended for the south side to minimize the building's energy consumption [29]. Therefore, in the energy-saving retrofit of existing buildings, if the budget is limited, you can prioritize the replacement of the window with the best orientation.

The research above shows that the continuous emergence of novel window technologies and materials has promoted global carbon reduction efforts. However, when retrofitting existing building windows, regional variations and long-term costs need to be further studied because the economic benefits of window technology vary from region to region, and detailed cost-benefit analysis is carried out for specific local climate and economic conditions. Assessing the economic benefits between the energy savings and replacement costs of advanced window technologies to understand their true economic impact over the building's life cycle is critical to making informed decisions about window retrofitting. The purpose of this study was to investigate the effects of different window types on building energy consumption and economic payback periods in four different climate zones in China. This paper utilizes EnergyPlus to establish energy simulation models for large office buildings in these four climate zones. Furthermore, it examines various window options based on their prices and associated energy consumption costs, ultimately identifying the most cost-effective window selection for office buildings in each climate zone. This research provides valuable insights into the economic implications of window retrofitting, emphasizing the importance of region-specific analyses.

## 2. Methodology

The research workflow of this paper is shown in Figure 1. The first step was establishing baseline energy models for large office buildings in different climate zones. According to the "Uniform Standard for the design of civil buildings (GB 50352-2019) [30] in China, China contains five major climate zones. In this study, large office buildings in Harbin (severe cold), Beijing (cold), Changsha (hot summer and cold winter), and Shenzhen (hot summer and warm winter) were selected as research objects. The baseline building models for new buildings were generated using AutoBPS. AutoBPS is a building energy modeling tool developed by Hunan University in Changsha, China [31]. AutoBPS can generate prototype building energy models using OpenStudio and EnergyPlus for different combinations of climate zone, year built, and building type [32]. AutoBPS can modify various parameter scenarios and generate different building models according to different specifications and standards [33,34], and can be used for retrofit analysis [35], demand response analysis [36], photovoltaic generation analysis [37], and thermal storage system analysis [38].

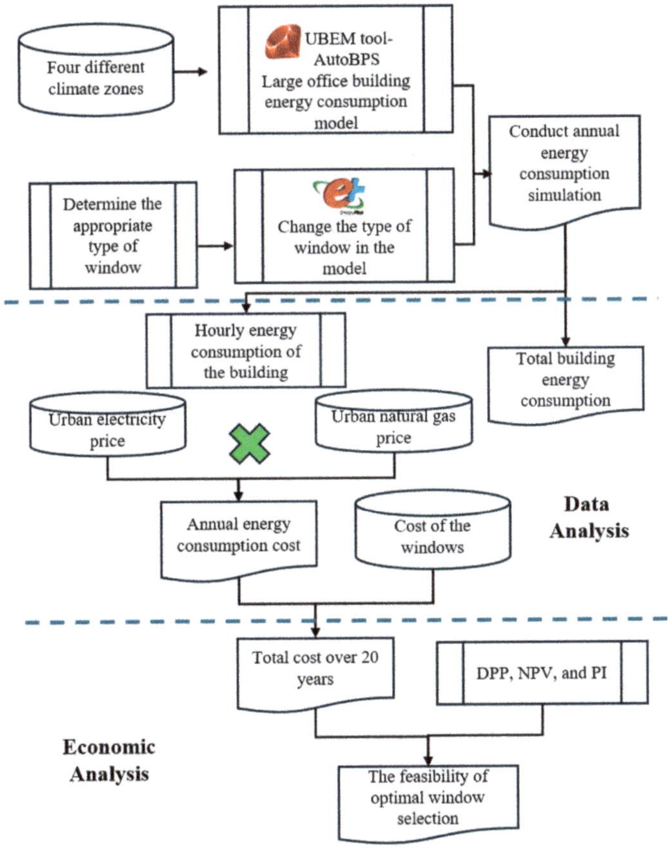


**Figure 1.** Workflow of the paper.

The second step was to develop the retrofit energy models for each window replacement scenario for each climate zone. Based on market research and inquiries, five different window replacement scenarios were selected for each climate zone. The window construction in EnergyPlus was modified to simulate the energy consumption of each window replacement scenario of each climate zone. The hourly energy consumption data of the baseline and retrofit models were used for the retrofit analysis.

The third step was to perform an economic analysis of five window replacement scenarios for each climate zone. The annual energy cost of the building was calculated based on city-specific electricity and natural gas prices, combined with the cost of the glass windows. The total cost of the building over 20 years in different climatic zones was then calculated, considering both the capital value and the time value of money. An economic performance analysis model was established to reflect the economic feasibility of window replacement. The feasibility of optimal window selection in the four regions was analyzed using DPP, NPV, and PI. The total 20-year cost of the building in this paper referred to the sum of the building's energy cost and the cost of windows, ignoring other costs such as building maintenance cost.

### 2.1. Climate Zones and Weather Data

In China, a major climate classification is for the thermal design of buildings, mainly concerned with the conduction of heat gain/loss through the building envelope and the corresponding thermal insulation issues [39]. It has five major climatic types: severe cold, cold, hot summer and cold winter, mild and hot summer, and warm winter.

The energy consumption of buildings in mild climate zones was less sensitive to the building envelope system. Therefore, four cities were selected to represent the other four climate zones, including Harbin (severe cold), Beijing (cold), Changsha (hot summer and cold winter), and Shenzhen (hot summer and warm winter). Their specific information is shown in Table 1. Geographical coordinates of the regions.

**Table 1.** Geographical coordinates of the regions.

| City | Thermal Climate Zone | | Longitude | Latitude | HDD18 | CDD26 | Solar Energy Resource Area | Climatic Regions |
|---|---|---|---|---|---|---|---|---|
| | GB50176-2016 [40] | ASHRAE Standard 169-2021 [41] | | | | | | |
| Harbin | 1C | 7 | 126.53 | 45.8 | 5032 | 14 | II | Severe cold (SC) |
| Beijing | 2A | 4 | 116.4 | 39.9 | 2699 | 94 | II | Cold (C) |
| Changsha | 3B | 3 | 112.93 | 28.23 | 1466 | 230 | III | Hot summer and cold winter (HC) |
| Shenzhen | 4A | 2 | 114 | 22.53 | 223 | 374 | III | Hot summer and warm winter (HW) |

The weather data in the EnergyPlus Weather (EPW) format were obtained from the official website (https://energyplus.net/weather accessed on 20 July 2024) for the cities.

### 2.2. Generation of Baseline Energy Models for Large Office Buildings

The case study large office prototype building has a total floor area of 43,330.86 $m^2$, with a north-south orientation, comprising twenty above-ground floors and one basement floor. The length is 57.83 m, and the width is 35.68 m. The floor-to-floor height is 3.96 m, and the total building height is 81.7 m. The building has a rectangular geometry with a window–wall ratio of 0.3881. As shown in Figure 2, the top, first, and basement floors were modeled in detail, while the middle floor was created with a multiplier of 18 to represent the other 18 floors to speed up the simulation. The geometrical specifications of the large office building remain consistent across the four regions studied.

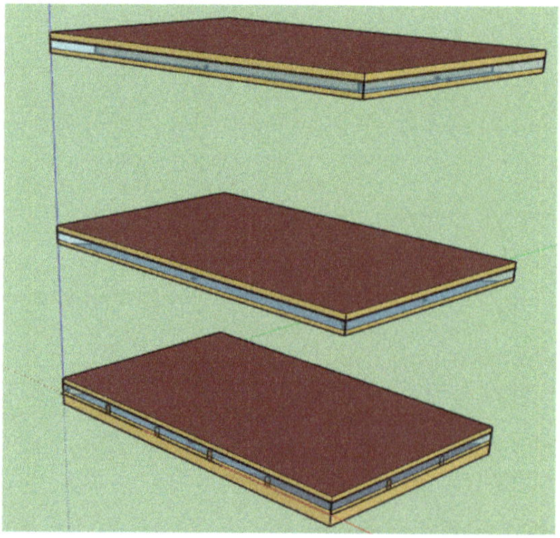

**Figure 2.** Baseline energy model of a large office building.

The heating, ventilation, and HVAC systems consist of chillers and cooling towers for cooling, the water-cooled centrifugal chiller nominal capacity is 1.17 MW and natural gas boilers for space heating and service hot water systems. The operational schedules of HVAC systems, occupants, lights, and plug loads are established according to the Chinese Standard GB50189-2015 [42]. Table 2 presents the major parameters of the building envelope and internal loads, including the U-factor and SHGC of exterior windows. Table 3 shows the window composition of the baseline model for the four cities.

**Table 2.** Major parameters of the building envelope and internal loads.

| Parameters | Harbin | Beijing | Changsha | Shenzhen |
|---|---|---|---|---|
| Exterior wall heat transfer coefficient (W/m$^2$·K) | 0.35<br>(≤0.43) | 0.45<br>(≤0.50) | 0.60<br>(≤0.60) | 0.80<br>(≤0.80) |
| Roof heat transfer coefficient (W/m$^2$·K) | 0.25<br>(≤0.35) | 0.40<br>(≤0.45) | 0.40<br>(≤0.40) | 0.50<br>(≤0.50) |
| Window U-value/SHGC | 2.208/0.217<br>(≤2.3/-) | 2.262/0.422<br>(≤2.4/≤0.48) | 2.493/0.356<br>(≤2.6/≤0.40) | 2.888/0.316<br>(≤3.0/≤0.35) |
| People density (person/m$^2$) | 0.05<br>(0.1) | 0.05<br>(0.1) | 0.05<br>(0.1) | 0.05<br>(0.1) |
| Equipment power density (W/m$^2$) | 15<br>(15) | 15<br>(15) | 15<br>(15) | 15<br>(15) |
| Lighting power density (W/m$^2$) | 8.93<br>(0.90) | 8.93<br>(0.90) | 8.93<br>(0.90) | 8.93<br>(0.90) |
| Cooling/heating temperature setpoint (°C) | 26/20<br>(26/20) | 26/20<br>(26/20) | 26/20<br>(26/20) | 26/20<br>(26/20) |
| Chiller COP | 5.40<br>(5.40) | 5.50<br>(5.50) | 5.60<br>(5.60) | 5.70<br>(5.70) |
| Boiler heating efficiency | 0.9<br>(0.9) | 0.9<br>(0.9) | 0.9<br>(0.9) | 0.9<br>(0.9) |

Note: The value in () is the standard value according to the Chinese Standard GB50189-2015.

**Table 3.** Window composition of the baseline model for the four cities.

| City | Glass Type |
|---|---|
| Harbin | REF C TINT HI 6 mm glass + 13 mm air + 6 mm clear glass |
| Beijing | ECREF-2 COLORED 6 mm glass + 13 mm argon gas + 6 mm clear glass |
| Changsha | REF D TINT 6 mm glass + 13 mm air + 6 mm clear glass |
| Shenzhen | REF B CLEAR LO 6 mm glass + 13 mm argon gas + 6 mm clear glass |

Note: the order of glass type marking is from outer layer to inner layer; REF B is titanium coating; REF C is pewter coating; REF D is tin-oxide coating; TINT is tinted with inorganic materials to increase absorption; LO, HI is low- and high-transmittance coating, respectively; ECREF is electrochromic glass that darkens by becoming more reflective; COLORED is the colored (darkest) state of electrochromic glass.

In order to compare the authenticity of the baseline model, eight public building windows were collected in four different climate zones from other existing research papers. Their window U-values and SHGC values were selected, and similar values were found in the EnergyPlus parameter file as simulation parameters. The specific parameters are shown in Table 4.

The main goal of this study is to evaluate the energy-saving potential and economic feasibility of building window retrofits in China's climate region. Although the window parameters of Seoul and Phoenix are not directly applicable to building energy analysis in China, their wide performance range (including high SHGC value and low U-value windows) provides a data reference for this study to expand the comparative analysis of window performance. To ensure the applicability of the study, the main simulation results and conclusions are based on window parameters that meet the Chinese design standards.

**Table 4.** Collected cases' window parameter information.

| Code | City | Building Types | U-Value (W/(m$^2$·K)) | SHGC (-) | Solar Energy Resource Area | Climatic Regions |
|---|---|---|---|---|---|---|
| Case1 [39] | Harbin | office | 2.556 | 0.609 | II | (SC) |
| Case2 [43] | Shenyang | office | 5.067 | 0.4 | II | (SC) |
| Case3 [39] | Beijing | office | 2.888 | 0.316 | II | (C) |
| Case4 [44] | Seoul | daycare centers | 3.835 | 0.768 | II | (C) |
| Case5 [45] | Changsha | office | 5.894 | 0.861 | III | (HC) |
| Case6 [46] | Wuxi | four-star resort hotel | 3.835 | 0.768 | III | (HC) |
| Case7 [47] | Phoenix | office | 2.511 | 0.704 | III | (HW) |
| Case8 [39] | Hongkong | office | 4.513 | 0.781 | III | (HW) |

### 2.3. Window Type Selection

The shape coefficient is the ratio of the exterior area of a building to the volume enclosed by the exterior area. The shape coefficient of the case study building is 0.11 m$^{-1}$. Based on the window–wall ratio and shape coefficient of the building, the exterior window requirement of the Chinese Standard GB55015-2021 [48] for the four climate zones is listed in Table 5. Ten different glass windows were selected from the EnergyPlus database, and their U-value and SHGC are shown in Table 6. Figure 3 shows the U-value and SHGC of the ten glass windows and the requirements of the four climate zones. As shown in Table 5, for the Harbin region, window types W1, W2, W3, and W6 were chosen for energy simulation; for the Beijing region, window types W3, W4, W5, and W7 were chosen; for the Changsha region, window types W5, W6, W9, and W10 were chosen; and for the Shenzhen region, window types W6, W7, W8, and W9 were chosen. The energy consumption simulations were carried out accordingly for each region. This study did not consider window shading devices (such as roller shades or blinds) and their opening and closing schedules. Future research could further refine the impact of shading devices on energy consumption.

**Table 5.** Exterior window requirement of GB55015-2021 for the case study building.

| Parameters | Harbin | Beijing | Changsha | Shenzhen |
|---|---|---|---|---|
| U-value (W/(m$^2$·K)) | ≤2.1 | ≤2.0 | ≤2.2 | ≤2.5 |
| SHGC (-) | - | ≤0.40 | ≤0.35 | ≤0.30 |

**Table 6.** U-value and SHGC values for glazed windows.

| Name | Glass Type | U-Value ((W/(m$^2$·K)) | SHGC | Harbin | Beijing | Changsha | Shenzhen |
|---|---|---|---|---|---|---|---|
| W1 | 6 low-transmittance Low-E + 13 air + 6 transparent | 1.761 | 0.568 | X | | | |
| W2 | 3 low-transmittance Low-E + 6 air + 3 clear + 6 air + 3 clear | 1.525 | 0.472 | X | | | |
| W3 | 6 low-transmittance low-film66 + 13 air + 6 transparent + 13 air + 6 transparent | 1.22 | 0.363 | X | X | | |
| W4 | 6 low-transmittance low-film66 + 6 air + 6 transparent + 6 air + 6 transparent | 1.712 | 0.358 | | X | | |
| W5 | 6 tri-silver Low-E + 12 air + 6 transparent | 1.63 | 0.296 | | X | X | |
| W6 | 6 tri-silver low-film66 + 13 air + 6 transparent + 13 air + 6 transparent | 1.22 | 0.254 | | | X | X |

**Table 6.** *Cont.*

| ame | Glass Type | U-Value ((W/(m$^2$·K)) | SHGC | Harbin | Beijing | Changsha | Shenzhen |
|---|---|---|---|---|---|---|---|
| W7 | 6 tri-silver low-film33 + 13 air + 6 clear + 13 air + 6 clear | 1.196 | 0.154 | X | X | | X |
| W8 | 6 highly transparent heat reflective + 13 air + 6 transparent | 2.416 | 0.234 | | | | X |
| W9 | 6 colored heat absorption + 13 air + 6 transparent | 1.761 | 0.169 | | | X | X |
| W10 | 6 tri-silver low-film66 + 6 air + 6 clear + 6 air + 6 clear | 1.712 | 0.265 | | | X | |

Note: all glass and inter-layer air thickness units are "mm"; film66 and film33 are coated polyester films with a nominal visible transmittance of 66 percent and 33 percent; "X" represents the selection.

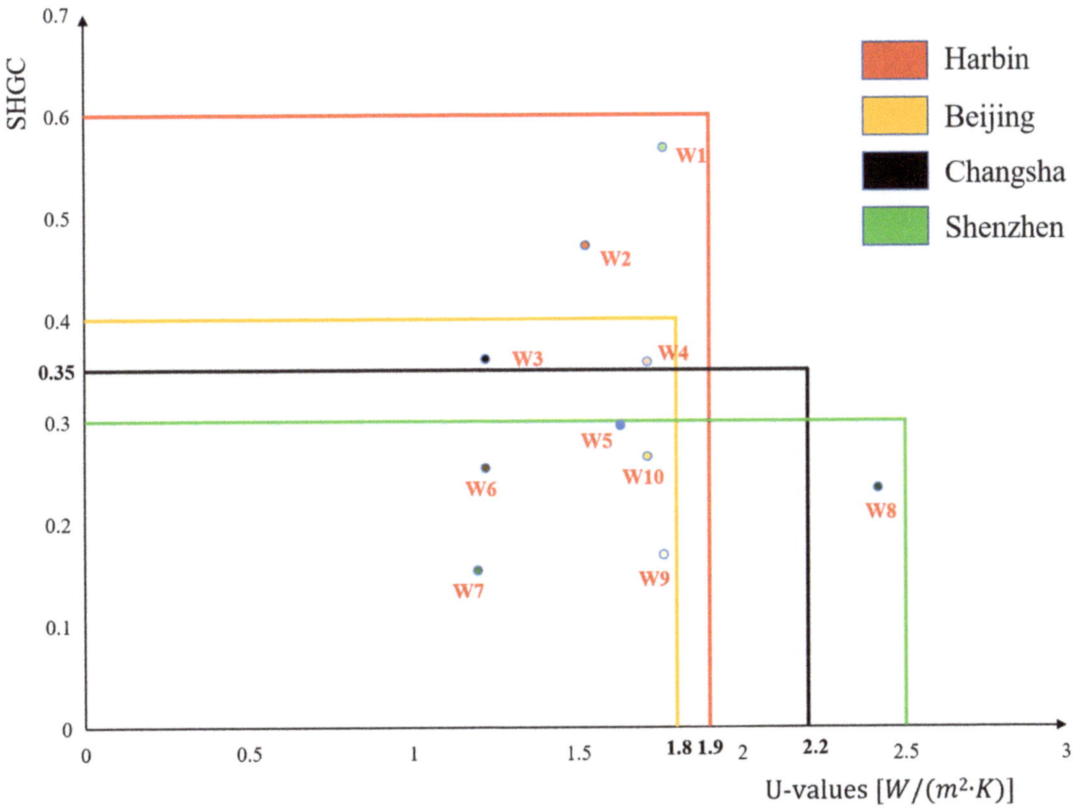


**Figure 3.** Relationship between U-values and SHGC-values of glazed windows.

This study did not consider window shading devices (such as roller shades or blinds) and their opening and closing schedules. Future research could further refine the impact of shading devices on energy consumption.

## 2.4. Cost and Energy Costs

### 2.4.1. Initial Total Investment of Window

In China, the price of windows includes glass and labor. The maintenance cost of windows during use is low, and the recycling efficiency of windows after scrapping is not high. Therefore, this paper does not consider the maintenance cost of windows during use and the recycling efficiency after scrapping. Through the investigation of 6 glass manufacturers, the unit area prices of ten types of windows in the four cities were obtained, as shown in Table 7. The window price in Harbin is 412–610 CNY/m$^2$, 527–680 CNY/m$^2$ in Beijing, 522–624 CNY/m$^2$, and 786–1103 CNY/m$^2$ in Shenzhen. The total area of windows in this building is 5929.56 square meters. According to the price data in Table 7, the total cost is the initial investment cost of the window retrofit.

**Table 7.** Price per unit area of windows.

| Name | Harbin (CNY/m$^2$) | Beijing (CNY/m$^2$) | Changsha (CNY/m$^2$) | Shenzhen (CNY/m$^2$) |
|---|---|---|---|---|
| W1 | 412 | - | - | - |
| W2 | 455 | - | - | - |
| W3 | 479 | 527 | - | - |
| W4 | - | 540 | - | - |
| W5 | - | 544 | 522 | - |
| W6 | 590 | - | 603 | 847 |
| W7 | - | 680 | - | 1103 |
| W8 | - | - | - | 906 |
| W9 | - | - | 624 | 1059 |
| W10 | - | - | 550 | - |

Note: "-" indicates no query.

### 2.4.2. Total Energy Cost

The research object of this paper is public buildings, the main energy sources of which are electricity and natural gas. Table 8 presents the industrial and commercial time-of-use electricity tariffs for the four cities, as obtained from the official websites of their respective provincial and municipal development and reform commissions. The four regions have different off-peak hours, mid-peak hours, on-peak hours, and critical peak hours.

**Table 8.** Electricity tariffs for commercial and industrial use in four cities.

| City | Critical Peak Period (CNY/kWh) | On Peak Period (CNY/kWh) | Mid-Peak Period (CNY/kWh) | Off-Peak Period (CNY/kWh) |
|---|---|---|---|---|
| Harbin | 1.3294 | 1.1181 | 0.7660 | 0.4139 |
| Beijing | 1.3029 | 1.1572 | 0.8333 | 0.5499 |
| Changsha | 1.4576 | 1.2223 | 0.7813 | 0.3403 |
| Shenzhen | 1.4259 | 1.1462 | 0.7615 | 0.2770 |

Note: data from China State Grid's latest electricity price in August 2024.

In northern China, natural gas is the main heating energy source; the heating season is from November 15 to March 15 each year. The calorific value of natural gas is about 35.53 MJ/m$^3$, and Table 9 shows the natural gas prices in the four cities queried on the official website of each city by China's Development and Reform Commission.

The total energy cost M is calculated by Equation (1):

$$C_{energy} = \sum_{i=1}^{8760} P_h \times Q_h \tag{1}$$

where

$P_h$ is the energy price corresponding to moment h;
$Q_h$ is the energy consumption corresponding to the moment h, calculated from the hour-by-hour electric energy consumption and hour-by-hour natural gas energy consumption generated by EnergyPlus.

**Table 9.** Industrial natural gas prices in four cities.

| City | Prices (CNY/m$^3$) |
|---|---|
| Harbin | 3.48 |
| Beijing | 2.45 |
| Changsha | 3.853 |
| Shenzhen | 4.3 |

In this study, the energy-saving potential of the baseline, the collected cases, and the retrofit case with the lowest energy cost were analyzed, after which the energy-saving rate could be obtained using Equation (2).

$$Energy - saveing\ rate = \left(1 - \frac{EC_y}{EC_y}\right) \times 100\% \quad (2)$$

where

$EC_y$ is the annual energy consumption cost for the retrofit cases;
$EC_y$ is the annual energy consumption cost for the baseline or collected cases.

### 2.5. Economic Indicators

There are many evaluation indexes of project economic benefit, such as payback period, return on investment, net present value of investment, etc. Each index reflects the economy of the project from an important angle. Whether the time value of money is considered in the evaluation index system of investment projects, the evaluation index of economic effect can be divided into static index and dynamic index. Static evaluation indicators include static investment payback period and investment return rate, and the calculation of indicator data does not consider the time value of funds. The calculation of dynamic evaluation indicators needs to consider the time value of funds, which are generally net present value, internal rate of return, investment present value rate, and so on. This study conducted a feasibility analysis in four regions using DPP, NPV, and PI as evaluation metrics to assess the economic viability of window replacement in buildings.

#### 2.5.1. Cumulative Discounted Cash Flow Payback Period (DPP)

After replacing the windows, the total operating cost of the building ($C_{total}$) includes the initial investment cost ($C_{initial}$) and the lifetime energy cost, calculated over a 20-year service life. For details, see Equations (3) and (4).

$$C_{initial} = P_w \times S_w \quad (3)$$

$$C_{total} = C_{initial} + C_{energy} \times T \quad (4)$$

where

Pw = price of ten different glass windows;
Sw = total window area of the building;
T = 20 years.

A simple payback period (PP) can serve as a basic index for evaluating a project. However, it does not account for the time value of money, may not accurately reflect the project's actual cash flow, and cannot truly represent the real financial situation. Therefore, the DPP, which is the time required to accumulate discounted cash flows to return the initial investment, is used. We used Formula (5) to calculate the annual cash inflow amount, calculated the discounted cash flow using Equation (6), and determined the investment return time using Equations (7) and (8); DPP occurs in the year when $V_t$ passes from negative to positive.

$$C_t = EC_y - EC_y \tag{5}$$

$$PV = \frac{C_t}{(1+i)^t} \tag{6}$$

when t = 0

$$V_0 = -C_{initial} \tag{7}$$

when t = 1, 2, 3 . . .T

$$V_t = V_0 + \frac{C_t}{(1+i)^t} \tag{8}$$

where

$V_t$ = unrecovered initial investment;
$C_t$ = net cash inflow during the period t, That is the energy cost that can be saved by replacing windows in each city;
i = discounted rate;
t = period in which cash inflows are related.

### 2.5.2. Net Present Value (NPV)

NPV is commonly used in projects to assess the benefits and costs of projects [49]. If the NPV of a project is positive, it indicates that the project can recover the investment. The greater the NPV, the earlier the investment cost recovery and the faster the return on investment [37,50]. NPV can be calculated by Equation (9).

$$NPV = \sum_{t=1}^{T} \frac{C_t}{(1+i)^t} - C_{initial} \tag{9}$$

where

$C_t$ = net cash inflow during the period t; t is the number of time periods;
i = discounted rate.

### 2.5.3. Profitability Index (PI)

The profitability index (PI) is a ratio of cash inflow's present value and cash outflow's present value from a project. PI measures the financial attractiveness of a project, with higher values corresponding to more attractive projects [51], and can be calculated using Equation (10).

$$PI = \frac{\sum_{t=1}^{T} \frac{C_t}{(1+i)^t}}{C_{initial}} \tag{10}$$

where

$C_t$ = net cash inflow during the period t;
t = the number of time periods.

## 3. Simulation Results

Considering the different climate zones in China, this paper established baseline energy models for large office buildings in four climate zones. Then, retrofit models were developed in each climate zone, considering five scenarios. Finally, this paper performed an economic analysis of five window replacement scenarios using three economic indicators.

### 3.1. Baseline Simulation Results

The energy consumption of the baseline and collected cases across four cities was simulated using the EnergyPlus software. It was difficult to quantitatively analyze the simulation results due to the lack of real data. The baseline was set based on the Chinese design standards, the Changsha prototype building model, and information on case windows collected from the literature. Governments require the Chinese design standards in any climate zone. The reasonableness of the prototype buildings was verified by comparing the total energy consumption represented by the prototype buildings with the local statistical yearbook [52]. The case window information was real data rather than hypotheses. Therefore, the results of the baseline were reasonable. The results for monthly electricity consumption are presented in Figure 4, while Figure 5 shows the monthly natural gas consumption. As observed in Figure 4, with the exception of Case 3 in Beijing, where monthly electricity consumption closely mirrors that of the baseline, the electricity consumption in all the other cases significantly exceeds the baseline. This indicates that public buildings constructed in compliance with newer standards consume less electricity, and replacing exterior window materials with more efficient options contributes to energy savings.

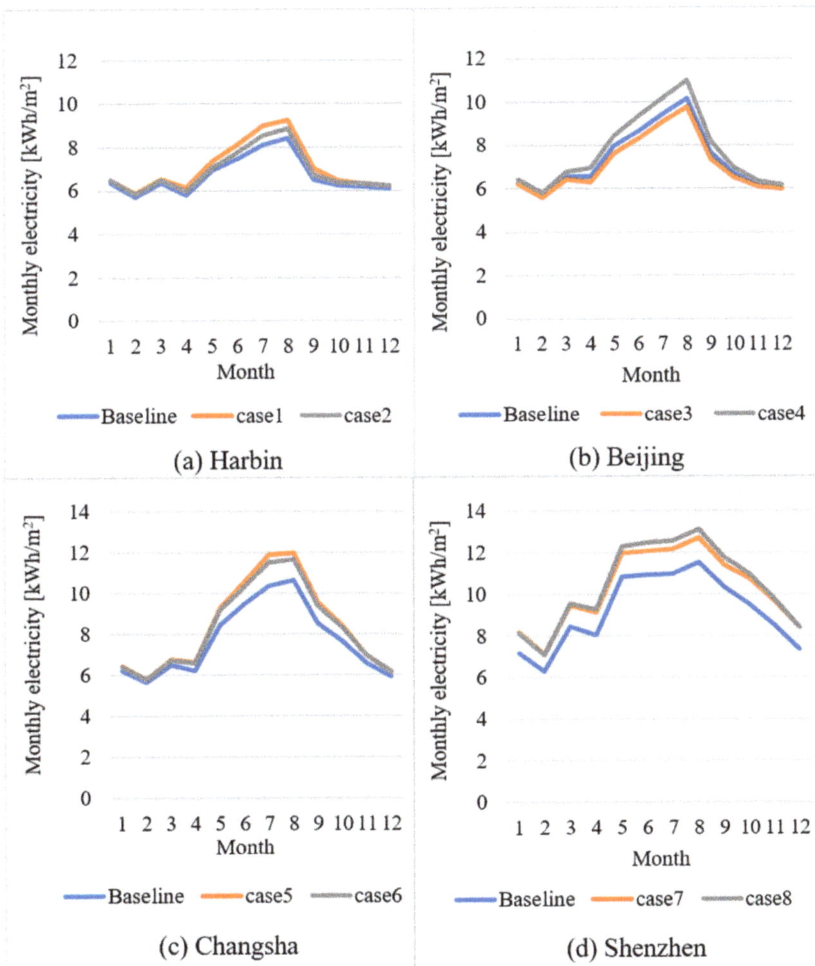


**Figure 4.** Monthly comparison of electricity consumption in four cities.

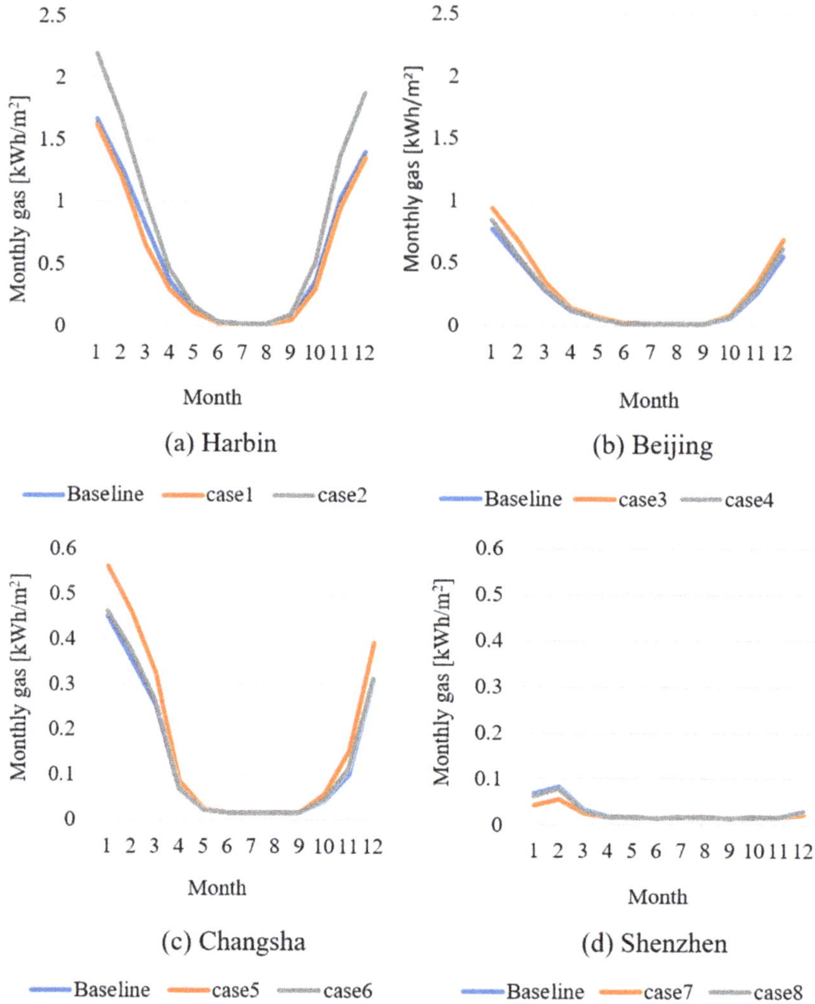


**Figure 5.** Monthly comparison of gas consumption in four cities.

In Figure 5, the monthly natural gas consumption of Harbin Case 1, Shenzhen Case 7, and Shenzhen Case 8 is lower than the baseline. Referring to Table 4, it is notable that the SHGC values for these cases are higher than those of the baseline and other cases within the same climate zone. We hypothesize that this is due to the specific climate conditions in Harbin, which is located in a severely cold zone with high heating demand in winter. A higher U-value generally leads to greater heat loss, but the increased SHGC in Case 1 may allow for higher solar heat gains during the day, partially offsetting the heat loss and reducing heating demand, particularly in winter. Similarly, Shenzhen, located in a subtropical climate zone where cooling is the primary energy demand, benefits from windows with higher SHGC values, as they increase solar heat gain and reduce the cooling load. Figure 6 illustrates the annual energy consumption across the four cities. With the exception of Case 3 in Beijing, where energy use is slightly lower than the baseline, the baseline scenario has the lowest total energy consumption.

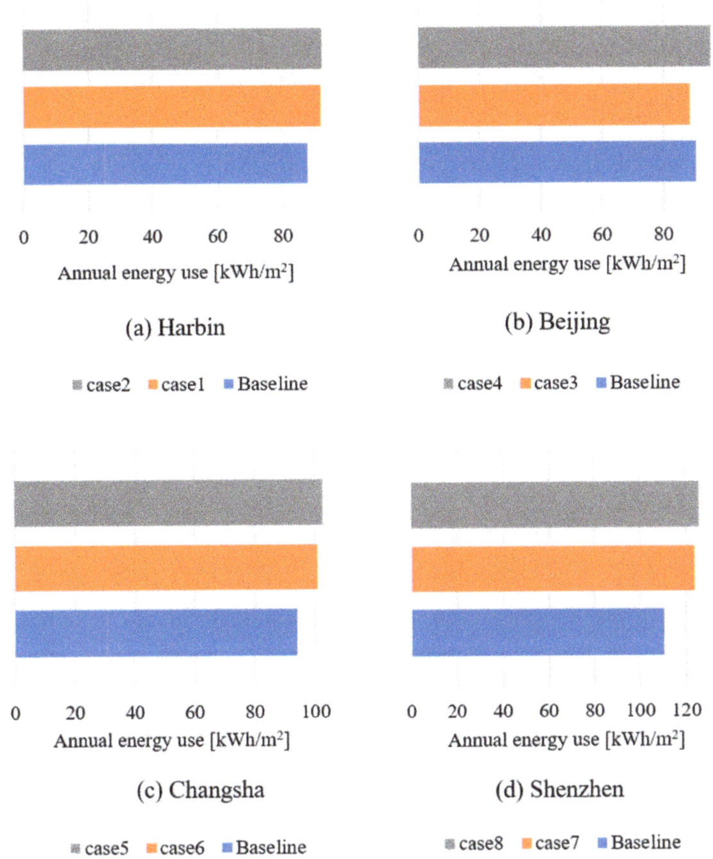


**Figure 6.** Annual energy use in four cities.

### *3.2. Retrofit Case Simulation Results*

The annual energy cost of each retrofit case was calculated based on the energy price and electricity price structure (see Tables 8 and 9, for specific data). The results are summarized in Table 10, which lists the annual energy costs of the retrofit options in the four cities. The window type with the lowest annual energy cost was selected for the four different regions from Table 7 for the next analysis.

**Table 10.** Cost of energy consumption per square meter per year.

| Name | Harbin (RMB/m$^2$) | Beijing (RMB/m$^2$) | Changsha (RMB/m$^2$) | Shenzhen (RMB/m$^2$) |
|---|---|---|---|---|
| W1 | 100.14 | - | - | - |
| W2 | 98.70 | - | - | - |
| W3 | 97.38 | 87.80 | - | - |
| W4 | - | 88.06 | - | - |
| W5 | - | 88.42 | 87.57 | - |
| W6 | 99.89 | - | 87.25 | 103.81 |
| W7 | - | 87.55 | - | 95.07 |
| W8 | - | - | - | 102.80 |
| W9 | - | - | 87.02 | 101.36 |
| W10 | - | - | 89.99 | - |

Note: "-" indicates no simulation.

After simulating the retrofit model with replaced windows, the energy-saving effect was evaluated and compared with the baseline and collection cases, and the energy-saving potential was analyzed. The results shown in Figure 7 show that retrofitting exterior windows can produce significant energy-saving effects in both the baseline and collection cases. In Changsha (HC) and Shenzhen (HW), replacing exterior windows with high-performance materials can significantly save energy, with Changsha (HC) achieving a reduction of 10.95% and Shenzhen (HW) achieving a higher reduction of 20.01%.

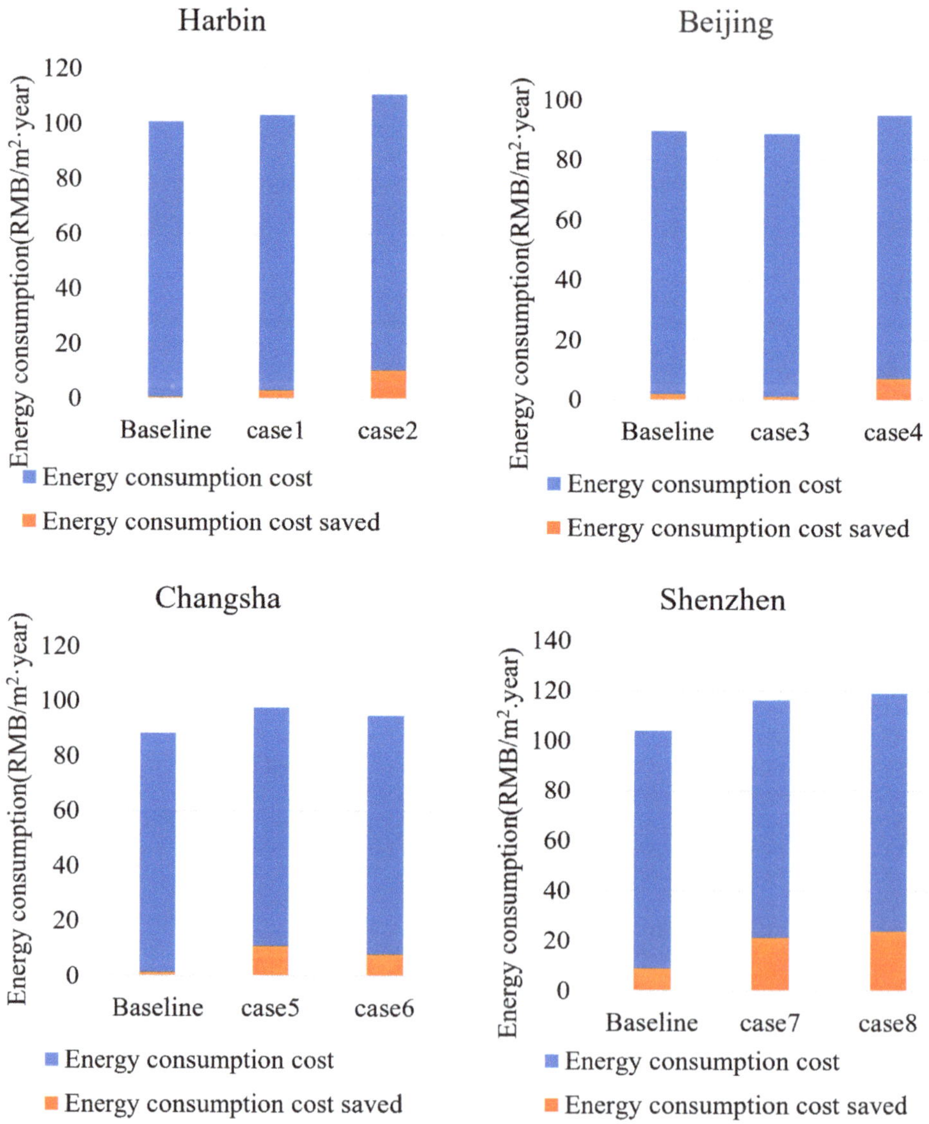


**Figure 7.** Comparative analysis of energy-saving effects.

### 3.3. Economic Feasibility Analysis

In previous studies [33,53], the service life of windows was often set at 20 years, and related research, such as window retrofit, was carried out based on this. So, the window life cycle is assumed to be 20 years in this study. The economic viability of optimal window selection is assessed using three key financial metrics: the DPP, NPV, and PI. Table 7 presents the retrofit costs for various window options, excluding the minimum maintenance costs and the residual value at disposal.

Using Equations (3) and (4) and the data provided in Tables 3 and 4, the total cost of the renovated building over the 20-year service life of the windows was calculated for different window options across the four cities. The results, summarized in Table 11, identify the most cost-effective window selections for each city: Harbin (W1), Beijing (W3), Changsha (W5), and Shenzhen (W6).

**Table 11.** Total 20-year cost for different windows in four cities.

| Harbin | Total Cost (CNY/m$^2$) | Beijing | Total Cost (CNY/m$^2$) | Changsha | Total Cost (CNY/m$^2$) | Shenzhen | Total Cost (CNY/m$^2$) |
|---|---|---|---|---|---|---|---|
| W1 | 2414.73 | W3 | 2283.02 | W5 | 2273.47 | W6 | 2923.1 |
| W2 | 2429.09 | W4 | 2301.12 | W6 | 2348.08 | W7 | 3004.37 |
| W3 | 2426.69 | W5 | 2312.35 | W9 | 2364.31 | W8 | 2961.97 |
| W6 | 2587.9 | W7 | 2430.94 | W10 | 2349.74 | W9 | 3086.23 |

This study adopts a discount rate of 3.576% per year based on the data published by the International Monetary Fund on 25 September 2024. This discount rate ensures that the economic analysis reflects the current financial conditions in China. The results of the DPP were calculated using Equations (5)–(7), with detailed values presented in Table 12. Similarly, the NPV of the optimal window selection was determined using Equation (8) and is summarized in Table 13. For the baseline scenario, the DPP for window replacements exceeds the service life of the windows, resulting in negative NPV values. In the case models, the results vary across different cities and building types. In Harbin and Beijing, for buildings constructed before 2015, while the replacement of high-performance windows offers energy-saving potential, the high initial costs lead to unfavorable economic outcomes. Specifically, in Cases 1 and 3, the DPP exceeds the service life of the windows, and the NPV remains negative. In Changsha and Shenzhen, high-performance window replacements in pre-2015 buildings demonstrate both energy-saving and economic feasibility. The payback periods are within the service life of the windows, with values of 10.16 years (Case 5), 15.66 years (Case 6), 14.14 years (Case 7), and 11.42 years (Case 8), respectively. These results highlight the greater economic viability of window retrofits in warmer climates with higher cooling energy demands.

**Table 12.** The DPP of the optimal window selections in the four different cities.

| City | Selected Window | Retrofitting Buildings | DPP (Year) |
|---|---|---|---|
| Harbin | W1 | Baseline | >20 |
| | | Case1 | >20 |
| | | Case2 | 6.60 |
| Beijing | W3 | Baseline | >20 |
| | | Case3 | >20 |
| | | Case4 | 15.66 |
| Changsha | W5 | Baseline | >20 |
| | | Case5 | 10.16 |
| | | Case6 | 15.66 |
| Shenzhen | W6 | Baseline | >20 |
| | | Case7 | 14.14 |
| | | Case8 | 11.42 |

**Table 13.** The NPV of the optimal window selections in the four different cities.

| City | Selected Window | Retrofitting Buildings | Returns Within the Service Life (RMB) | Initial Investment (RMB) | NPV (RMB) |
|---|---|---|---|---|---|
| Harbin | W1 | Baseline | $38.60 \times 10^4$ | $244.30 \times 10^4$ | $-205.70 \times 10^4$ |
| | | Case1 | $182.10 \times 10^4$ | | $-62.20 \times 10^4$ |
| | | Case2 | $64.97 \times 10^4$ | | $405.34 \times 10^4$ |
| Beijing | W3 | Baseline | $114.50 \times 10^4$ | $312.49 \times 10^4$ | $-197.98 \times 10^4$ |
| | | Case3 | $58.30 \times 10^4$ | | $-254.19 \times 10^4$ |
| | | Case4 | $439.85 \times 10^4$ | | $127.36 \times 10^4$ |
| Changsha | W5 | Baseline | $44.11 \times 10^4$ | $309.52 \times 10^4$ | $-265.41 \times 10^4$ |
| | | Case5 | $620.91 \times 10^4$ | | $311.38 \times 10^4$ |
| | | Case6 | $443.41 \times 10^4$ | | $133.89 \times 10^4$ |
| Shenzhen | W6 | Baseline | $72.07 \times 10^3$ | $502.23 \times 10^4$ | $-495.03 \times 10^4$ |
| | | Case7 | $762.71 \times 10^4$ | | $260.48 \times 10^4$ |
| | | Case8 | $921.16 \times 10^4$ | | $418.93 \times 10^4$ |

According to the results of the NPV, the retrofitting buildings with positive values were screened out and further analyzed through the PI. The PI is calculated using Equation (9), as shown in Table 14. From the results in Table 14, it can be seen that the PI values of the retrofit models screened by the NPV are all more than one, indicating that the replacement of high-performance windows has good profitability.

**Table 14.** The PI of the optimal window selections in the four different cities.

| City | Window | Retrofitting Buildings | PI |
|---|---|---|---|
| Harbin | W1 | Case2 | 2.66 |
| Beijing | W3 | Case4 | 1.41 |
| Changsha | W5 | Case5 | 2.01 |
| | | Case6 | 1.43 |
| Shenzhen | W6 | Case7 | 1.52 |
| | | Case8 | 1.83 |

Based on the evaluation of the above three economic indicators, Harbin (W1), Beijing (W3), Changsha (W5), and Shenzhen (W6) have energy-saving potential and are economically feasible in the energy-saving retrofit of buildings constructed before 2015 in the four cities.

## 4. Discussion

This study evaluates the energy efficiency and economic feasibility of replacing exterior windows with high-performance materials in public buildings in four cities. The analysis results indicate that window characteristics are a key determinant of the effectiveness of such energy efficiency retrofits and vary significantly between climate zones, which is consistent with previous studies of the same type [54]. The key findings, window selection recommendations, and future research directions are highlighted below.

(1) Window characteristics and retrofit effectiveness

Buildings constructed before implementing modern energy standards (e.g., GB50189-2015) typically realize more significant energy savings and economic benefits when retrofitted with high-performance windows. This trend is particularly evident in Changsha and Shenzhen, where older buildings with lower base thermal performance realized significant energy consumption reductions, resulting in shorter payback periods and NPV. The longer the time lag between building construction and implementing the latest standards, the lower the performance of their original windows, and the more significant the energy savings resulting from the retrofit.

In contrast, the existing windows with lower U-values and SHGC values had lower economic returns, even though some energy savings can be realized. For example, in Harbin and Beijing, the high cost of replacing otherwise relatively efficient windows resulted in payback periods that exceeded the useful life of the windows, leading to negative net present values. In such cases, alternative strategies such as improved wall or roof insulation may be more effective.

(2) Window selection strategies for climate zones

This study's important contribution is providing actionable guidelines for exterior window selection for different window characteristics and climate zones. For older buildings in hot summer climate zones (e.g., Changsha and Shenzhen), windows with high SHGC can significantly improve cooling energy efficiency, consistent with the performance characteristics of the Phoenix data. In contrast, in cold climate zones (e.g., Harbin), windows with low heat transfer coefficients (U-values) (similar to the Seoul data) are more effective in reducing heating energy consumption. For new construction, window selection should be carefully weighed against alternative retrofit strategies to ensure affordability.

This recommendation for window selection by climate zone and window characteristics provides additional refinement to the existing energy efficiency retrofit studies and directly aligns with the goal of this study to optimize exterior window retrofit strategies.

(3) Policy implications

The findings emphasize the importance of developing targeted retrofit programs and climate-responsive policies. In regions with a large share of older buildings, such as Changsha and Shenzhen, energy efficiency retrofits can be promoted through phased incentives (e.g., window replacement subsidies or loan discounts) to maximize energy and economic benefits. In regions with predominantly new buildings, promoting other retrofit measures, such as insulation improvements, may be more appropriate. Policymakers should also consider window retrofit guidelines for different climate zones to optimize energy savings and affordability.

(4) Limitations and future research directions

There are some limitations to this study. For example, the effects of other envelope factors, such as wall or roof insulation, were not considered, and these factors may work together with window retrofits to affect the overall energy savings. In addition, although international data were used to extend the performance characteristics, the main analysis was still dominated by the Chinese context, and the international applicability of the study's findings needs to be further explored. Future research could delve deeper into the synergistic effects of integrated retrofit strategies for windows, walls, and roofs and the retrofit effects under different global standards.

Internal shading devices (e.g., curtains and shades) and external shading devices (e.g., roller shades and blinds) play an important role in building energy performance. According to Raimundo et al. [55], these shading devices are effective in reducing solar heat gain in the summer, thus reducing cooling demand and assisting heating in the winter by optimizing solar gain. In addition, the study by Steen Englund et al. [56] verified the significant impact of shading device use and opening/closing strategies on building energy consumption through actual measurements and simulations, especially in buildings with high variations in occupancy behavior. These studies show that the proper configuration and operational management of shading devices can significantly improve energy efficiency in different climatic conditions and building types. Therefore, future research should further focus on shading device design optimization and intelligent control strategies to maximize building energy efficiency.

## 5. Conclusions

This study investigated the energy efficiency and economic feasibility of high-performance window replacement in different climate zones and building types. The results highlight the importance of tailoring retrofit strategies to a building's specific climate, economy, and built climate, economy, and environment. The following are the main conclusions drawn from the analysis:

(1) Economic evaluation beyond U-values and SHGC values

The economic performance of window replacement is not only affected by the thermal transfer coefficient (U-value) and SHGC, although these indicators are critical. Rather than pursuing the minimum U-value and SHGC values, the study identified a balance that both meets energy conservation standards and optimizes economic outcomes. Moderate U-values and SHGC values are generally more cost-effective, especially in the climate zones studied.

(2) Economic feasibility in Changsha and Shenzhen

High-performance windows are both energy-efficient and economically viable in Changsha and Shenzhen. The discounted payback period in these cities is within the useful life of the window, and the NPV is positive. These results show that window retrofits can achieve dual goals under specific climate conditions and building characteristics: improving sustainability and achieving a return on investment.

(3) Policy recommendations for older buildings

Retrofit programs targeting older buildings have the greatest potential to achieve significant energy savings and economic returns. In areas such as Changsha and Shenzhen, where much of the building stock predates modern standards, policymakers could prioritize retrofitting these buildings. Incentive mechanisms, such as subsidies or financial support for window replacement, could encourage building owners to adopt energy-efficient upgrades, helping to align personal economic interests with broader sustainable development goals.

**Author Contributions:** Conceptualization, S.L. and Y.C.; Methodology, S.L. and Y.C.; Software, S.L., Z.G. and Y.C.; Validation, S.L., F.E.M.G., J.Y. and Y.C.; Formal analysis, S.L., J.Y. and Y.C.; Investigation, S.L., J.Y. and K.Z.; Resources, S.L. and K.Z.; Data curation, S.L., Z.G. and Y.C.; Writing—original draft, S.L.; Writing—review & editing, S.L. and Y.C.; Visualization, S.L., Z.G. and K.Z.; Supervision, F.E.M.G. and Y.C.; Project administration, S.L., F.E.M.G. and Y.C.; Funding acquisition, S.L. All authors have read and agreed to the published version of the manuscript.

**Funding:** This research was supported by "A Project Supported by Scientific Research Fund of Hunan Provincial Education Department No. 23C0409".

**Data Availability Statement:** Data are contained within the article.

**Acknowledgments:** During the preparation of this work, the authors used Grammarly and ChatGPT-4o to improve readability and detect spelling/grammar mistakes. After using this tool/service, the authors reviewed and edited the content as needed, and take full responsibility for the content of the publication.

**Conflicts of Interest:** The authors declare no conflict of interest.

## References

1. Tong, D.; Zhang, Q.; Zheng, Y.; Caldeira, K.; Shearer, C.; Hong, C.; Qin, Y.; Davis, S.J. Committed Emissions from Existing Energy Infrastructure Jeopardize 1.5 °C Climate Target. *Nature* **2019**, *572*, 373–377. [CrossRef]
2. Ahmed, A.; Ge, T.; Peng, J.; Yan, W.C.; Tee, B.T.; You, S. Assessment of the Renewable Energy Generation towards Net-Zero Energy Buildings: A Review. *Energy Build.* **2022**, *256*, 111755. [CrossRef]
3. Zhou, Y.; Fan, F.; Liu, Y.; Zhao, S.; Xu, Q.; Wang, S.; Luo, D.; Long, Y. Unconventional Smart Windows: Materials, Structures and Designs. *Nano Energy* **2021**, *90*, 106613. [CrossRef]
4. Gaetani, I.; Hoes, P.J.; Hensen, J.L.M. Occupant Behavior in Building Energy Simulation: Towards a Fit-for-Purpose Modeling Strategy. *Energy Build.* **2016**, *121*, 188–204. [CrossRef]
5. Uddin, M.M.; Ji, J.; Wang, C.; Zhang, C. Building Energy Conservation Potentials of Semi-Transparent CdTe Integrated Photovoltaic Window Systems in Bangladesh Context. *Renew. Energy* **2023**, *207*, 512–530. [CrossRef]
6. Yin, H.; Zhou, X.; Zhou, Z.; Liu, R.; Mo, X.; Chen, Z.; Yang, E.; Huang, Z.; Li, H.; Wu, H.; et al. Switchable Kirigami Structures as Window Envelopes for Energy-Efficient Buildings. *Research* **2023**, *6*, 0103. [CrossRef]
7. Altun, A.F. Determination of Optimum Building Envelope Parameters of a Room Concerning Window-to-Wall Ratio, Orientation, Insulation Thickness and Window Type. *Buildings* **2022**, *12*, 383. [CrossRef]

8. Kirankumar, G.; Saboor, S.; Vali, S.S.; Mahapatra, D.; Talanki Puttaranga Setty, A.B.; Kim, K.H. Thermal and Cost Analysis of Various Air Filled Double Glazed Reflective Windows for Energy Efficient Buildings. *J. Build. Eng.* **2020**, *28*, 101055. [CrossRef]
9. RollCar, M.K.; Anderson, A.M.; Mangu, S.T.; Hajjaj, Z.; Capron, M. Aesthetic Aerogel Window Design for Sustainable Buildings. *Sustainability* **2022**, *14*, 2887. [CrossRef]
10. Nundy, S.; Mesloub, A.; Alsolami, B.M.; Ghosh, A. Electrically Actuated Visible and Near-Infrared Regulating Switchable Smart Window for Energy Positive Building: A Review. *J. Clean. Prod.* **2021**, *301*, 126854. [CrossRef]
11. Zhang, S.; Hu, W.; Li, D.; Zhang, C.; Ar, M. Energy Ef Fi Ciency Optimization of PCM and Aerogel-Fi Lled Multiple Glazing Windows. *Energy* **2021**, *222*, 119916. [CrossRef]
12. Fu, J.; Feng, C.; Liao, Y.; Mao, M.; Liu, H.; Liu, K. Broadband Light Management in Hydrogel Glass for Energy Efficient Windows. *Front. Optoelectron.* **2022**, *15*, 33. [CrossRef] [PubMed]
13. Shi, S.; Zhu, N.; Wu, S.; Song, Y. Evaluation and Analysis of Transmitted Daylight Color Quality for Different Colored Semi-Transparent PV Glazing. *Renew. Energy* **2024**, *222*, 119826. [CrossRef]
14. Abraham, E.; Cherpak, V.; Senyuk, B.; ten Hove, J.B.; Lee, T.; Liu, Q.; Smalyukh, I.I. Highly Transparent Silanized Cellulose Aerogels for Boosting Energy Efficiency of Glazing in Buildings. *Nat. Energy* **2023**, *8*, 381–396. [CrossRef]
15. Mendes, V.F.; Cruz, A.S.; Gomes, A.P.; Mendes, J.C. A Systematic Review of Methods for Evaluating the Thermal Performance of Buildings through Energy Simulations. *Renew. Sustain. Energy Rev.* **2024**, *189*, 113875. [CrossRef]
16. Chen, X.; Qiu, Y.; Wang, X. A Systematic Review of Research Methods and Economic Feasibility of Photovoltaic Integrated Shading Device. *Energy Build.* **2024**, *311*, 114172. [CrossRef]
17. Farzaneh, S.; Motlagh, M.; Sohani, A.; Saghafi, M.D.; Sayyaadi, H.; Nastasi, B. The Road to Developing Economically Feasible Plans for Green, Comfortable and Energy Efficient Buildings. *Energies* **2021**, *14*, 636. [CrossRef]
18. Hussain, S.N.; Ghosh, A. Highly Insulated Hydrogen Powered Vacuum Gasochromic Smart Windows for Energy-Efficient Buildings. *J. Build. Eng.* **2024**, *97*, 110855. [CrossRef]
19. Gao, Y.; Zheng, Q.; Jonsson, J.C.; Lubner, S.; Curcija, C.; Fernandes, L.; Kaur, S.; Kohler, C. Parametric Study of Solid-Solid Translucent Phase Change Materials in Building Windows. *Appl. Energy* **2021**, *301*, 117467. [CrossRef]
20. Zhang, Y.; Tennakoon, T.; Chan, Y.H.; Chan, K.C.; Fu, S.C.; Tso, C.Y.; Yu, K.M.; Huang, B.L.; Yao, S.H.; Qiu, H.H.; et al. Energy Consumption Modelling of a Passive Hybrid System for Office Buildings in Different Climates. *Energy* **2022**, *239*, 121914. [CrossRef]
21. Ma, Y.; Li, D.; Yang, R.; Zhang, S.; Arıcı, M.; Liu, C.; Zhang, C. Energy and Daylighting Performance of a Building Containing an Innovative Glazing Window with Solid-Solid Phase Change Material and Silica Aerogel Integration. *Energy Convers. Manag.* **2022**, *271*, 116341. [CrossRef]
22. Nicoletti, F.; Kaliakatsos, D.; Ferraro, V.; Cucumo, M.A. Analysis of the Energy and Visual Performance of a Building with Photochromic Windows for a Location in Southern Italy. *Build. Environ.* **2022**, *224*, 109570. [CrossRef]
23. Ali, H.; Hashlamun, R. Envelope Retrofitting Strategies for Public School Buildings in Jordan. *J. Build. Eng.* **2019**, *25*, 100819. [CrossRef]
24. Chiradeja, P.; Thongsuk, S.; Ananwattanaporn, S.; Ngaopitakkul, A.; Yoomak, S. A Study on Transparent Type Envelope Material in Terms of Overall Thermal Transfer, Energy, and Economy for an Office Building Based on the Thai Building Energy Code. *Sustainability* **2023**, *15*, 10435. [CrossRef]
25. Liu, J.; Yang, R.; Zhang, J.; Tao, Q.; Li, A.; Liu, Z.; Su, Y.; Liu, Y. Dual-Function Smart Windows with Dynamic and Fast Thermal Response for Building Energy-Saving/Storage. *Sol. Energy Mater. Sol. Cells* **2023**, *249*, 112048. [CrossRef]
26. Heydari, A.; Sadati, S.E.; Gharib, M.R. Effects of Different Window Configurations on Energy Consumption in Building: Optimization and Economic Analysis. *J. Build. Eng.* **2021**, *35*, 102099. [CrossRef]
27. Kowalczyk, Z.; Tomasik, M. Economic and Energy Analysis of the Operation of Windows in Residential Buildings in Poland. *Energies* **2023**, *16*, 6810. [CrossRef]
28. Moghaddam, S.A.; Mattsson, M.; Ameen, A.; Akander, J.; Da Gameiro Silva, M.; Simões, N. Low-Emissivity Window Films as an Energy Retrofit Option for a Historical Stone Building in Cold Climate. *Energies* **2021**, *14*, 7584. [CrossRef]
29. Foroughi, R.; Asadi, S.; Khazaeli, S. On the Optimization of Energy Efficient Fenestration for Small Commercial Buildings in the United States. *J. Clean. Prod.* **2021**, *283*, 124604. [CrossRef]
30. *GB 50352-2019*; Uniform Standard for Design of Civil Buildings. China Architecture & Building Press: Beijing, China, 2019.
31. Deng, Z.; Chen, Y.; Yang, J.; Causone, F. AutoBPS: A Tool for Urban Building Energy Modeling to Support Energy Efficiency Improvement at City-Scale. *Energy Build.* **2023**, *282*, 112794. [CrossRef]
32. Yang, J.; Zhang, Q.; Peng, C.; Chen, Y. AutoBPS-Prototype: A Web-Based Toolkit to Automatically Generate Prototype Building Energy Models with Customizable Efficiency Values in China. *Energy Build.* **2024**, *305*, 113880. [CrossRef]
33. Zhao, W.; Deng, Z.; Ji, Y.; Song, C.; Yuan, Y.; Wang, Z.; Chen, Y. Analysis of Peak Demand Reduction and Energy Saving in a Mixed-Use Community through Urban Building Energy Modeling. *Energies* **2024**, *17*, 1214. [CrossRef]
34. Song, C.; Deng, Z.; Zhao, W.; Yuan, Y.; Liu, M.; Xu, S.; Chen, Y. Developing Urban Building Energy Models for Shanghai City with Multi-Source Open Data. *Sustain. Cities Soc.* **2024**, *106*, 105425. [CrossRef]
35. Ji, Y.; Li, G.; Su, F.; Chen, Y.; Zhang, R. Retrofit Analysis of City-Scale Residential Buildings in the Hot Summer and Cold Winter Climate Zone. *Energies* **2023**, *16*, 6152. [CrossRef]

36. Peng, C.; Chen, Z.; Yang, J.; Liu, Z.; Yan, D.; Chen, Y. Assessment of Electricity Consumption Reduction Potential for City-Scale Buildings under Different Demand Response Strategies. *Energy Build.* **2023**, *297*, 113473. [CrossRef]
37. Ren, Z.; Chen, Y.; Song, C.; Liu, M.; Xu, A.; Zhang, Q. Economic Analysis of Rooftop Photovoltaics System under Different Shadowing Conditions for 20 Cities in China. In *Building Simulation*; Tsinghua University Press: Beijing, China, 2023. [CrossRef]
38. Su, F.; Wang, Z.; Yuan, Y.; Song, C.; Zeng, K.; Chen, Y.; Zhang, R. Enhanced Operation of Ice Storage System for Peak Load Management in Shopping Malls across Diverse Climate Zones. *Sustainability* **2023**, *15*, 14759. [CrossRef]
39. Yang, L.; Lam, J.C.; Tsang, C.L. Energy Performance of Building Envelopes in Different Climate Zones in China. *Appl. Energy* **2008**, *85*, 800–817. [CrossRef]
40. *GB50176-2016*; Code for Thermal Design of Civil Building. China Architecture & Building Press: Beijing, China, 2016.
41. *ANSI/ASHRAE 169-2021*; Climatic Data for Building Design Standards. American Society of Heating, Refrigerating and Air-Conditioning Engineers, Inc.: Washington, DC, USA, 2021.
42. *GB 50189–2015*; Ministry of Housing and Urban-Rural Development, Design Standard for Energy Efficiency of Public Buildings. China Architecture & Building Press: Beijing, China, 2015.
43. Huang, J.; Lv, H.; Gao, T.; Feng, W.; Chen, Y.; Zhou, T. Thermal Properties Optimization of Envelope in Energy-Saving Renovation of Existing Public Buildings. *Energy Build.* **2014**, *75*, 504–510. [CrossRef]
44. Hee Chung, M.; Eun Kim, S.; Woo Song, Y.; Chul Park, J. Strategies for Improving Impact of Energy Renovation: A Case Study on Korean Daycare Centers. *Energy Build.* **2023**, *284*, 112844. [CrossRef]
45. Wang, P.; Gong, G.; Wang, Y.; Li, L. Thermodynamic Investigation of Building Integrated Energy Efficiency for Building Retrofit. *Energy Build.* **2014**, *77*, 139–148. [CrossRef]
46. Xiang, Q.C.; Feng, X.P.; Jia, X.Y.; Cai, L.; Chen, R. Reducing Carbon Dioxide Emissions through Energy-Saving Renovation of Existing Buildings. *Aerosol Air Qual. Res.* **2019**, *19*, 2732–2745. [CrossRef]
47. Aksamija, A. Impact of Retrofitting Energy-Efficient Design Strategies on Energy Use of Existing Commercial Buildings: Comparative Study of Low-Impact and Deep Retrofit Strategies. *J. Green Build.* **2017**, *12*, 70–88. [CrossRef]
48. *GB55015-2021*; General Code for Energy Efficiency and Renewable Energy Application in Buildings. China Architecture & Building Press: Beijing, China, 2021.
49. Knoke, T.; Gosling, E.; Paul, C. Use and Misuse of the Net Present Value in Environmental Studies. *Ecol. Econ.* **2020**, *174*, 106664. [CrossRef]
50. Chiradeja, P.; Ngaopitakkul, A. Energy and Economic Analysis of Tropical Building Envelope Material in Compliance with Thailand's Building Energy Code. *Sustainability* **2019**, *11*, 6872. [CrossRef]
51. Ahsan Kabir, M.; Hasan, M.M.; Hossain, T.; Ahnaf, A.; Monir, H. Sustainable Energy Transition in Bangladeshi Academic Buildings: A Techno-Economic Analysis of Photovoltaic-Based Net Zero Energy Systems. *Energy Build.* **2024**, *312*, 114205. [CrossRef]
52. Deng, Z.; Chen, Y.; Yang, J.; Chen, Z. Archetype Identification and Urban Building Energy Modeling for City-Scale Buildings Based on GIS Datasets. *Build. Simul.* **2022**, *15*, 1547–1559. [CrossRef]
53. Chen, D.; He, Z.; Liu, Y.; Tao, C.; Wang, H.; Lu, B.; Li, J.; Lu, Y. Influence of Transparent Envelope Parameters on Office Building Energy Use Intensity in the Different Thermal Climate Regions of China. *Processes* **2023**, *11*, 2908. [CrossRef]
54. Chen, S.; Guan, J.; Levine, M.D.; Xie, L.; Yowargana, P. Elaboration of Energy Saving Renovation Measures for Urban Existing Residential Buildings in North China Based on Simulation and Site Investigations. *Build. Simul.* **2013**, *6*, 113–125. [CrossRef]
55. Raimundo, A.M.; Saraiva, N.B.; Dias Pereira, L.; Rebelo, A.C. Market-Oriented Cost-Effectiveness and Energy Analysis of Windows in Portugal. *Energies* **2021**, *14*, 3720. [CrossRef]
56. Englund, J.S.; Cehlin, M.; Akander, J.; Moshfegh, B. Measured and Simulated Energy Use in a Secondary School Building in Sweden-a Case Study of Validation, Airing, and Occupancy Behaviour. *Energies* **2020**, *13*, 2325. [CrossRef]

buildings

Article

# Is the Energy Transition of Housing Financially Viable? Unlocking the Potential of Deep Retrofits with New Business Models

**Ezio Micelli [1,*], Giulia Giliberto [2] and Eleonora Righetto [3]**

1 Department of Architecture and Art, Università Iuav di Venezia, 30135 Venice, Italy
2 Department of Civil, Environmental Engineering and Architecture, University of Cagliari, 09123 Cagliari, Italy; giulia.giliberto@unica.it
3 Department of Management Engineering and Real Estate Economics, University of Padua, 36100 Vicenza, Italy; eleonora.righetto.2@phd.unipd.it
* Correspondence: micelli@iuav.it

**Abstract:** The transition to energy-efficient buildings is a priority of the European EPBD (Energy Performance Building Directive) and requires deep retrofits to reduce consumption and emissions. However, their financial viability remains underexplored. This research assesses the financial feasibility of deep retrofit interventions through innovative business models, focusing on the Managed Energy Services Agreement (MESA), which is considered the most effective for residential buildings. Additionally, we integrate off-site production from the Energiesprong model, which optimizes costs and time through long-term contracts and industrialized retrofit technologies. The analysis targets two investment profiles—owner/tenant and developer/entrepreneur—in Italian urban contexts with different market dynamics. A static analysis evaluates retrofits based on existing costs and technologies, while a dynamic analysis considers future profitability improvements because of cost reductions enabled by off-site production. The results indicate that, under current conditions, residential retrofitting is not financially sustainable without public subsidies. However, cost reductions driven by off-site technologies improve profitability, making large-scale retrofits feasible. Moreover, real estate market characteristics affect financial sustainability: in smaller cities, deeper cost reductions are necessary for retrofit interventions to become viable.

**Keywords:** business models; energy efficiency retrofit; green premium price; Energiesprong; building energy efficiency; value proposition; off-site construction; learning function

## 1. Introduction

Climate change is an unavoidable issue because of its impact on ecological and social systems [1,2]. Globally, the building stock contributes 38% of $CO_2$ emissions and 36% of greenhouse gas emissions, and the expected growth in urbanization will exacerbate these figures [3].

The European Union stands out as a regulatory leader in the energy transition. Since the Paris Agreement, reducing greenhouse gas emissions has been the basis of major public policies [4]. However, to date, 75% of buildings in Europe are still energy inefficient [5].

The European Union has introduced specific directives over the years to reduce greenhouse gas emissions. The most recent of these, the Energy Performance of Buildings

Directive, approved on 15 April 2024, sets out measures to improve the energy performance of buildings, with the final goal of decarbonizing the building stock by 2050 [6].

In addition to aligning with global climate targets, the directives represent an opportunity for the development of green buildings with environmental and economic benefits [7,8]. Residential buildings with high energy performance contribute substantially to optimizing the consumption of energy resources and reducing greenhouse gas emissions [9,10]. Reduced energy consumption, in turn, translates into lower costs for households and businesses, a factor that the real estate market reflects as an increase in property value [11–13].

However, the transition to energy-efficient buildings requires significant investments. Under current market conditions, some research argues that the costs of retrofitting existing building stock exceed the market benefits, making financial support from the public sector essential [14–16].

On the other hand, numerous studies have highlighted the crucial role of business models (BM) in promoting and disseminating technological innovations [17–19] and energy transitions [20,21], which can effectively integrate economies of scale and learning in production processes [22,23].

Among the various BMs studied in the literature, the Managed Energy Service Agreement (MESA) model is particularly effective. Its distinguishing features include the guarantee of energy performance of retrofit measures and the assumption of responsibility for payment of the energy bill through an energy supply contract, thus managing the entire energy provision and procurement process for the customer [24].

Among the variants of the Managed Energy Service Agreement (MESA) business model, the Energiesprong proposal emerges as a particularly efficient solution, as it guarantees net-zero energy performance and industrialized production processes capable of increasing the long-stagnant productivity of the construction industry [25,26].

The research aims to assess whether deep energy retrofit interventions in the residential sector, through the adoption of the MESA business model and enhanced by off-site production processes, a distinctive feature of Energiesprong's approach, can deliver a value proposition that exceeds the retrofit costs, making them financially viable.

The analysis focuses on two urban markets in Northern Italy, which are representative of real estate markets differing in dynamism and values. It examines the value proposition linked to the improvement of buildings' intrinsic characteristics and their energy performance. Through a dynamic analysis, the research considers the profitability of retrofit interventions in terms of cost reductions resulting from economies of scale and learning effects.

The financial feasibility of such interventions is crucial to make them feasible at a large scale, with positive effects on reducing energy consumption [27], increasing living comfort [28], and the value of buildings.

The decision to focus on Italy as a case study is driven by economic and environmental considerations. From an economic perspective, the nation's building stock, comprising approximately 25 million dwellings, constitutes over 50% of the total cost savings of Italian households [29,30]. From an environmental perspective, according to the EU's Directive 2024/1275, the building stock is inefficient mainly, with approximately 75% of dwellings falling into the least efficient energy classes [6].

The article is organized into five sections. The first section considers the main theoretical references regarding the value proposition and business models adopted in the field of real estate retrofitting, with particular attention to the MESA model. The second section examines the research methodology linking investment perspectives to the positions of the stakeholders involved. The third section presents data and financial models, detailing cost and revenue items for each business model. The fourth section presents the results of the research, while the fifth section provides a discussion of the evaluations carried out.

## 2. Literature on Value Proposition and Business Models for Residential Retrofit

The energy transformation of the building stock is a crucial element in achieving the European Green Deal target of zero net emissions by 2050 [31–33].

Environmental sustainability is becoming increasingly important in the field of property valuation, with a growing number of studies analyzing the influence of energy efficiency on property values and the impact of Energy Performance Certificate (EPC) policies on consumer choices [34].

Improving energy performance represents a significant investment for households and businesses, but it is also an important feature of house value formation. The price premium of properties with better energy performance has been verified and estimated in numerous international studies [12,35–40].

The appreciation of energy-efficient properties is related to two factors. First, properties with superior energy efficiency benefit from lower operating costs, which, when capitalized, result in a recognizable price premium in the market [41,42]. Second, increasingly stringent energy performance regulations may limit the ability to sell or rent non-compliant properties, leading the market to reward properties in the most efficient energy classes.

The empirical estimation of price premiums related to the energy characteristics of dwellings has mainly been conducted at the level of individual countries or cities. In Europe, significant research has focused on markets in the Netherlands [43], Ireland [44], Denmark [45], the United Kingdom [46], Wales [12], and Spain [47].

In Italy, further studies have examined the impact of energy performance on the market value of residential properties in different cities [48–52]. Research conducted by Ruggeri et al. [53] and Micelli et al. [54] revealed an average price premium in northern Italy of 14% in metropolitan cities and 28% in medium-sized cities when comparing properties from Class G to Class A on the EPC scale. Research conducted by the Bank of Italy on a national scale found an average increase of 25% in the value of energy-efficient dwellings, with significant regional variations: the price premium varies from 12% in more temperate climatic zones to 37% in colder ones, a variation attributed to the higher energy upgrading costs required in areas with a harsh climate [55].

The increase in property values thus represents an opportunity for households and developers. The adoption of appropriate BMs becomes crucial to capture the energy price premium at costs that make retrofits of existing residential stock financially viable [17,19].

The efficiency of BMs depends on the specificities related to the application area and process considered [56,57]. These particularities are reflected in the core components, which include the value proposition, supply chain, customer interface, and financial model [58–62].

Most studies that analyze BMs for residential retrofit focus on the production processes of individual companies, neglecting to consider the entire process and the specific challenges of retrofitting residential buildings [63–68]. Moschetti e Brattebø [69] and Brown [24] comprehensively review the existing academic and grey literature, identifying the main barriers to retrofitting residential buildings.

To date, the atomized market model is still dominant in the retrofit sector. This model involves individual interventions and technologies installed by separate contractors. As a result, retrofits are generally incomplete and inefficient, as the customer must manage the entire project individually, coordinating all parties involved in the process. This increases the potential for abandonment of the initiative with outcomes of limited efficiency and effectiveness [70].

On the front of the models considered most effective, the MESA model contemplates a single specialized company capable of managing the entire retrofit process, from design to verification of energy cost savings. In contrast to the atomized model, in the MESA model, the supplier deals directly with energy bills and monitors consumption, passing on the financial cost savings achieved to the customer (Figure 1).

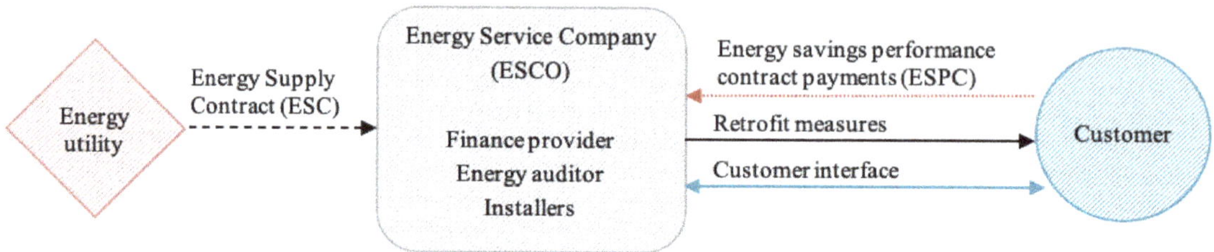


**Figure 1.** Managed energy services agreement (MESA).

The MESA model combines transformative interventions with a significant set of services for the benefit of ownership and can both take over existing technologies and promote innovation in the specific area of technological and production solutions [71].

Among the various articulations of the MESA model, the solutions promoted by Energiesprong in several European countries are distinguished as advanced solutions for the technologies employed. Initially developed in the Netherlands, Energiesprong functions as a market intermediary, promoting innovation in business models dedicated to residential retrofitting. The Energiesprong BM is based on a net-zero energy performance contract, an integrated and industrialized supply chain, a single customer interface, and a financial model based on increasing the value of the house and the cost savings related to energy performance improvements [72].

Energiesprong is distinguished by its industrial character, which is based on the off-site production of prefabricated components. This approach has significant effects on reducing the time and cost of interventions [26]. Energiesprong envisages rapid, efficient, and scalable whole-home energy upgrades that can achieve high standards of quality and environmental sustainability at decreasing costs due to learning and scale economies [73–75].

Off-site production, in addition to reducing installation time to less than a week in some cases, is distinguished by the possibility of carrying out interventions without requiring the temporary relocation of tenants during the works, thus minimizing disruption and eliminating associated costs. This approach encompasses the prefabrication of elements such as insulated facades and modules integrated with heating systems and renewable energy production [25,70,76–79].

Technical innovation is integrated with new models of economic and financial feasibility based on the concept of Total Cost of Ownership (TCO) and the guarantee of long-term performance [80].

The Energiesprong model, in addition to guaranteeing a 30-year net zero energy balance, significantly improves the intrinsic characteristics of buildings, resulting in an aesthetic and functional impact—termed "*kerbside appeal*"—that can stimulate a growing demand for building retrofits [24].

This is of particular importance, as benefits related to living comfort, such as more efficient thermal insulation and superior indoor air quality, are perceived by owners to be of similar importance to cost savings related to energy consumption [81]. The heightened emphasis on indoor comfort has become particularly evident in the post-pandemic era, as many households have expressed an increasing propensity to invest in enhancing the thermal and acoustic conditions of their dwellings despite the substantial costs associated with such interventions [82].

In addition, the possibility of customizing interventions and improving a building's overall aesthetics makes energy retrofits of relevant interest. Aesthetic features have been shown to have a significant impact on the marketability of energy-efficient buildings, thereby enhancing their desirability and consequently stimulating demand [83].

Under current market conditions, some research argues that the costs of upgrading existing assets outweigh the market benefits of improving their energy efficiency [84,85]. In light of the current market dynamics and the values involved, the necessity for financial support from the public sector is evident [14–16].

The high number of residential buildings requiring retrofit interventions makes it desirable to invest in new technologies and new production processes to increase productivity in the construction sector and to reduce the costs of interventions to make them more economically feasibility [86–89]. This aspect is crucial in a traditionally low-tech and often innovation-resistant sector such as the construction industry [65,68].

For these reasons, the research carries out a dynamic analysis concerning the possible evolutionary path of the retrofit sector to determine the financial feasibility conditions of retrofit interventions for the two investment profiles and urban and territorial contexts characterized by different real estate dynamics.

Assuming the adoption of a MESA BM enriched with advanced technological solutions capable of reducing production time and costs, the retrofit could, over time, become economically feasible and not require any public contribution for its implementation.

The research proposes an evaluation of the economic feasibility of the retrofit by integrating three key components for the first time: the value proposition linked to the energy efficiency of the building, the MESA business model, and the optimization of the production chain through the economies of scale and learning inherent in the variant promoted by the Energiesprong model.

## 3. Methodology

The methodology we adopted aimed to offer the financial feasibility of retrofit interventions on the existing residential stock, the main recipient of European energy efficiency regulations and, in Italy, largely characterized by modest energy efficiency [90]. According to data provided by the ENEA [91], approximately 4 million residential buildings in Italy, representing 68.6% of the total, are classified within the least efficient EPC energy Classes E/F/G.

The research considers two investment profiles, representative of the main actors involved in the energy retrofit of the residential sector: the owner–tenant, a person who carries out retrofit work on their home to improve its energy and aesthetic performance, and the entrepreneur–developer, an economic operator who carries out the work to increase the value of the property and then to capitalize on it by selling it.

The two profiles analyzed consider distinct time horizons determined by the different purposes pursued. The first profile, associated with the owner–tenant, considers a long-term period of twenty years. This is linked to the desire to capture all the economic and financial benefits deriving from the retrofit intervention, taking full advantage of the guarantee period provided by Energiesprong. Conversely, the second profile, typified by the real estate developer, is associated with a more limited time horizon, spanning a mere two years.

On the supply side, the methodological hypothesis is that both profiles are aimed at companies adopting the BM MESA, enriched by the use of off-site technologies to optimize property transformation processes [92,93].

The assessment of the economic feasibility of the retrofit intervention was carried out using discounted cash flow analysis (DCFA), which is the main procedure for evaluating real estate investments in both academic and professional practice [94–96]. Discounted cash flow analysis allows the determination of the net present value (NPV) of the investment by discounting future cash flows over a given time horizon, considering in a single model the costs of implementing the retrofit (hard costs, which include materials, equipment rental, and transportation), the associated operating costs (soft costs, such as design, consultancy, contingencies, and administrative fees), and the expected value from the sale of the property.

Data on revenues, costs, and discount rates were obtained through market analysis and consultation with the main reference manuals specific to retrofits in the real estate sector. The costs of retrofits considered in the valuation model are derived from the Energiesprong France report [97], which provides a detailed analysis of the current costs of residential retrofits carried out according to the Energiesprong model.

The reason to refer to data from the French context is motivated by the lack of clear and reliable estimates of the specific costs of retrofits carried out according to the Energiesprong model in the Italian market, together with information on the evolution of production volumes and cost trends.

Energiesprong renovations are characterized by an off-site production process and the use of prefabricated components, resulting in relatively stable costs internationally. This differentiates them from traditional building renovations, whose costs are strongly influenced by on-site variables, with significant local fluctuations. In the two cities under analysis, building costs are higher than the Italian national average and in line with French costs, which are slightly higher than Italian costs [98].

Therefore, we assume here the full transferability of the cost values and technological processes to the Italian context because of the great similarity between France and Italy in terms of industrial structure, building stock characteristics, and the substantial convergence of construction and renovation costs in the two countries [99].

To ensure accurate comparability, the cost values in the report have been appropriately updated using the French Total Cost Index (CCI) for the period 2021–2023 to put these data in context with the current economic landscape [100].

The report also includes projections of possible cost reductions as the number of renovated buildings increases because of economies of scale and learning. The process has been widely studied in the literature. Since Wright's [101] study of airplanes, the systematic relationship between unit production costs and cumulative output has been recognized: unit costs tend to fall by a constant factor each time cumulative output doubles [102]. This relationship, known as the learning curve, has been widely used to predict cost reductions from the adoption of innovative technologies. A significant example in this field is the market for photovoltaic panels, whose prices have decreased by 99.6% since 1976 [103,104].

The benefits associated with retrofitting interventions are economic and financial, differentiated according to the profiles studied. For the owner–tenant, the economic benefit relates to the increase in the asset value of the property because of the improved energy performance and maintenance status of the dwelling unit, while the financial benefit includes the reduction in energy costs and the elimination of the costs associated with temporary relocation during retrofitting interventions.

The valuation of the price premium is based on the hedonic pricing model, a widely used method for assessing the impact of EPC certification on property values. The underlying assumption is that the value of an asset, in this case, a property unit, depends on its characteristics. Consequently, the method is used to determine the contribution of positional and performance characteristics to the overall value of the property [105,106].

The analysis is based on a representative sample of a given geographical area, which allows the application of multivariate regression analysis to the observed market values [107]. The evaluations obtained make it possible to assess the increase in value of an energy-efficient property compared with a less efficient one.

The financial benefits to the owner–tenant fall into two main categories. The first is the cost savings made by eliminating the costs of moving and paying rent for a replacement property. The Energiesprong model allows retrofits to be carried out while keeping the owner or tenant in the home during the retrofit, eliminating the need for temporary relocation. In addition to eliminating the costs associated with temporary relocation to another dwelling, continuity of housing also supports the emotional and social well-being of residents during the retrofit process [25].

The second category of financial benefits relates to the reduction in energy and maintenance expenses over the 20 years following the intervention. More precisely, energy expenses are reduced to zero thanks to the net-zero nature of the Energiesprong interventions, while maintenance expenses are postponed until the end of the 20-year warranty period offered by Energiesprong, which covers both the energy and structural performance of the building [108–110].

For the entrepreneur–developer, the valuation is made on the assumption that the retrofitting intervention is carried out on a property owned by him/her, without tenants, to sell it two years after the investment. In this case, the only benefit of the intervention is financial, and it consists of the cash flow generated by the sale of the property at a higher value as a function of the improved energy characteristics and the overall upgrading of the property.

Table 1 delineates the benefits of the retrofit intervention for both investment profiles, classifying them into economic benefits related to changes in asset value and financial benefits related to direct and immediate cash flows.

**Table 1.** Economic and financial benefits of retrofitting with off-site technologies according to the Energiesprong model for the two investment profiles.

| | Owner–Tenant | Entrepreneur–Developer |
|---|---|---|
| Economic benefits | Δ price premium from G and habitable to Class A and renovated | - |
| Financial benefits | - | Δ price premium from G and habitable to Class A and renovated |
| | Avoided cost of energy consumption | - |
| | Avoided the cost of moving | - |
| | Avoided the cost of rent for the alternative accommodation | - |
| | Avoided cost of extraordinary maintenance | - |

The price premium associated with energy retrofitting turns out to be very different in cities of different sizes, real estate market dynamics [54], and climatic environments [55]. It seems crucial to carry out a more spatially sensitive comparative investment analysis. The selected cities are Milan and Udine, both located in Northern Italy but with different urban dimensions and market dynamics.

Milan, a metropolitan city with 1.3 million inhabitants, is recognized as the main economic center of Italy. The city has a highly dynamic real estate market with positive growth indicators, both economic and social [111]. On the other hand, Udine, with 98,040 inhabitants, is a representative city of medium-sized Italian cities, where more than 40% of the population lives [112].

According to the climatic classification of Italian municipalities established by current legislation [113], both cities fall within climate zone E, the area on which this research focuses. Zone E is characterized by a continental climate known to be one of the harshest in Europe [114], with a number of degree-days between 2101 and 3000. Buildings in the cities of this zone are characterized by high energy demand, resulting in significant annual costs for the inhabitants, which are only lower than in the Alpine area in the north of the country.

## 4. Data Selection and Source

The development of the DCFA model requires the identification of the financial costs and benefits of retrofits. For the developer-investor, the change in value associated with improved energy and intrinsic property characteristics is the most important benefit.

The price premium associated with superior energy efficiency was estimated using a dataset of offer prices. There are several reasons for using asking prices rather than actual sales and purchase values. First, asking prices ensure greater data availability and wider access to information on energy performance certificates (EPCs). In Italy, the collection of data on actual transactions is often hampered by legal constraints related to privacy and significant restrictions on access to such information [107]. It is no coincidence that the scientific literature has consequently used asking prices because of their availability, as evidenced by numerous studies that have confirmed their usefulness for analyzing real estate market trends [115] and for investigations very similar to the one developed in this research [53–55].

Asking prices were collected from Immobiliare.it [116], the leading real estate asking platform in Italy, which provides a large database of detailed information from both real estate agencies and private sellers [117].

Data collection followed a systematic process, with random sampling within the administrative boundaries of the urban areas of the cities under research. A total of 873 asking prices were collected for Milan and 404 for Udine. For each asking price sampled, eight variables were collected relating to the geographical location and the intrinsic, typological, and technological characteristics of the property. This information was collected in July 2023.

The descriptive and frequency statistics show a homogeneous picture in terms of typological characteristics and maintenance status (Appendix A—Tables A1 and A2). The data analysis revealed a predominant dwelling in the sample, located in a semi-central area with little infrastructure. The predominant dwelling consists of flats of approximately 70 square meters, with only one bathroom, which can be classified as ordinary class properties belonging to Class G on the EPC scale.

The methodology used to determine the price premium involves the use of the hedonic price model. The regression used to estimate the hedonic price can take several functional forms [118,119]. The literature does not provide clear guidance on the most appropriate functional form to represent the complexity of value formation. In this research, we choose to use a semilogarithmic function, which has been widely adopted for its properties, including its ability to reduce the problem of heteroscedasticity [120] and to highlight the nonlinear relationship between property prices and the characteristics that explain their value [121].

The regression model uses the natural logarithm of the unit market value as the dependent variable. Formally, the model used is as follows:

$$P_i = \beta_0 + \sum_{i=1}^{I} \beta_i X_i + e_i \tag{1}$$

where:

- $P_i$ is the natural logarithm of the price of a dwelling expressed in EUR/sqm;
- $\beta_0$ is the constant of the model;
- $\beta_i$ represents the marginal price of the characteristic;
- $X_i$ is the numerical value of the observed variables, including EPC;
- $e_i$ represents a random error.

The regression models for Milan and Udine show a coefficient of determination ($R^2$) of 62.6% and 42.9% and an adjusted $R^2$ of 62.3% and 42.0%. The statistical reliability of the models was further confirmed by the F-test, whose $p$-value was less than 0.005 in both cities (Table 2).

**Table 2.** Models.

| Model | $R^2$ | Rq Adjusted | F | df1 [a] | df2 [b] | $p$-Value |
|---|---|---|---|---|---|---|
| Milan | 0.626 | 0.623 | 192.859 | 7 | 807 | <0.001 |
| Udine | 0.429 | 0.420 | 49.7 | 6 | 397 | <0.001 |

[a] df1: regression; [b] df2: residual.

Table 3 presents the results of the regression model, together with the analysis of the Variance Inflation Factor (VIF), which is used to assess the degree of multicollinearity between the independent variables. In both cases analyzed, the VIF values are close to one, indicating a low correlation between the variables and confirming their independence.

**Table 3.** Regression models.

| Predictors $x_i$ | $x_i$ | Milan | | Udine | |
|---|---|---|---|---|---|
| | | βi | VIF | βi | VIF |
| K | | 10.306 | - | 8.421 | - |
| Zone | 2 | −0.314 | 1242 | −0.041 | 1.11 |
| Proximity to infrastructure | 3 | −0.076 | 1061 | - | - |
| Typology | 1 | - | - | - | - |
| Property class | 3 | −0.17 | 1408 | −0.132 | 1.34 |
| Number of bathrooms | 1 | 0.064 | 2413 | 0.150 | 2.01 |
| Surface | 70 | −0.001 | 2496 | −0.002 | 1.96 |
| Energy class | 12 | −0.027 | 1409 | −0.037 | 2.13 |
| Maintenance status | 3 | −0.036 | 1381 | −0.103 | 2.29 |

According to the analysis presented in Table 4, the difference in value between a Class G property and a Class A property is estimated to be 21.90% for Milan and 38.26% for Udine.

For the owner–tenant, the increase in value has only an economic significance and is therefore not considered in the evaluation, while other benefits are considered in the evaluation of the investment: cost savings on energy costs, reduction in costs for extraordinary maintenance, and elimination of the costs of moving and renting alternative accommodation during an ordinary renovation.

**Table 4.** Difference in value for energy efficiency.

| | Milan | Udine |
|---|---|---|
| Vm ante unit: Class G and habitable | 4925 EUR | 1312 EUR |
| Vm post unit: Class A and renovated | 6003 EUR | 1814 EUR |
| Vm ante: Class G and habitable | 344,723 EUR | 91,844 EUR |
| Vm post: Class A and renovated | 420,204 EUR | 126,979 EUR |
| From G and habitable to Class A and renovated | 75,481 EUR<br>21.90% | 35,135 EUR<br>38.26% |

The cost of energy consumption (Ce) was determined concerning the typical building previously identified in the sample using the following formula:

$$Ce = (EPgl,nren \times Pg + EPgl,ren \times PUN) \times S \quad (2)$$

where:

- *Ce* is the cost of energy consumption;
- *EPgl,nren* is the overall non-renewable energy performance index (kWh/sqm year);
- *Pg* is the price of non-renewable energy (EUR/kWh);
- *EPgl,ren* is the overall renewable energy performance index (kWh/sqm year);
- *PUN* is the price of renewable energy (EUR/kWh);
- *S* is the surface area of the building unit (sqm).

Unit energy prices have been determined based on the average of historical prices recorded over the last 15 years to obtain a representative value of market dynamics and to reduce the influence of short-term economic fluctuations. The average cost of non-renewable energy (Pg) was determined from data provided by the Gestore dei Mercati Energetici [122], the main institution responsible for managing the Italian electricity market. The cost of renewable energy (PUN), on the other hand, was obtained from information published by the Regulatory Authority for Energy Networks and the Environment [123], which is responsible for regulating and controlling the energy and water sectors. In summary, the annual cost of the energy consumption of the typical building is 1745 EUR (Appendix—Table A3).

The retrofit intervention with off-site technology proposed by Energiesprong guarantees the tenants' continuity of living during all phases of the intervention. This feature eliminates the cost of moving, estimated at around 3000 EUR, as well as the costs associated with renting alternative accommodation. The avoided rental costs have been estimated over four months, the average duration of a complete renovation not carried out off-site. The amount of avoided costs depends on the local real estate market: 4150 EUR in Milan and 1500 EUR in Udine [116,124].

Further financial benefit comes from the possibility of postponing extraordinary maintenance costs thanks to the investment made and the 20-year guarantee on the technological and engineering features and performance offered by the Energiesprong model. Dividing these costs into annual installments allows a saving of 0.25% per year on construction costs [109].

The hard and soft costs associated with deep retrofitting are identical in the two assessments for the two investment profiles considered. The hard costs, represented by the expenses for implementing the retrofit using off-site technology, were estimated at 1150 EUR/sqm in 2021 in France [97]. The 21% increase recorded by the French Overall Cost Index (CCI) between 2021 and 2024 results in an updated renovation cost of 1400 EUR/sqm [100].

The soft costs, i.e., the additional costs associated with retrofitting, were estimated through a comparative analysis of their incidence in traditional renovation projects to the specificities of the off-site model of Energiesprong [109]. The analysis was necessary because the official reports of Energiesprong do not provide any information on this. The incidence of soft costs in the off-site retrofit is reduced due to greater efficiency and streamlining of the design and construction phases. Detailed values of financial benefits and intervention costs are given in Appendix A—Table A4.

The discount rate used for the economic evaluation is determined using the Weighted Average Cost of Capital (WACC), which weights the cost of debt and equity capital. The cost of debt capital is set at 3.06%, which is the average value of green mortgages offered by the main Italian banks [125] (Appendix A—Table A5). This value, which is lower than the rates applied to ordinary building interventions, reflects the favorable conditions offered by banking institutions to projects promoting energy efficiency and environmental sustainability.

The cost of equity capital varies according to the development scenario considered. For the owner–tenant, the financing structure makes greater use of debt capital than in conventional configurations, given the plausible propensity of households to maximize leverage by taking advantage of the favorable conditions offered by green mortgages. On the contrary, in the case of the developer, the financing structure adopts an ordinary structure with lower debt leverage and higher equity contribution because of the higher return expectations and risks typically associated with such projects [109,126]. Details on the WACC and its decomposition into its components can be found in Appendix A—Table A6.

The assessment of the economic feasibility of the retrofit intervention considers a time horizon of 2 years for the developer and 20 years for the owner–tenant.

The first assessment is based on current retrofit costs and technologies. It aims to determine the financial feasibility of the retrofit investment by the two investor profiles identified in the different urban contexts studied. A second assessment was carried out to verify how the economies of scale and learning associated with the increasing number of retrofits—a particular aspect of the BM Energiesprong examined in this research—affect the profitability of the investments.

## 5. Analysis and Results

The assessment of the financial feasibility of the retrofit intervention, based on the BM Energiesprong, was carried out on the previously defined dwelling. The retrofit intervention aims to improve the energy efficiency of a 70 sqm building in the case of the maximum performance gap to transform a building from energy Class G to Class A on the EPC scale while at the same time improving its intrinsic characteristics. The evaluation was carried out by discounting the costs (Table 5) and the financial and economic returns of the investment, considering the time perspective of each investment profile (Appendix A—Table A4).

**Table 5.** Unit and total costs, both undiscounted, of the retrofit intervention with off-site technologies according to the Energiesprong model for the dwelling in Milan and Udine for the two investment profiles.

| | Owner–Tenant | | Entrepreneur–Developer | |
|---|---|---|---|---|
| | Unit Value [EUR/sqm] | Total Value [EUR] | Unit Value [EUR/sqm] | Total Value [EUR] |
| Hard costs | 1400 | 98,000 | 1400 | 98,000 |
| Soft costs | 168 | 11,760 | 224 | 15,600 |
| Total costs | 1568 | 109,760 | 1624 | 113,600 |

The financial feasibility assessment was carried out by estimating the NPV of the investment. The results obtained indicate that, under current cost conditions, retrofit interventions have a negative NPV for both investment profiles and urban contexts analyzed.

For the owner–tenant profile, the values obtained show a similarity between the two cities analyzed. In Milan, the NPV shows a negative unit value of −890 EUR/sqm and a total of −62,000 EUR for the whole investment. Similarly, in Udine, the unit NPV is −930 EUR/sqm, and the total is −64,860 EUR.

The costs and benefits of the retrofitting intervention are largely independent of the specificities of the local real estate markets, except for the lost costs of renting an alternative accommodation during the intervention phases. In metropolitan cities such as Milan, rents are high, and their absence, even for a short period associated with the renovation of the property, reduces the investment deficit but does not significantly affect the final NPV.

The situation is different for the entrepreneur–developer who sells the dwelling after the retrofit. In this case, the NPV values for Milan and Udine show a significant divergence. In Milan, the NPV of the unit is negative at −536 EUR/sqm and a total of −37,500 EUR for the whole investment, while in Udine, the NPV of the unit is −1112 EUR/sqm and a total of −77,900 EUR for the dwelling.

This difference is due to the immediate capitalization at the time of sale of the price premium resulting from the improved energy and maintenance quality of the retrofit. In metropolitan cities, where property values are generally higher, the change in value because of the improvement in energy efficiency is also significantly higher. The NPV in the simulation carried out for Milan is, therefore, higher than that developed for the city of Udine for the same type of intervention and retrofitting costs.

At this point, it is possible to determine the public contributions necessary to make the investment financially viable. In the hypothesis of upfront contributions to the different investor profiles considered, the results elaborated based on the financial models' return mirror values regarding the financial performance considered previously.

For the owner–tenant, the financial contribution required to make the retrofit investment financially feasible amounts to 57% of the total investment value in Milan and 59% in Udine. These values, in line with those found for the NPV, indicate the need for a contribution of more than half of the investment for the retrofit with BM Energiesprong. The case of the entrepreneur–developer is different: the required upfront contribution is 27% in Milan, while in Udine, it is more than double, reaching 65% of the total investment (Figure 2). In Milan, the higher property values allow the developer to obtain a greater absolute price premium after retrofitting. This means that, for the same intervention cost, the resale value of the upgraded property covers a greater proportion of the investment. As a result, the financial gap to be filled by public support is smaller.

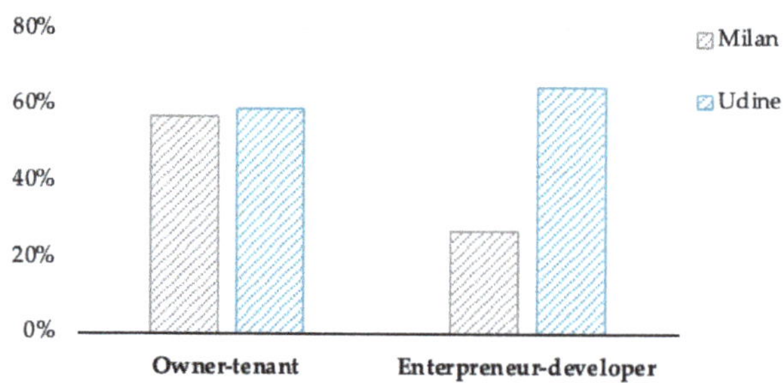


**Figure 2.** Amount of upfront financial contribution (%) compared with the total amount of financial investment required to make a retrofit financially feasible for the two investment profiles in Milan and Udine.

Udine, on the other hand, has a lower value and is a less dynamic market. Although the percentage increase in value from retrofitting is significant, the absolute gain is much smaller. This makes it more difficult for the developer to recoup the investment through resale, and therefore, a higher public contribution is needed to ensure financial feasibility.

At current cost levels, retrofits are not financially feasible for the profiles examined, and significant public intervention is required to attract the interest of owners and developers. However, the research aims to test whether the Energiesprong BM can be financially feasible in the future, taking into account the cost reductions associated with increasingly efficient off-site production, which is a distinctive aspect of the Energiesprong model.

The function of hard costs of retrofits as a function of production volume was developed using data provided by Energiesprong France, the only source that provides clear and reliable estimates of cost reductions per production volume for retrofits carried out according to the Energiesprong model [97].

As illustrated in Table 6, three distinct functional forms were examined, namely the Power Law Model [101], the Learning Curve Model [127], and the Logarithmic Cost Function [128]. These models are frequently employed to approximate cost as a function of production quantity, incorporating the concepts of economies of scale and learning.

**Table 6.** Comparison of three cost reduction models—Power Law, Learning Curve, and Logarithmic Cost Model—showing their mathematical formulation and typical industrial applications.

| Model | Functional Form | Typical Application |
|---|---|---|
| Power Law Model | $C(Q) = C_0 Q^b$ | Strong economies of scale, manufacturing, and industrial production [23,101] |
| Learning Curve Model | $C(Q) = C_0 Q^{log_2(1-r)}$ | Learning effect in production costs, renewable energy, and aerospace [127,129,130] |
| Logarithmic Cost Function | $C(Q) = a + b\, ln(Q)$ | Gradual cost reduction, industrial processes with limited scalability [128,131] |

A considerable amount of research in the industrial sector has revealed that the decline in cost is more likely to be a function of a logarithmic trend as opposed to a pure power function [128]. The logarithmic function is characterized by a more gradual decrease in production cost compared with the power function, which renders it more suitable for sectors where large-scale production does not result in significant cost reductions. Consequently, for building retrofit projects, where fixed costs and learning variables are less pronounced than in manufacturing, the logarithmic reduction may be more realistic.

The logarithmic function provides a useful representation of the cost reduction, which decelerates over time. The dynamics of economies of scale have a strong initial impact, which then tends to stabilize. In comparison to the conventional learning curve, which assumes a constant rate of reduction for each duplication of production, the logarithmic function facilitates the modeling of scenarios in which learning is non-uniform.

As outlined in the preceding paragraph, the relationship between cost and production volume is described in this case by the logarithmic functional form because of the reasons stated and the ability to interpolate data obtained from Energiesprong France [97] (Figure 3—see the black dots).

The cost function, which has been elaborated, is based on a limited number of surveys and thus suffers from a significant level of approximation. However, it represents, at present, the most accurate interpolation obtainable based on the extant empirical evidence of the Energiesprong off-site production model.

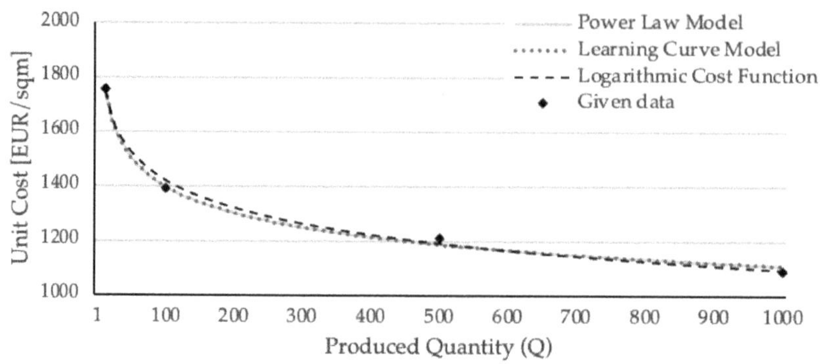


**Figure 3.** Cost reduction trends for different production quantities, comparing the Power Law Model, Learning Curve Model, and Logarithmic Cost Model. The Power Law model (light grey) predicts strong economies of scale, leading to rapid cost declines. The Learning Curve model (dotted grey line) incorporates experience-driven cost reductions commonly observed in high-tech and renewable energy industries. The Logarithmic Cost Model (dashed black line) represents industries where cost decreases occur more gradually over time. The black dots indicate the observed data points used to fit the models.

The estimated cost reduction is conservative when compared with values identified in previous studies on learning curves of other industrial products. Studies conducted on the production of photovoltaic panels, for example, have shown an average production cost reduction of between 20% and 32% for each doubling of production [103,104].

Based on the elaborated cost function, with a cumulative production of 50,000 units, the cost of a deep retrofit intervention would be around 500 EUR/sqm, an estimated 60% reduction compared with the current cost of 1400 EUR/sqm reported by Energiesprong France (Figure 4). The reference value is normalized to 100 for the current cost of 1400 EUR/sqm, while the initial intervention cost of 125 reflects an initial value of 1750 EUR/sqm recorded with lower production volumes.

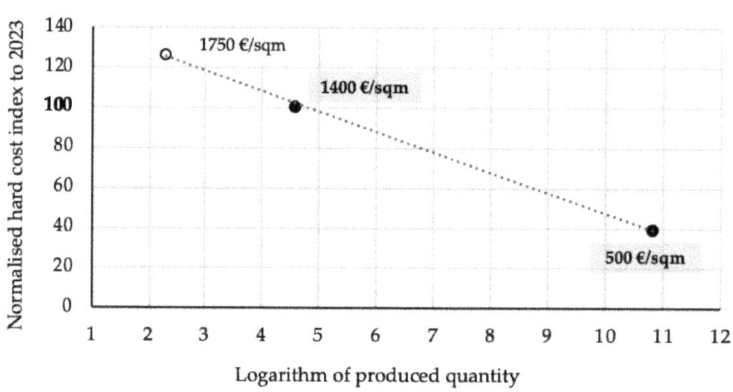


**Figure 4.** Unit cost function of retrofitting with off-site technologies as a function of the logarithm of the number of units produced, based on data from Energiesprong France.

The reduction function of the identified intervention costs, with unchanged soft costs, enables the correlation of the development of production with the financial models for the two identified profiles. The objective is to ascertain the cost threshold value for which the Energiesprong retrofit intervention is financially feasible without the necessity for public contribution.

The cost reduction leads to a progressive increase in the net present value (NPV) of the investment for both the considered profiles and the investigated urban contexts. For the owner–tenant, the feasibility of retrofit investments is ensured for cost values around 500 EUR/sqm in both high-property-value cities such as Milan and low-property-value cities such as Udine (see Figure 5).

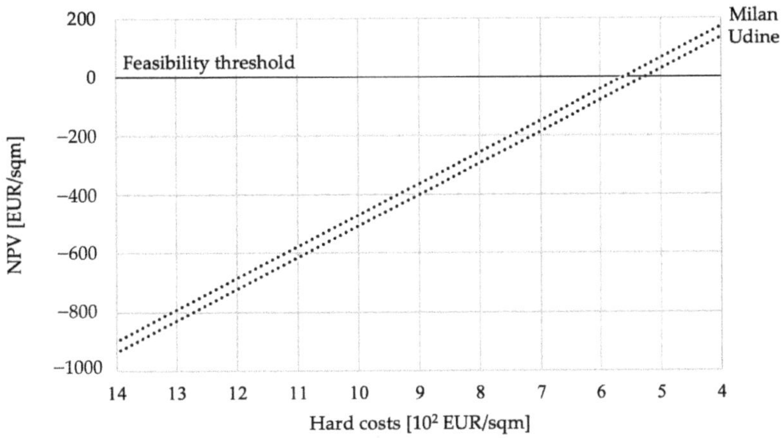


**Figure 5.** Variation in the unit NPV of the investment as a function of the reduction in the cost of the retrofit intervention using off-site technologies in Milan and Udine for the tenant owner.

Nevertheless, the conclusions drawn for the developer entrepreneur differ. In Milan, the investment in retrofitting is feasible for much lower values than in Udine. The immediate capture of the price premium value by the developer determines, in the simulation carried out on the values of Milan's real estate market, profitability conditions that ensure the feasibility of the retrofit investment with a hard cost of approximately 900 EUR/sqm (Figure 6).

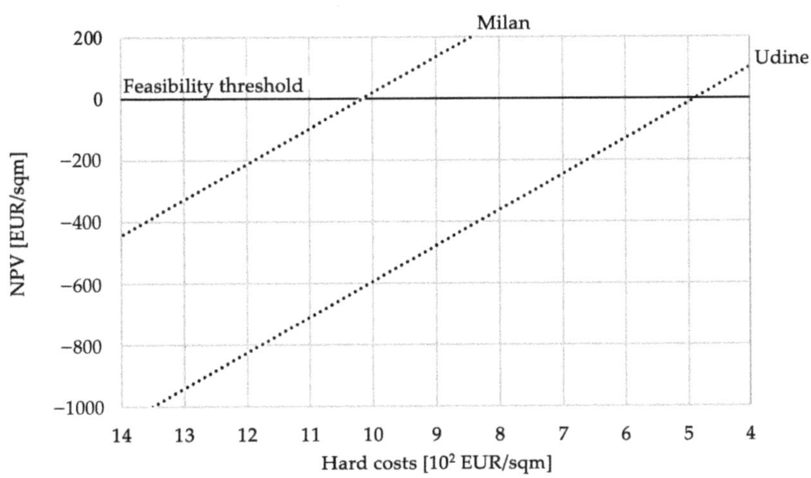


**Figure 6.** Variation in the unit NPV of the investment as a function of the reduction in the cost of the retrofit intervention using off-site technologies in Milan and Udine for the entrepreneur–developer.

In the case of lower price premiums, as is the case of the Udine real estate market, the conditions of feasibility of the investment are reached with hard costs of approximately 500 EUR/sqm (Figure 6).

## 6. Discussion of the Results

The results highlight some important aspects. First, evaluations conducted at current costs and technologies show that, to date, the economic and financial benefits of retrofitting are lower than the costs incurred. For the owner–tenant, about 60% of the investment is not compensated by the benefits, while for the developer, the share of the investment not recovered varies from 30% in metropolitan cities to 70% in medium-sized cities.

This finding aligns with the extant literature [83,84], which suggests that the financial benefits associated with enhancing the energy efficiency and inherent characteristics of residential properties, even when considering the full range between Class G and Class A properties, are inadequate to ensure the economic and financial feasibility of the investment.

The retrofit sector—strategic for the decarbonization of the built environment—is therefore not financially feasible and must rely on public subsidies to increase an otherwise inadequate return. Direct subsidies are a key mechanism, but alternative policy measures such as tax incentives, levies, an increased carbon tax, and the promotion of green mortgages can also play a crucial role in bridging the financial gap and incentivizing long-term investment in energy-efficient retrofits.

Nevertheless, the values that emerged from the research indicate that upfront contributions intended for the transition of the building stock should only cover a part of the retrofit investment and not its entirety. In some states, the public contribution has instead been integral and not partial [132]. The Italian legislator, for example, has in recent years supported a policy aimed at promoting energy efficiency investments by financing them in full, while an adequate return would have been achieved even with only a partial contribution to the total investment.

The second point that deserves to be discussed concerns the dynamic nature of the sector, examining the new technologies that characterize the MESA Energiesprong model.

The interest of this model lies in the possibility of considering a progressive reduction in the costs of retrofit interventions following a learning curve that has already characterized many other industrial sectors, including those related to retrofits, such as photovoltaic panels [104].

The analyses conducted, although based on data referring to only one state, France, and referring only to early trends, show an important trend of decreasing costs because of learning and scale economies linked to the development of new off-site production processes and industrialized construction. The reduction in hard costs assumed in the research because of economies of scale and learning is essentially half of the current construction cost. Consequently, retrofit investments become financially feasible, and no public subsidy is required for households and companies.

To achieve this objective, the major production of several tens of thousands of dwellings is expected compared with the current production volumes of industrialized construction [26]. While this production volume may appear ambitious, it should be noted that it represents a modest fraction of the overall building stock in need of modernization by the European deadlines set by the EPDB Directive for 2050. In a single state, such as Italy, the number of buildings in need of action exceeds 4 million, a number that grows further when considering the European building stock.

Cost reduction could have a significant impact on the affordability of retrofit interventions. As the analyses demonstrate, even with a reduction of a few tens of percentage points in the hard cost component alone, the feasibility of retrofit investments could be ensured without depending on any public contribution, at least in those areas where premium prices can make building retrofit investments more profitable.

In many states—and this is particularly the case in Italy—policies have adopted technology and construction costs as unchangeable data. As a result, the leverage used to make interventions financially feasible has been to contribute to households and support them to invest in upgrading their dwellings. The MESA Energiesprong model demonstrates how a different path is possible by supporting not only demand but also capable companies, thanks to the industrialization path of construction that allows an increase in productivity and results in lower unit costs.

Such a path of increasing the efficiency of the retrofit sector makes it possible to consider the gradual decrease in public support for the decarbonization of the existing building stock, with retrofitting becoming a distinct segment of the construction market, independent of public contributions. The development of off-site technologies on a European scale, rather than a national scale, renders it highly probable that construction costs will decline and the consequent emergence, progressively over time and depending on the specificities of territories and local real estate markets, of renewed conditions of economic feasibility of interventions to decarbonize the European residential stock.

Third, the financial feasibility of retrofit interventions is subject to variation according to the territorial context. In large cities characterized by larger real estate markets, the price premium derived from retrofit intervention at the time of sale, in absolute value, is higher. In contrast, cities with less substantial real estate markets encounter greater challenges in transitioning to energy-efficient assets, a problem that is especially pronounced in the context of private conveniences.

This conclusion raises a question regarding the distributive justice of the burdens associated with energy transition. The financial feasibility of interventions, given the same typology and costs, appears more onerous, in relative terms, in territories where the amount of the price premium is limited, as shown in a small–medium-sized city such as Udine compared with a metropolitan city such as Milan.

The energy transition does not invariably have a uniform effect on assets. Indeed, the transformation of building stock imposes a greater burden in relative terms on properties in cities already marked by lower market values. This dynamic deserves special attention, as it could fuel discontent and resistance to the transition itself, motivated more by economic aspects than theoretical or ideological reasons [133,134].

In addition to the disparities between cities of different ranks and characterized by important gaps in real estate values, there are also differences related to the climatic characteristics of the territories. Research shows how the price premium related to the technological upgrade of the property varies depending on the geographical area considered.

Public incentives and contributions should take into account the different prominence of energy costs and the price premium attributed to the characteristics of each property for the proper allocation of public resources [135]. The current policy of uniform national contributions is likely to demonstrate inefficiency and inadequacy in the distribution of public resources [53,55,136].

In the future, therefore, policies should provide incentive instruments appropriately tailored to the territorial context and characteristics of local markets to ensure the optimal allocation of available resources.

## 7. Conclusions and Policy Implications

The energy transition and decarbonization require a deep transformation of the building stock, in line with the net zero emissions targets set by the European Green Deal. The challenge is crucial for Italy, where 70% of the building stock belongs to the least efficient EPC energy classes.

The research analyses the financial feasibility of deep energy retrofit interventions in the residential sector, adopting the BM MESA, which is recognized as the most effective for its integrated service offering and simplifies the customer experience [24]. The proposed methodology integrates BM MESA principles with the off-site production processes typical of the Energiesprong offering.

The evaluations consider two investment profiles: an owner–tenant who renovates the dwelling for private use and an entrepreneur–developer who renovates the property to sell it, capitalizing on the increase in value. The analysis was conducted in two distinct urban contexts: Milan, a metropolitan city, and Udine, a small–medium-sized city.

The findings indicate that, despite the presence of relevant benefits represented by a price premium linked to superior energy efficiency, the renewed quality of the property, and the cost savings on energy costs, retrofits are not financially feasible with current costs and technologies without external upfront contributions.

The research indicates, however, that the development of off-site technologies, a distinctive aspect of the MESA Energiesprong model, may lead to a substantial reduction in hard costs due to learning economies and economies of scale. Despite the absence of fully reliable data on the reduction in costs as a function of production volumes, it is reasonable to assume that, particularly in large cities characterized by higher price premiums in absolute terms, the retrofit sector may eventually represent an autonomous and independent sector of the real estate construction industry without the public contribution being a diriment support.

Future research may cover several areas. First, the development of new players capable of giving substance to the MESA Energiesprong model appears to be of great importance. While the model has been delineated [24] and can be considered a virtuous and efficient evolution of the construction and real estate sector, its affirmation in concrete national markets remains to be demonstrated. Therefore, it appears central to verify its concrete development in the different national markets; otherwise, the virtuous and efficient development of the housing stock retrofit market will not be achieved.

Second, it appears necessary to pay the utmost attention to the evolution of retrofit costs operated with off-site production technologies, inseparable from the Mesa Energiesprong model considered in this research. Since these technologies, in any case, have a significant presence of the labor factor, it seems hardly credible to assume the same cost reductions as for buildings that are entirely mass-produced and, therefore, susceptible to the greatest economies of scale and learning. The development of off-site retrofit volumes will, therefore, provide crucial insight into the development of a strategic component of European construction.

Finally, a very important issue concerns public policies aimed at supporting the achievement of the EPBD 2050 in the building stock and housing stock. The research has highlighted the need to investigate not only demand-support policies, differentiating these for contexts characterized by very different conditions of the economic viability of the retrofit, but also those aimed at supply, supporting the use of technologies capable of ensuring increasing technical and economic efficiency to the processes of energy and real estate requalification of the residential building stock. The optimal combination of these policies could usefully contribute not only to the efficient use of public resources in the sector but also to the concrete achievement of European objectives on the decarbonization of the built environment.

**Author Contributions:** Conceptualization, E.M.; methodology, E.M.; validation, E.M., E.R. and G.G.; formal analysis, E.M., E.R. and G.G.; investigation, E.M., E.R. and G.G.; resources, E.M., E.R. and G.G.; data curation, E.M., E.R. and G.G.; writing—original draft preparation, E.M., E.R. and G.G.; writing—review and editing, E.M., E.R. and G.G.; visualization, E.M., E.R. and G.G.; supervision, E.M.; project administration, E.M. All authors have read and agreed to the published version of the manuscript.

**Funding:** This research received no external funding.

**Data Availability Statement:** Data presented in this study are available on request from the corresponding author.

**Conflicts of Interest:** The authors declare no conflicts of interest.

## Appendix A

**Table A1.** Frequency analysis.

| Variables | Categories | Milan | | Udine | |
|---|---|---|---|---|---|
| | | n | % | n | % |
| Zone | Central zone | 209 | 23.9 | 151 | 37.4 |
| | Semicentral zone | 224 | 25.7 | 160 | 39.6 |
| | Suburban zone | 440 | 50.4 | 93 | 23.0 |
| Proximity to infrastructure | Up to 200 m | 265 | 30.4 | | |
| | From 201 to 500 m | 518 | 59.3 | | |
| | Over 500 m | 90 | 10.3 | | |
| Typology | Villa | 0 | 0.0 | 24 | 5.9 |
| | Apartment | 873 | 100.0 | 380 | 94.1 |
| Property class | Luxury | 53 | 6.1 | 15 | 3.7 |
| | Prestigious | 469 | 53.7 | 152 | 37.6 |
| | Ordinary | 283 | 32.42.2 | 222 | 55.0 |
| | Economic | 19 | | 15 | 3.7 |
| Number of bathrooms | One bathroom | 489 | 56.0 | 179 | 44.3 |
| | Two bathrooms | 295 | 33.8 | 193 | 47.8 |
| | Three bathrooms | 86 | 9.9 | 32 | 7.9 |
| | Four bathrooms | 2 | 0.2 | - | - |
| Energy class | A4 | 7 | 0.8 | 21 | 5.2 |
| | A3 | 8 | 0.9 | 2 | 0.5 |
| | A2 | 12 | 1.4 | 5 | 1.2 |
| | A1 | 11 | 1.3 | 4 | 1.0 |
| | A+ | 6 | 0.7 | 8 | 2.0 |
| | A | 24 | 2.7 | 4 | 1.0 |
| | B | 22 | 2.5 | 26 | 6.4 |
| | C | 20 | 2.3 | 31 | 7.7 |
| | D | 87 | 10.0 | 52 | 12.9 |
| | E | 130 | 14.9 | 91 | 22.5 |
| | F | 182 | 20.8 | 91 | 22.5 |
| | G | 362 | 41.5 | 69 | 17.1 |
| Maintenance status | New—under construction | 48 | 5.5 | 31 | 7.7 |
| | Excellent—renovated | 346 | 39.6 | 116 | 28.7 |
| | Good—habitable | 357 | 40.9 | 234 | 57.9 |
| | Poor—to be renovated | 110 | 12.6 | 23 | 5.7 |

**Table A2.** Descriptive statistics.

| Variables | Milan | | | | | Udine | | | | |
|---|---|---|---|---|---|---|---|---|---|---|
| | n | Mean | S.D. | Min | Max | n | Mean | S.D. | Min | Max |
| Unit value | 873 | 6324.9 | 2999 | 1307 | 21,883 | 404 | 1655.4 | 649.8 | 523.0 | 4063 |
| Zone | 873 | 2.26 | 0.821 | 1 | 3 | 404 | 1.8564 | 0.765 | 1 | 2 |
| Proximity to infrastructure | 873 | 1.80 | 0.606 | 1 | 3 | 0 | - | - | - | - |
| Typology | 873 | 1.00 | 0.000 | 1 | 1 | 404 | 0.9406 | 0.237 | 0 | 1 |
| Property class | 824 | 2.33 | 0.628 | 1 | 4 | 404 | 2.5866 | 0.626 | 1 | 4 |
| Number of bathrooms | 873 | 1.54 | 0.680 | 0 | 4 | 404 | 1.6361 | 0.625 | 1 | 3 |

**Table A2.** *Cont.*

| Variables | Milan | | | | | Udine | | | | |
|---|---|---|---|---|---|---|---|---|---|---|
| | n | Mean | S.D. | Min | Max | n | Mean | S.D. | Min | Max |
| Energy class | 871 | 10.36 | 2.270 | 1 | 12 | 404 | 9.2995 | 2.756 | 1 | 12 |
| Maintenance status | 861 | 2.61 | 0.778 | 1 | 4 | 404 | 2.6163 | 0.711 | 1 | 4 |
| Surface | 873 | 108.09 | 57.371 | 19 | 482 | 404 | 128.73 | 66.716 | 23 | 490 |

**Table A3.** Total energy consumption cost for a residential property in climate zone E, with floor area between 50 and 100 sqm and energy Class G.

| **Energy Consumption** *Ec = (EPgl,nren × Pg + EPgl,ren × PUN) × S* | |
|---|---|
| EPgl,nren | 313.6 kWh/sqm·y [1] |
| EPgl,ren | 5.8 kWh/sqm·y [1] |
| Pg | 0.078 EUR/kWh [2] |
| PUN | 0.081 EUR/kWh [3] |
| Surface (S) | 70 sqm |
| Ec | 1745 EUR·y |

[1] [137]; [2] [123]; [3] [122].

**Table A4.** Financial benefits and total costs, both undiscounted, of the retrofit intervention with off-site technologies according to the Energiesprong model for the dwelling in Milan and Udine for the two investment profiles.

| | Owner–Tenant | | | | Entrepreneur–Developer | | | |
|---|---|---|---|---|---|---|---|---|
| | Milan | | Udine | | Milan | | Udine | |
| | [EURsqm] | [€] | [EUR/sqm] | [€] | [EUR/sqm] | [€] | [EUR/sqm] | [EUR] |
| Hard costs | 1400 | 98,000 | 1400 | 98,000 | 1400 | 98,000 | 1400 | 98,000 |
| Soft costs | 168 | 11,760 | 168 | 11,760 | 224 | 15,600 | 224 | 15,600 |
| Total costs | 1568 | 109,760 | 1568 | 109,760 | 1624 | 113,600 | 1624 | 113,600 |
| Δ price premium from G to A | 1078 | 75,481 | 502 | 35,135 | 1078 | 75,481 | 502 | 35,135 |
| Avoided cost of rent for the accommodation | 59 | 4147 | 21 | 1496 | - | - | - | - |
| Avoided cost of moving | 43 | 3000 | 43 | 3000 | - | - | - | - |
| Avoided cost of energy consumption | 499 | 34,907 | 499 | 34,907 | - | - | - | - |
| Avoided cost of extraordinary maintenance | 70 | 4870 | 70 | 4879 | - | - | - | - |
| Financial benefits | 1749 | 122,405 | 1134 | 79,409 | 1078 | 75,481 | 502 | 35,135 |

**Table A5.** Fixed rates for 20-year green mortgages offered by major Italian banks in 2022 [125].

| Bank | 20-Year Green Mortage Fixed Rate [%] |
|---|---|
| Casa Banca delle Terre Venete Credito Cooperativo | 2.70 |
| Banca delle Terre Venete Credito Cooperativo—Gruppo Bancario Cooperativo Iccrea | 2.70 |
| Banca Monte dei Paschi di Siena | 3.50 |
| Banca Sella—Gruppo Banca Sella | 2.99 |
| Banco BPM—Gruppo Banco BPM | 3.40 |
| Banco di Desio e della Brianza—Gruppo Banco Desio | 3.40 |
| Banco di Sardegna—Gruppo BPER | 2.80 |
| Bcc Milano—Gruppo Bancario Cooperativo Iccrea | 3.50 |
| BNL—Gruppo BNP Paribas | 2.80 |
| BPER Banca—Gruppo BPER | 3.06 |
| CrediFriuli—Gruppo Bancario Cooperativo Iccrea | 2.70 |
| Crédit Agricole Italia—Gruppo bancario Crédit Agricole Italia | 2.60 |
| ING | 3.20 |
| Intesa Sanpaolo—Gruppo Intesa Sanpaolo | 3.20 |
| UniCredit | 3.40 |
| Average rate | 3.06 |

**Table A6.** Financing structure and calculation of WACC (Weighted Average Cost of Capital) for the two investments in Milan and Udine.

| | Owner–Tenant | | Entrepreneur–Developer | |
|---|---|---|---|---|
| | Equity [%] | Debt [%] | Equity [%] | Debt [%] |
| Financial structure | 20 | 80 | 30 | 70 |
| Capital weight | 3.81 [a] | 3.06 | 8.00 | 3.06 |
| WACC | 3.21 | | 4.54 | |

[a] Italian BTPs—May 2024.

## References


1. Calvin, K.; Dasgupta, D.; Krinner, G.; Mukherji, A.; Thorne, P.W.; Trisos, C.; Romero, J.; Aldunce, P.; Barrett, K.; Blanco, G.; et al. *IPCC, 2023: Climate Change 2023: Synthesis Report. Contribution of Working Groups I, II and III to the Sixth Assessment Report of the Intergovernmental Panel on Climate Change*; Core Writing Team, Lee, H., Romero, J., Eds.; IPCC: Geneva, Switzerland, 2023. [CrossRef]
2. United Nations. Annual Report 2022. United Nations Climate Change, Bonn. 2022. Available online: https://unfccc.int/sites/default/files/resource/UNClimateChange_AnnualReport_2022.pdf (accessed on 1 January 2025).
3. GlobalABC. 2021 Globalabc Annual Narrative Report. 2022. Available online: https://globalabc.org/index.php/resources/publications/2021-globalabc-annual-narrative-report (accessed on 1 January 2025).
4. UNEP—United Nations Environment Programme. Global Status Report for Buildings and Construction: Towards a Zero-Emission, Efficient and Resilient Buildings and Construction Sector. Nairobi. 2021. Available online: https://globalabc.org/ (accessed on 18 July 2024).
5. European Commission. REfurbishment Decision Making Platform Through Advanced Technologies for Near Zero Energy BUILDing Renovation. 2024. Available online: https://cordis.europa.eu/project/id/768623/reporting (accessed on 19 June 2024).
6. European Parliament and Council. *Directive (EU) 2024/1275 of the European Parliament and of the Council of 24 April 2024 on the Energy Performance of Buildings*; European Parliament and Council: Strasbourg, France, 2024.
7. Ngwakwe, C.C. Environmental and economic benefits of compliance to green building. *Int. J. Green Econ.* **2019**, *13*, 288–299. [CrossRef]
8. Ries, R.; Bilec, M.M.; Gokhan, N.M.; Needy, K.L. The Economic Benefits of Green Buildings: A Comprehensive Case Study. *Eng. Econ.* **2006**, *51*, 259–295. [CrossRef]
9. Araújo, C.; Almeida, M.; Bragança, L.; Barbosa, J.A. Cost–benefit analysis method for building solutions. *Appl. Energy* **2016**, *173*, 124–133. [CrossRef]
10. D'Alpaos, C.; Bragolusi, P. Buildings energy retrofit valuation approaches: State of the art and future perspectives. *Valori Valutazioni* **2018**, *2018*, 79–94.
11. Dell'anna, F.; Bottero, M. Green premium in buildings: Evidence from the real estate market of Singapore. *J. Clean. Prod.* **2021**, *286*, 125327. [CrossRef]
12. Fuerst, F.; McAllister, P.; Nanda, A.; Wyatt, P. Energy performance ratings and house prices in Wales: An empirical study. *Energy Policy* **2016**, *92*, 20–33. [CrossRef]
13. Leskinen, N.; Vimpari, J.; Junnila, S. A Review of the Impact of Green Building Certification on the Cash Flows and Values of Commercial Properties. *Sustainability* **2020**, *12*, 2729. [CrossRef]
14. Bragolusi, P.; D'Alpaos, C. The valuation of buildings energy retrofitting: A multiple-criteria approach to reconcile cost-benefit trade-offs and energy savings. *Appl. Energy* **2022**, *310*, 118431. [CrossRef]
15. Kerr, N.; Gouldson, A.; Barrett, J. The rationale for energy efficiency policy: Assessing the recognition of the multiple benefits of energy efficiency retrofit policy. *Energy Policy* **2017**, *106*, 212–221. [CrossRef]
16. Rosenow, J.; Kern, F.; Rogge, K. The need for comprehensive and well targeted instrument mixes to stimulate energy transitions: The case of energy efficiency policy. *Energy Res. Soc. Sci.* **2017**, *33*, 95–104. [CrossRef]
17. Baden-Fuller, C.; Haefliger, S. Business Models and Technological Innovation. *Long Range Plan.* **2013**, *46*, 419–426. [CrossRef]
18. Cluett, R.; Amann, J. *Multiple Benefits of Multifamily Energy Efficiency for Cost-Effectiveness Screening*; American Council for an Energy-Efficient Economy: Washington, DC, USA, 2015.
19. Teece, D.J. Business Models, Business Strategy and Innovation. *Long Range Plan.* **2010**, *43*, 172–194. [CrossRef]
20. Bidmon, C.M.; Knab, S.F. The three roles of business models in societal transitions: New linkages between business model and transition research. *J. Clean. Prod.* **2018**, *178*, 903–916. [CrossRef]
21. Bolton, R.; Hannon, M. Governing sustainability transitions through business model innovation: Towards a systems understanding. *Res. Policy* **2016**, *45*, 1731–1742. [CrossRef]

22. Belova, N.; Sbitneva, D. Business model of construction organization management under conditions of new industrial basis. *SHS Web Conf.* **2018**, *44*, 00015. [CrossRef]
23. Nagy, B.; Farmer, J.D.; Bui, Q.M.; Trancik, J.E. Statistical Basis for Predicting Technological Progress. *PLoS ONE* **2013**, *8*, e52669. [CrossRef]
24. Brown, D. Business models for residential retrofit in the UK: A critical assessment of five key archetypes. *Energy Effic.* **2018**, *11*, 1497–1517. [CrossRef]
25. EnergieSprong. Transition Zero. 2015. Available online: https://cdn.prod.website-files.com/59944999990f53000134107e/5bc8766a33d973180b34d58c_ESUK-Transition_Zero_document.pdf (accessed on 1 January 2025).
26. Energiesprong. Available online: https://energiesprong.org/ (accessed on 1 January 2025).
27. Sovacool, B.K. Fuel poverty, affordability, and energy justice in England: Policy insights from the Warm Front Program. *Energy* **2015**, *93*, 361–371. [CrossRef]
28. Willand, N.; Ridley, I.; Maller, C. Towards explaining the health impacts of residential energy efficiency interventions—A realist review. Part 1: Pathways. *Soc. Sci. Med.* **2015**, *133*, 191–201. [CrossRef]
29. Banca d'Italia, Eurosistema, e Istat. La Ricchezza dei Settori Istituzionali in Italia 2005–2021. Available online: https://www.bancaditalia.it (accessed on 1 January 2025).
30. Dalla Zuanna, G.D.; Weber, G. *Cose da Non Credere: Il Senso Comune Alla Prova dei Numeri*; Laterza: Bari, Italy, 2012.
31. Commissione Europea, Comunicazione della Commissione al Parlamento Europeo, al Consiglio, al Comitato Economico e Sociale Europeo e al Comitato delle Regioni. 2019. Available online: https://eur-lex.europa.eu/legal-content/IT/TXT/?uri=COM:2019:640:FIN (accessed on 1 January 2025).
32. Haines, A.; Scheelbeek, P. European Green Deal: A major opportunity for health improvement. *Lancet* **2020**, *395*, 1327–1329. [CrossRef]
33. Pierrehumbert, R. There is no Plan B for dealing with the climate crisis. *Bull. At. Sci.* **2019**, *75*, 215–221. [CrossRef]
34. Owen, A.; Middlemiss, L.; Brown, D.; Davis, M.; Hall, S.; Bookbinder, R.; Brisbois, M.C.; Cairns, I.; Hannon, M.; Mininni, G. Who applies for energy grants? *Energy Res. Soc. Sci.* **2023**, *101*, 103123. [CrossRef]
35. Cajias, M.; Piazolo, D. Green performs better: Energy efficiency and financial return on buildings. *J. Corp. Real Estate* **2013**, *15*, 53–72. [CrossRef]
36. Cerin, P.; Hassel, L.G.; Semenova, N. Energy Performance and Housing Prices. *Sustain. Dev.* **2014**, *22*, 404–419. [CrossRef]
37. Copiello, S.; Donati, E. Is investing in energy efficiency worth it? Evidence for substantial price premiums but limited profitability in the housing sector. *Energy Build.* **2021**, *251*, 111371. [CrossRef]
38. Marmolejo-Duarte, C.; Chen, A. The Uneven Price Impact of Energy Efficiency Ratings on Housing Segments. Implications for Public Policy and Private Markets. *Sustainability* **2019**, *11*, 372. [CrossRef]
39. Micelli, E.; Gilberto, G.; Righetto, E.; Tafuri, G. The economic value of sustainability. Real estate market and energy performance of homes: Il Valore Economico della Sostenibilità. Mercato Immobiliare e Prestazioni Energetiche delle Abitazioni. *Valori Valutazioni* **2023**, *34*, 3–16. [CrossRef]
40. Zhang, L.; Li, Y.; Stephenson, R.; Ashuri, B. Valuation of energy efficient certificates in buildings. *Energy Build.* **2018**, *158*, 1226–1240. [CrossRef]
41. Aydin, E.; Brounen, D.; Kok, N. The capitalization of energy efficiency: Evidence from the housing market. *J. Urban Econ.* **2020**, *117*, 103243. [CrossRef]
42. Eichholtz, P.; Kok, N.; Quigley, J.M. The Economics of Green Building. *Rev. Econ. Stat.* **2013**, *95*, 50–63. [CrossRef]
43. Brounen, D.; Kok, N. On the economics of energy labels in the housing market. *J. Environ. Econ. Manag.* **2011**, *62*, 166–179. [CrossRef]
44. Hyland, M.; Lyons, R.C.; Lyons, S. The value of domestic building energy efficiency—Evidence from Ireland. *Energy Econ.* **2013**, *40*, 943–952. [CrossRef]
45. Jensen, O.M.; Hansen, A.R.; Kragh, J. Market response to the public display of energy performance rating at property sales. *Energy Policy* **2016**, *93*, 229–235. [CrossRef]
46. Fuerst, F.; McAllister, P.; Nanda, A.; Wyatt, P. Does energy efficiency matter to home-buyers? An investigation of EPC ratings and transaction prices in England. *Energy Econ.* **2015**, *48*, 145–156. [CrossRef]
47. de Ayala, A.; Galarraga, I.; Spadaro, J.V. The price of energy efficiency in the Spanish housing market. *Energy Policy* **2016**, *94*, 16–24. [CrossRef]
48. Bisello, A.; Antoniucci, V.; Marella, G. Measuring the price premium of energy efficiency: A two-step analysis in the Italian housing market. *Energy Build.* **2020**, *208*, 109670. [CrossRef]
49. Del Giudice, V.; Massimo, D.E.; Salvo, F.; De Paola, P.; De Ruggiero, M.; Musolino, M. Market Price Premium for Green Buildings: A Review of Empirical Evidence. Case Study. In *New Metropolitan Perspectives*; Bevilacqua, C., Calabrò, F., Della Spina, L., Eds.; Smart Innovation, Systems and Technologies; Springer International Publishing: Cham, Switzerland, 2021; pp. 1237–1247. [CrossRef]

50. Mangialardo, A.; Micelli, E.; Saccani, F. Does Sustainability Affect Real Estate Market Values? Empirical Evidence from the Office Buildings Market in Milan (Italy). *Sustainability* **2019**, *11*, 12. [CrossRef]
51. Morano, P.; Rosato, P.; Tajani, F.; Di Liddo, F. An Analysis of the Energy Efficiency Impacts on the Residential Property Prices in the City of Bari (Italy). In *Green Energy and Technology*; Springer: Cham, Switzerland, 2020; pp. 73–88. [CrossRef]
52. Dell'anna, F. Machine learning framework for evaluating energy performance certificate (EPC) effectiveness in real estate: A case study of Turin's private residential market. *Energy Policy* **2025**, *198*, 114407. [CrossRef]
53. Ruggeri, A.G.; Gabrielli, L.; Scarpa, M.; Marella, G. What Is the Impact of the Energy Class on Market Value Assessments of Residential Buildings? An Analysis throughout Northern Italy Based on Extensive Data Mining and Artificial Intelligence. *Buildings* **2023**, *13*, 2994. [CrossRef]
54. Micelli, E.; Giliberto, G.; Righetto, E.; Tafuri, G. Urban Disparities in Energy Performance Premium Prices: Towards an Unjust Transition? *Land* **2024**, *13*, 224. [CrossRef]
55. Loberto, M.; Mistretta, A.; Spuri, M. The Capitalization of Energy Labels into House Prices. Evidence from Italy. *SSRN J.* **2024**. [CrossRef]
56. Richter, M. Utilities' business models for renewable energy: A review. *Renew. Sustain. Energy Rev.* **2012**, *16*, 2483–2493. [CrossRef]
57. Steinberger, J.K.; van Niel, J.; Bourg, D. Profiting from negawatts: Reducing absolute consumption and emissions through a performance-based energy economy. *Energy Policy* **2009**, *37*, 361–370. [CrossRef]
58. Boons, F.; Montalvo, C.; Quist, J.; Wagner, M. Sustainable innovation, business models and economic performance: An overview. *J. Clean. Prod.* **2013**, *45*, 1–8. [CrossRef]
59. Brege, S.; Stehn, L.; Nord, T. Business models in industrialized building of multi-storey houses. *Constr. Manag. Econ.* **2014**, *32*, 208–226. [CrossRef]
60. Johnson, M.W.; Christensen, C.M.; Kagermann, H. Reinventing your business model. *Harv. Bus. Rev.* **2008**, *86*, 50–59.
61. Osterwalder, A.; Pigneur, Y.; Tucci, C.L. Clarifying Business Models: Origins, Present, and Future of the Concept. *Commun. Assoc. Inf. Syst.* **2005**, *16*, 1. [CrossRef]
62. Suikki, R.M.; Goman, A.M.J.; Haapasalo, H.J.O. A framework for creating business models a challenge in convergence of high clock speed industry. *Int. J. Bus. Environ.* **2006**, *1*, 211–233. [CrossRef]
63. Foxon, T.J.; Bale, C.S.E.; Busch, J.; Bush, R.; Hall, S.; Roelich, K. Low carbon infrastructure investment: Extending business models for sustainability. *Infrastruct. Complex.* **2015**, *2*, 4. [CrossRef]
64. Gauthier, C.; Gilomen, B. Business Models for Sustainability: Energy Efficiency in Urban Districts. *Organ. Environ.* **2016**, *29*, 124–144. [CrossRef]
65. Hall, S.; Roelich, K. Business model innovation in electricity supply markets: The role of complex value in the United Kingdom. *Energy Policy* **2016**, *92*, 286–298. [CrossRef]
66. Hannon, M.J.; Bolton, R. UK Local Authority engagement with the Energy Service Company (ESCo) model: Key characteristics, benefits, limitations and considerations. *Energy Policy* **2015**, *78*, 198–212. [CrossRef]
67. Mahapatra, K.; Gustavsson, L.; Haavik, T.; Aabrekk, S.; Svendsen, S.; Vanhoutteghem, L.; Paiho, S.; Ala-Juusela, M. Business models for full service energy renovation of single-family houses in Nordic countries. *Appl. Energy* **2013**, *112*, 1558–1565. [CrossRef]
68. Winther, T.; Gurigard, K. Energy performance contracting (EPC): A suitable mechanism for achieving energy savings in housing cooperatives? Results from a Norwegian pilot project. *Energy Effic.* **2017**, *10*, 577–596. [CrossRef]
69. Moschetti, R.; Brattebø, H. Sustainable Business Models for Deep Energy Retrofitting of Buildings: State-of-the-art and Methodological Approach. *Energy Procedia* **2016**, *96*, 435–445. [CrossRef]
70. Brown, D.; Kivimaa, P.; Sorrell, S. *How Can Intermediaries Promote Business Model Innovation: The Case of 'Energiesprong' Whole-House Retrofits in the United Kingdom (UK) and the Netherlands*; SPRU Working Paper Series, Art. f2018–19; University of Sussex Business School: Falmer, UK, 2018. Available online: https://ideas.repec.org//p/sru/ssewps/2018-19.html (accessed on 11 January 2025).
71. Brown, D.; Hall, S.; Martiskainen, M.; Davis, M.E. Conceptualising domestic energy service business models: A typology and policy recommendations. *Energy Policy* **2022**, *161*, 112704. [CrossRef]
72. Sovacool, B.K.; Turnheim, B.; Martiskainen, M.; Brown, D.; Kivimaa, P. Guides or gatekeepers? Incumbent-oriented transition intermediaries in a low-carbon era. *Energy Res. Soc. Sci.* **2020**, *66*, 101490. [CrossRef]
73. Delpont, S. Industrialisation de la rénovation énergétique. *Ann. Des Mines Réalités Ind.* **2022**, *2022*, 35–38. [CrossRef]
74. Massa, L.; Tucci, C.L. Business model innovation. In *The Oxford Handbook of Innovation Management*; Oxford Academic: Oxford, UK, 2013; Volume 20, pp. 420–441.
75. Micelli, E.; Mangialardo, A. Recycling the City New Perspective on the Real-Estate Market and Construction Industry. In *Smart and Sustainable Planning for Cities and Regions: Results of SSPCR 2015*; Bisello, A., Vettorato, D., Stephens, R., Elisei, P., Eds.; Green Energy and Technology; Springer International Publishing: Cham, Switzerland, 2017; pp. 115–125. [CrossRef]

76. Friedler, C.; Kumar, C.; Reinventing retrofit: How to scale up home energy efficiency in the UK. Green Alliance. 2019. Available online: https://green-alliance.org.uk/publication/reinventing-retrofit-how-to-scale-up-home-energy-efficiency-in-the-uk/ (accessed on 20 August 2024).
77. Pellegrino, M.; Wernert, C.; Chartier, A. Social Housing Net-Zero Energy Renovations with Energy Performance Contract: Incorporating Occupants' Behaviour. *Urban Plan.* **2022**, *7*, 5–19. [CrossRef]
78. Pellegrino, M.; Building Sector Stakeholders Facing the Challenge of Scalingup high-Performance Energy Renovation: The case of the Energiesprong Approach in the Netherlands and France. Revue Internationale d'Urbanisme. 2019. Available online: https://hal.science/hal-02937949v1/document (accessed on 8 March 2025).
79. Nesticò, A.; De Mare, G.; Aurigemma, I. Off-Site Construction. The Economic Analyses for the Energy Requalification of the Existing Buildings. In *Values and Functions for Future Cities*; Mondini, G., Oppio, A., Stanghellini, S., Bottero, M., Abastante, F., Eds.; Springer International Publishing: Cham, Switzerland, 2020; pp. 447–462. [CrossRef]
80. Ellram, L.M. Total cost of ownership: An analysis approach for purchasing. *Int. J. Phys. Distrib. Logist. Manag.* **1995**, *25*, 4–23. [CrossRef]
81. Ferreira, M.; Almeida, M. Benefits from Energy Related Building Renovation Beyond Costs, Energy and Emissions. *Energy Procedia* **2015**, *78*, 2397–2402. [CrossRef]
82. Berto, R.; Tintinaglia, F.; Rosato, P. How much is the indoor comfort of a residential building worth? A discrete choice experiment. *Build. Environ.* **2023**, *245*, 110911. [CrossRef]
83. Aydin, Y.C.; Mirzaei, P.A.; Akhavannasab, S. On the relationship between building energy efficiency, aesthetic features and marketability: Toward a novel policy for energy demand reduction. *Energy Policy* **2019**, *128*, 593–606. [CrossRef]
84. Morano, P.; Tajani, F.; Di Liddo, F.; Amoruso, P. A Feasibility Analysis of Energy Retrofit Initiatives Aimed at the Existing Property Assets Decarbonisation. *Sustainability* **2024**, *16*, 3204. [CrossRef]
85. Streicher, K.N.; Mennel, S.; Chambers, J.; Parra, D.; Patel, M.K. Cost-effectiveness of large-scale deep energy retrofit packages for residential buildings under different economic assessment approaches. *Energy Build.* **2020**, *215*, 109870. [CrossRef]
86. Nesticò, A.; Moffa, R. Economic analysis and operational research tools for estimating productivity levels in off-site construction. *Valori Valutazioni* **2018**, *20*, 107–128.
87. Cachon, G.P.; Harker, P.T. Competition and Outsourcing with Scale Economies. *Manag. Sci.* **2002**, *48*, 1314–1333. [CrossRef]
88. Pellegrino, R.; Costantino, N. An empirical investigation of the learning effect in concrete operations. *Eng. Constr. Arch. Manag.* **2018**, *25*, 342–357. [CrossRef]
89. Tagliaro, C.; Pomè, A.P.; Bellintani, S.; Ciaramella, A. *La Digitalizzazione del Settore Immobiliare e Delle Costruzioni: Nuove Tecnologie e Soluzioni per la Gestione Dell'ambiente Costruito*; FrancoAngeli: Milan, Italy, 2024.
90. Guerrieri, G. *Il Mercato Della Casa. Domanda, Offerta, Tassazione e Spesa Pubblica*; Carocci: Rome, Italy, 2022.
91. ENEA. Certificazione Energetica Degli Edifci Italiani. Available online: https://pnpe2.enea.it/statistiche (accessed on 25 July 2024).
92. Fylan, F.; Glew, D.; Smith, M.; Johnston, D.; Brooke-Peat, M.; Miles-Shenton, D.; Fletcher, M.; Aloise-Young, P.; Gorse, C. Reflections on retrofits: Overcoming barriers to energy efficiency among the fuel poor in the United Kingdom. *Energy Res. Soc. Sci.* **2016**, *21*, 190–198. [CrossRef]
93. Michelsen, C.; Rosenschon, S.; Schulz, C. Small might be beautiful, but bigger performs better: Scale economies in "green" refurbishments of apartment housing. *Energy Econ.* **2015**, *50*, 240–250. [CrossRef]
94. Christersson, M.; Vimpari, J.; Junnila, S. Assessment of financial potential of real estate energy efficiency investments–A discounted cash flow approach. *Sustain. Cities Soc.* **2015**, *18*, 66–73. [CrossRef]
95. IPD; KTI. Property Valuation in the Nordic Countries. 2012. Available online: https://kti.fi/wp-content/uploads/Property-valuation-in-the-Nordic-countries.pdf (accessed on 1 January 2025).
96. Shapiro, E.; Mackmin, D.; Sams, G. *Modern Methods of Valuation*, 12th ed.; Estates Gazette: London, UK, 2019. [CrossRef]
97. EnergieSprong France. *Observatoire Coûts, Qualité et Impact des Rénovations EnergieSprong*; Energiesprong: Paris, France, 2021.
98. Eurostat Database. 2018. Available online: https://ec.europa.eu/eurostat/data/database (accessed on 30 March 2025).
99. Burroni, L.; Gherardini, A.; Scalise, G. Policy Failure in the Triangle of Growth: Labour Market, Human Capital, and Innovation in Spain and Italy. *South Eur. Soc. Politics* **2019**, *24*, 29–52. [CrossRef]
100. Insee. Indice du Coût de la Construction des Immeubles à Usage D'habitation (ICC). Available online: https://www.insee.fr/fr/statistiques/serie/000008630 (accessed on 25 July 2024).
101. Wright, T.P. Factors affecting the cost of airplanes. *J. Aeronaut. Sci.* **1936**, *3*, 122–128. [CrossRef]
102. Thompson, P. The Relationship between Unit Cost and Cumulative Quantity and the Evidence for Organizational Learning-by-Doing. *J. Econ. Perspect.* **2012**, *26*, 203–224. [CrossRef]
103. de La Tour, A.; Glachant, M.; Ménière, Y. Predicting the costs of photovoltaic solar modules in 2020 using experience curve models. *Energy* **2013**, *62*, 341–348. [CrossRef]

104. Lafond, F.; Bailey, A.G.; Bakker, J.D.; Rebois, D.; Zadourian, R.; McSharry, P.; Farmer, J.D. How well do experience curves predict technological progress? A method for making distributional forecasts. *Technol. Forecast. Soc. Chang.* **2018**, *128*, 104–117. [CrossRef]
105. Lancaster, K.J. A New Approach to Consumer Theory. *J. Political Econ.* **1966**, *74*, 132–157. [CrossRef]
106. Rosen, S. Hedonic Prices and Implicit Markets: Product Differentiation in Pure Competition. *J. Politic. Econ.* **1974**, *82*, 34–55. [CrossRef]
107. Michelangeli, A. Il metodo dei prezzi edonici per la costruzione di indici dei prezzi per il mercato immobiliare. In *Principi Metodologici per la Costruzione di Indici dei Prezzi del Mercato Immobiliare*; Del Giudice, V., D'Amato, M., Eds.; Maggioli: Rimini, Italy, 2008.
108. Energiesprong Italia. Introduzione a EnergieSprong. *Edera, Milano.* 2024. Available online: https://drive.google.com/drive/u/1/folders/1jVocyxki8l5fIkHLI3S8UtH7rXTYamEQ (accessed on 25 July 2024).
109. Morri, G.; Benedetto, P.; Mazzocco, A.; Caselli, S. *Valutazione Immobiliare. Metodologie e Casi*; EGEA: Milan, Italy, 2017.
110. Orefice, M.; Orefice, L. *Estimo Civile*; De Agostini: Rome, Italy, 2014.
111. Micelli, E.; Righetto, E. How do metropolitan cities evolve after the 2008/2012 crisis and the Covid-19 pandemic? An analysis from real estate market values: Come evolvono le città metropolitane dopo crisi e pandemia Covid-19? Un'analisi a partire dai valori del mercato immobiliare. *Valori Valutazioni* **2023**, *31*, 49–67. [CrossRef]
112. Eurostat. Our Lives in the City. Available online: https://ec.europa.eu/eurostat/web/products-eurostat-news/-/EDN-20180207-1 (accessed on 25 July 2024).
113. Presidente della Repubblica Italiana, d.P.R. n. 412—Regolamento Recante Norme per la Progettazione, L'installazione, L'esercizio e la Manutenzione Degli Impianti Termici Degli Edifici ai Fini del Contenimento dei Consumi di Ener 1993. Available online: https://www.bosettiegatti.eu/info/norme/statali/1993_0412.htm (accessed on 12 January 2025).
114. Tsikaloudaki, K.; Laskos, K.; Bikas, D. On the Establishment of Climatic Zones in Europe with Regard to the Energy Performance of Buildings. *Energies* **2012**, *5*, 32–44. [CrossRef]
115. Curto, R.; Fregonara, E.; Semeraro, P. Asking prices vs market prices: An empirical analysis. *Territ. Ital.* **2012**, *12*, 53–72.
116. Immobiliare.it. Dati dei Prezzi Degli Immobili in Italia. Available online: https://www.immobiliare.it (accessed on 1 January 2025).
117. Guglielminetti, E.; Loberto, M.; Zevi, G.; Zizza, R. *Living on My Own: The Impact of the COVID-19 Pandemic on Housing Preferences*; Bank of Italy Occasional Paper, Fasc. 627; Bank of Italy: Rome, Italy, 2021.
118. Kain, J.F.; Quigley, J.M. Measuring the value of house quality. *J. Am. Stat. Assoc.* **1970**, *65*, 532–548. [CrossRef]
119. Yayar, R.; Demir, D. Hedonic estimation of housing market prices in Turkey. *Erciyes Üniversitesi İktisadi İdari Bilim. Fakültesi Derg.* **2014**, *43*, 67–82. [CrossRef]
120. Ottensmann, J.; Payton, S.; Man, J. Urban Location and Housing Prices within a Hedonic Model. *J. Reg. Anal. Policy* **2008**, *38*, 19–35.
121. Simonotti, M. I prezzi marginali impliciti delle risorse immobiliari. In *Estimo ed Economia Ambientale*; Angeli: Milan, Italy, 1993.
122. Gestore Mercati Energetici. Prezzi Medi Energie. Available online: https://www.mercatoelettrico.org/it-it/ (accessed on 1 January 2025).
123. ARERA. Andamento del Prezzo del Gas Naturale per un Consumatore Domestico Tipo in Regime di tutela. Available online: https://www.arera.it (accessed on 1 March 2024).
124. Agenzia delle Entrate. Osservatorio del Mercato Immobiliare, Banca Dati delle Quotazioni Immobiliari. Available online: http://www.agenziaentrate.gov.it (accessed on 1 March 2023).
125. Assofin. Osservatorio sul Credito Immobiliare. Available online: https://www.assofin.it/ (accessed on 1 March 2024).
126. Baum, A. *Real Estate Investment: A Strategic Approach*, 4th ed.; Routledge: London, UK, 2015. [CrossRef]
127. Argote, L.; Epple, D. Learning Curves in Manufacturing. *Science* **1990**, *247*, 920–924. [CrossRef]
128. Hirschmann, W.B. Profit from the Learning Curve. *Harv. Bus. Rev.* **1964**, *42*, 125–139.
129. Nemet, G.F. Beyond the learning curve: Factors influencing cost reductions in photovoltaics. *Energy Policy* **2006**, *34*, 3218–3232. [CrossRef]
130. Yeh, S.; Rubin, E.S. A review of uncertainties in technology experience curves. *Energy Econ.* **2012**, *34*, 762–771. [CrossRef]
131. Thompson, P. Chapter 10—Learning by Doing. In *Handbook of the Economics of Innovation*; Hall, B.H., Rosenberg, N., Eds.; Elsevier: Amsterdam, The Netherlands, 2010; Volume 1, pp. 429–476. [CrossRef]
132. de Blasio, G.; Fiori, R.; Lavecchia, L.; Loberto, M.; Michelangeli, V.; Padovani, E.; Pisano, E.; Rodano, M.L.; Roma, G.; Rosolin, T.; et al. *Improving the Energy Efficiency of Homes in Italy: The State of the Art and Some Considerations for Public Intervention*; Fasc. 845; Banca d'Italia: Rome, Italy, 2024.
133. Dijkstra, L.; Poelman, H.; Rodríguez-Pose, A. The geography of EU discontent. *Reg. Stud.* **2020**, *54*, 737–753. [CrossRef]
134. Rodríguez-Pose, A. The revenge of the places that don't matter (and what to do about it). *Camb. J. Reg. Econ. Soc.* **2018**, *11*, 189–209. [CrossRef]
135. Amstalden, R.W.; Kost, M.; Nathani, C.; Imboden, D.M. Economic potential of energy-efficient retrofitting in the Swiss residential building sector: The effects of policy instruments and energy price expectations. *Energy Policy* **2007**, *35*, 1819–1829. [CrossRef]

136. Fabbri, K.; Gaspari, J. Mapping the energy poverty: A case study based on the energy performance certificates in the city of Bologna. *Energy Build.* **2021**, *234*, 110718. [CrossRef]
137. Sistema Informativo Sugli Attestati di Prestazione Energetica (SIAPE). Indice di Prestazione ed Emissioni. SIAPE. Available online: https://siape.enea.it/ (accessed on 2 August 2024).

*buildings*

MDPI

*Article*

# Building Energy Assessment of Thermal and Electrical Properties for Compact Cities: Case Study of a Multi-Purpose Building in South Korea

**Jaeho Lee and Jaewan Suh ***

Department of Electrical Engineering, College of Engineering, Incheon National University, Incheon 22035, Republic of Korea; mpforurbanenergy@inu.ac.kr
* Correspondence: jwsuh@inu.ac.kr

**Abstract:** This study conducts a simulation-based assessment of a recently commissioned office building in the Republic of Korea, representing a typical public office facility. The building was modeled using EnergyPlus 23.1.0 after construction, although no validation was performed due to the absence of metered consumption data. Previous approaches relying on simplified methods such as the Radiant Time Series (RTS), which neglect dynamic building behavior, have often led to overestimated cooling and heating loads. This has emerged as a major obstacle in designing energy-efficient buildings within the context of compact and smart cities pursuing carbon neutrality. Consequently, the trend in building performance analysis is shifting toward dynamic simulations and digital twin-based design methodologies. Furthermore, electrification of buildings without adequate thermal load assessment may also contribute to overdesign, irrespective of urban environmental characteristics. From an urban planning standpoint, there is a growing need for performance criteria that reflect occupant behavior and actual usage patterns. However, dynamics-based building studies remain scarce in the Republic of Korea. In this context, the present study demonstrates that passive design strategies, implemented through systematic changes in envelope materials, HVAC operational standards, and compliance with ASHRAE 90.1 criteria, can significantly enhance thermal comfort and indoor air quality. The simulation results show that energy consumption can be reduced by over 36.21% without compromising occupant health or comfort. These findings underscore the importance of thermal load understanding prior to electrification and highlight the potential of LEED-aligned passive strategies for achieving high-performance, low-energy buildings.

**Keywords:** radiant time series (RTS); heat ventilation air conditioning (HVAC); leadership in energy and environmental design (LEED); dedicated outdoor air system (DOAS); passive design

## 1. Introduction

Owing to the global demand to achieve carbon neutrality and concerns about energy security, efforts to achieve carbon neutrality have become paramount. A recent report by the World Nuclear Association (WNA) reveals that fossil fuel-based generation resources accounted for 40% of carbon emissions as of May 2021, with densely populated cities contributing 60% [1]. As one of the measures, smart cities oriented towards introducing energy-saving opportunities have been actively and voluntarily developed in member states of the European Union (EU) [2,3]. However, voluntary participation in smart city initiatives cannot be achieved solely through the compulsion of authorities, such as the forced curtailment of energy demand by sacrificing occupants' comfort. Thus, leadership initiatives in smart cities have asserted the need to develop balanced energy-saving strategies in the building sector without compromising comfort [4–6]. However, because these measures have focused on building operations as active measures when a building is commissioned, they have not considered the potential greenhouse gas (GHG) emissions from the lifecycle perspective of the building.

In the United States, building operations contribute to 40% of the total energy consumption, and the building energy sector is responsible for 29% of all GHG emissions. Before the emergence of attention to climate change due to GHG emissions, such emissions from direct energy consumption were considered [7,8].

However, the trends towards considering GHG emissions have begun shifting to include GHG emissions from the perspective of the entire life cycle, considering the production of building materials [9,10]. Therefore, the calculation of carbon emissions includes emissions during the use of a building and emissions generated during the design and construction processes before commercial operation.

Consequently, GHG emissions from buildings contribute to 48% of the total annual emissions in the United States, with 36% originating from building operations and 12% from construction. These emissions fall under the category of embodied carbon emissions per building, as outlined by Shingler [11].

As the focus on sustainable development increases in consideration of energy-saving opportunities from passive design factors in buildings, Leadership in Energy and Environmental Design (LEED) criteria have emerged as generalizable standards for building operations throughout the entire life cycle [12,13]. A major reason for designating LEED criteria in simulation environments is their alignment with the environmental and energy requirements of smart cities [14,15]. Table 1 compares the registration criteria of LEED with the principles of smart cities. It assesses energy efficiency, environmental impact, and ethical benefits for residents and neighbors. This analysis combines the planning philosophies of sustainability and smart city concepts [16,17].

**Table 1.** Comparisons of design criteria of LEED and smart cities.

| Criteria | LEED Criteria | Smart Cities |
|---|---|---|
| Infrastructure | Flexible | High technology required |
| Definition | Clear | Varies upon organizations |
| Recognition | Clear by credits | Subtle and esoteric |
| Primary Purpose | Sustainability | Smartness |

As evidenced by the numerous smart city projects worldwide, many buildings have obtained LEED certification. The city of Songdo, Republic of Korea, near the metropolitan city of Incheon, is actively participating in the state-launched smart city initiative and has drawn considerable attention, particularly in the green building industry, owing to its high concentration of LEED certifications [18]. Because the ultimate design motto of a smart city revolves around sustainable infrastructure for coexistence, where smartness does not necessarily imply sustainability, the design motto of a LEED city was deemed appropriate for the current simulation [19,20].

In addition to cases reported in the Republic of Korea, LEED cities have been developed in Brazil, Russia, India, and China (BRICS) to participate in sustainable development [21,22]. Currently, such federations are known as nations disinterested in achieving carbon neutrality owing to greater dependency upon the generation resources by fossil fuels by considering nation-varying economic reasons [23]. However, introducing a standardized framework for passive building designs could contribute to the requirements for thermal load in terms of building operations, regardless of energy security [24].

In addition to the passive factors that exacerbate reduction opportunities for GHG and energy consumption, initiatives such as the green button focused on active measures have been promoted in very recent periods [25]. One of the significant factors in green button initiatives to minimize is the energy consumption of heating, ventilation, and air conditioning (HVAC) operations [26,27]. However, because of their highly nonlinear characteristics and complicated operational mechanisms, volunteers find it challenging to develop and diagnose measures applicable to their buildings [28]. Thus, precise energy consumption assessments must be conducted using a simulation tool that can provide component-wise responses [29]. However, these initiatives also focus on active measures that control electrical loads when the building is commissioned.

Moreover, despite efforts towards the promotion of voluntary participation in active measurements by controlling the electrical load resources by providing additional incentives proportional to its curtailed kilowatt capacity, the stakeholders responsible for the consultancy of the potential curtailment capability have been registering the availability of curtailment capacity by considering only the nominal capacity of the devices on the nameplate [30,31].

General operations of HVAC with constant air volume (CAV) systems, commonly installed owing to their cost, have demonstrated that the electrical load is highly dependent on thermal energy consisting of human activities, ventilation requirements, and weather conditions, which are passive factors [32,33]. Hence, exhuming the energy-saving potential without a precise diagnosis of the system is virtually impossible, and the passive factors in the building have been considered as influential factors for the thermal gains and losses that determine HVAC efficiency [34,35]. Therefore, control logic and diagnosis in steps capable of incorporating passive and active measures must be developed [36,37]. EnergyPlus 23.1.0 is considered a reliable engine that can provide solutions of the desired quality. In contrast, the widely utilized radiant time sites (RTSs) cannot perform precise simulations based on the structural environment of the actual site [38,39].

Moreover, most previous studies focused on the operational logic of HVAC with variable air volume (VAV) systems, whereas HVAC with CAV systems is dominant in land markets because of its cost efficiency. Furthermore, previous studies related to energy-saving potential from the perspective of thermal and electrical energy have dealt with the analysis in a separate manner, without a system-wide integrative and interactive analysis of the design phases and components of HVAC. For instance, Abuseif et al. [40], Song et al. [41], Edun et al. [42], Chen et al. [43], and Navia-Osorio et al. [44] reported the effect of changing the roof envelope as a passive measure and observed its effects on the electrical load. However, critical design parameters as passive measures in a building are not restricted by envelopes alone.

From the perspective of active measures, Queiroz et al. [45], Zhang et al. [46] and Xie et al. [47] studied ventilation strategies and their effects on thermal load calculations. However, these analyses were limited to the effects of ventilation on active measures, without considering passive measures.

The adaptation of digital-twin-driven building designs that consider dynamics has recently attracted interest, whereas the majority of previous design methods were based on steady-state or RTS-driven approaches. Thus, this paper presents the effectiveness of a building design method that incorporates dynamic characteristics during the initial design phases but considers potential effects during the operational phase of the building.

The major contributions of this paper are summarized as follows:

- Technical necessities and contributions from digital-twin-based building designs from the perspective of urban planning are discussed.
- A methodology for assessing passive and active energy curtailment capability in design phases is proposed.
- The impact of changes in design criteria on occupant thermal comfort.
- $CO_2$ discomfort standards are analyzed.

From a governmental standpoint, where generalizable and standardized instruction dissemination is prioritized, the use of simplified cooling load estimation methods remains a preferred approach for designers due to its intuitive applicability. One such method, the Radiant Time Series (RTS) method developed by ASHRAE, serves as a conventional baseline in estimating peak cooling loads. The RTS method distributes internal and external heat gains over time using standardized profiles, offering a straightforward and consistent framework for early-stage design.

However, reliance on this method has led to systematic overestimations of building system capacities, particularly in cases where energy-efficient design strategies are employed. This is primarily due to the RTS method's inherent limitations in capturing the dynamic and transient behaviors of real buildings. In contrast, simulation tools such as EnergyPlus incorporate real-time control strategies, thermal mass interactions, and ventilation dynamics, offering a more nuanced and temporally resolved representation of thermal loads. Therefore, the comparative analysis between RTS-based estimations and EnergyPlus-based simulations reveals critical discrepancies, emphasizing the need for dynamic modeling in accurately assessing energy performance.

Table 2 provides comparative analysis between the previous studies and the proposed study that were performed from the perspective of the whole-building analysis. The proposed study focuses on a university campus building in South Korea, providing insight into the energy dynamics of education facilities. In contrast, Camblong et al. [48] analyze a real residential or mixed-use building in Europe, while Pérez-Carramiñana et al. [49] examine a real office building in Spain. This variation in building typologies enables a diverse understanding of energy consumption patterns across different functional categories.

**Table 2.** Comparison of the Previous Studies on the Influences from the Standards.

| Category | Proposed Analysis | Camblong et al. [48] | Pérez-Carramiñana et al. [49] |
|---|---|---|---|
| Case Basis | University campus building | Real building | Real office building |
| Energy Focus | Electricity + Thermal | Photovoltaics + EMS | Heating energy consumption and thermal comfort |
| Analysis Tools | EnergyPlus, OpenStudio | Rule-based EMS | Simulation-based assessment |
| Policies/Standards | LEED v4.1, ASHRAE 90.1 | PV self-consumption regulations | Spanish energy-saving standards |
| Comfort Metrics | PMV, RH, $CO_2$ | Electricity use optimization | PMV and temperature satisfaction |
| Climate Zone | ASHRAE 4C (Incheon) | European temperate climate | Dry Mediterranean climate |

From the perspectives of load characteristics, evaluating both electricity and thermal energy requirement, particularly emphasizing HVAC performance. Camblong et al. [48] focus on photovoltaic integration and EMSs (Energy Management Systems) to enhance self-consumption. Pérez-Carramiñana et al. [49] primarily address heating energy and thermal comfort, especially under regulatory changes triggered by geopolitical events.

A key strength of the proposed analysis is its use of EnergyPlus and OpenStudio, industry-standard simulation tools that offer detailed thermodynamic modeling. Camblong et al. [48] apply a rule-based EMS, while Pérez-Carramiñana et al. [49] adopt a simulation-based assessment framework tailored to policy evaluation. Each tool is selected to suit the study's objectives that are control logic vs. performance benchmarking vs. code compliance.

The proposed study adheres to LEED v4.1 [50] and ASHRAE 55 [51], ensuring international standardization and comparability. Camblong et al. [48] implement PV self-consumption regulations, likely aligned with EU directives. Pérez-Carramiñana et al. [49] evaluate the Spanish energy-saving standards (2022), revealing how national regulations impact comfort and performance. This policy diversity illustrates how energy performance is shaped by both global and local frameworks.

Comfort in the proposed analysis is quantified using PMV (Predicted Mean Vote), RH (Relative Humidity), and $CO_2$ concentration, covering both thermal and indoor air quality. Camblong et al. [48] focus on electricity optimization, indirectly improving comfort by stabilizing supply. Pérez-Carramiñana et al. [49] specifically assess PMV and temperature satisfaction, linking regulatory heating limitations to occupant well-being.

The three studies collectively contribute to the evolving discourse on building energy efficiency. While differing in scope and tools, they all emphasize the importance of integrating local climate conditions, occupant comfort, and regulatory compliance into performance-based energy design. The proposed analysis uniquely bridges passive and active design strategies using LEED-compliant simulations, providing a robust and transferable framework for education and public buildings across temperate climates.

This paper presents a structured analysis using EnergyPlus that considers the criteria set by LEED and the coordination of a HVAC system with CAV and a dedicated outdoor air system (DOAS), considering the characteristics of multipurpose buildings capable of generalization. The remainder of this paper is organized as follows: Section 2 describes the applicable LEED protocols in the present case study and the simulation environment for each case. Section 3 presents and discusses the results. Finally, Section 4 concludes the paper.

## 2. Materials and Methods

Assessments of building energy consumption generally follow the protocols set by either the American Society of Heating, Refrigerating, and Air-Conditioning Engineers (ASHRAE) or specific LEED versions. LEED protocols are often derived from the ASHRAE standards, as in LEED v4.1, which uses ASHRAE 90.1-2016 as a benchmark for evaluating building performance improvements.

### 2.1. LEED Criteria and Design Strategies

LEED Building Design and Construction (BD+C) is primarily applied during the design and construction phase of new buildings or major renovations, focusing on proactive integration of sustainability from the outset. In contrast, LEED Existing Building Operation and Maintenance (EBOM) targets existing buildings that have been in operation for at least two years, emphasizing performance-based assessments and ongoing operational improvements with minimal structural alterations that are typically under 50%.

As shown in Table 3, each protocol encompasses distinct eligibility criteria, intervention strategies, and performance metrics that must be considered independently when evaluating sustainability outcomes. EBOM was considered eligible for this simulation because it aligns with the qualification criteria. Table 3 shows the overlaps between BD+C and EBOM criteria. Key aspects such as Indoor Air Quality (IAQ), Thermal comfort (TC), and Energy Efficiency (EE) under indoor environmental quality (IEQ) are consistent across both protocols, adhering to the baseline standard 90.1-2016. This case study focused on modifications in the envelope properties from the initial design, without considering the properties of ASHRAE 90.1-2016, IAQ, TC, and EE, following the specified design densities. EBOM primarily aims to enhance IAQ by setting minimum standards to improve occupant health and comfort. For the subject building, compliance with the ASHRAE Standard 62.1-2016 [52] is required in the states or equivalent CEN standards elsewhere. This study selected the ASHRAE Standard 62.1-2016, aligning with the building envelope changes considered.

**Table 3.** Criteria for LEED BD+C and EBOM applicable to commercial buildings.

| Criteria | BD+C | EBOM | References |
|---|---|---|---|
| Protocol eligibility | Design phase | Operations period > 2 years | [53–55] |
| Design change | Design phase | Alterations under 50% | |
| Energy and atmosphere | Grid harmonization/Enhanced/Fundamental refrigerant management/<br>Energy performance/Energy efficiency best management practices | | |
| Materials and resources | Construction and demolition waste management/<br>Building life-cycle impact reduction | | |
| Indoor environmental quality | Minimum indoor air quality/Environment tobacco smoke control/<br>Indoor environmental quality performance | | |

The design flow considered in the simulation is presented in Table 4. It contrasts a standard reference case with three distinct scenarios: active, passive, and integrated measures, all set against a backdrop of consistent initial conditions, such as operation under a CAV system and fixed utility tariffs.

**Table 4.** Types of the LEED Protocols Applicable in the Proposed Design [56,57].

| Case Environment | Protocol | Replacement | Description of Replacement |
|---|---|---|---|
| Case 1 | Reference design | No-change | Specification in Tables 5 and 6 as current material properties |
| Case 2 | LEED v4.1 BD+C/Baseline by ASHRAE 90.1-2016 | External envelope | Material property and air infiltration rate per area referenced by baseline 90.1-2016 |
| | LEED v4.1 BD+C/Baseline by ASHRAE 90.1-2016 | | |
| Case 3 | Grid harmonization/Enhanced and fundamental refrigerant management/ | Energy efficiency | Design densities of electric equipment and occupants after space types |
| Case 4 | LEED v4.1 EBOM (Commissioning) | Indoor air quality | Minimum OA requirement for each person specified by EBOM |
| | | Thermal comfort | Range of TC specified by EBOM |
| Case 5 | Whole-in-One (Commissioning) | All changes adapted | Replacement in case environment from Cases 2 to 5 |

**Table 5.** Representative Material Properties of the Building Envelope of the Test Site [58].

| Space | Material (Type) | Thickness (mm) | Conductivity (W/m·K) | Density (kg/m$^3$) | Specific Heat (J/kg·K) |
|---|---|---|---|---|---|
| Exterior Wall | PF | 100 | 0.189 | 39 | 1400 |
| | Gypsum | 25 | 0.260 | 784.90 | 830 |
| Interior Wall | Gypsum | 200 | 0.35 | 1000 | 1090 |
| Exterior Door | Wood | 25 | 0.15 | 608 | 1630 |
| Interior Door | | | | | |
| Exterior Window (Aluminum Frame) | 1st Frame (Low-E) | 6 | 1.414 | - | |
| | Argon | 16 | | | |
| | 2nd Frame | 6 | | | |
| Exterior Roof | Concrete w/o Frame | 150 | 1.6 | 2200 | 1000 |
| | Concrete w/Frame | 200 | 1.6 | 2200 | 1000 |
| | Extrusion Protection | 180 | 0.027 | 35 | 837 |
| Interior Ceiling | Gypsum | 9.5 | 0.260 | 784.90 | 830 |
| | Air | 281 | - | - | - |
| | Gypsum | 9.5 | 0.260 | 784.90 | 830 |
| Interior Floor | Cement Mortar w/ Carpet Tiles | 100 | 1.4 | 2000 | 1000 |

**Table 6.** Electric and light load density in a reference case [58].

| Space Type | Occupant (ppl/m$^2$) | Light Load (W/m$^2$) | Equipment (W/m$^2$) | Ventilation (CMH/ppl) |
|---|---|---|---|---|
| Office | 0.2 | 15 | 15 | 36 |
| Lecture Room | 0.5 | 30 | 40 | 36 |
| Classroom | 0.5 | 15 | 10 | 36 |
| Corridor | 0.1 | 20 | 0 | 36 |
| Library | 0.2 | 15 | 15 | 36 |
| Conference Room | 0.5 | 15 | 10 | 36 |
| Coffee Station | 0.2 | 25 | 0 | 36 |
| Elevator | 0.1 | 20 | 0 | 36 |
| Computer Laboratory | 0.5 | 15 | 10 | 36 |
| Server Room | 0.1 | 20 | 70 | 36 |
| Entrance | 0.1 | 20 | 0 | 36 |
| Restroom | 0.1 | 20 | 0 | 36 |

The set-point controller for each thermal zone was designed for zone-average conditioning because of the technical characteristics of CAV-HVAC. The temperature ranges for the reference cases were designed to comply with the local recommendations of the government. For the reference case, the winter temperature was set to be between 17 and 23 °C, whereas the summer values were between 26 and 28 °C. Additionally, a night cycle operation schedule was applied in the simulation. The climate environment was considered accordingly to the conditions represented in Appendix A.

Relevant thermal comfort set-points for the breathing zones considered in the cases in compliance with LEED v4.1 are based on a Predicted Mean Vote (PMV) range of −0.5 to 0.5, with relative humidity (RH) levels selected for occupant health. Air velocity in breathing zones was set to 0.2 m/s, accounting for occupant clothing schedules that were 0.5 in summer and 1.0 in winter. The metabolic rates of the occupants from 1.0 to 1.3 met were considered suitable for office work and low-demand physical activity. Temperatures for cold and heat stress thresholds were considered to be at 22 °C, at which elderly occupants would begin feeling discomfort, and 27 °C, which is the initial temperature for the stress chart.

### 2.2. Building Envelope

The subject education hall in Siheung, Republic of Korea, part of the university campus in the present simulation, was commissioned in 2020. Figure 1a shows the education hall. The subject education hall consisted of 10 stories, an underground parking lot, and a centralized HVAC control room. The hall was constructed in compliance with the design regulations applicable for construction in the metropolitan area of Incheon, as specified by the Ministry of Land, Transportation, and Infrastructure (MOLIT) in the Republic of Korea, and a centralized HVAC control room. The hall was constructed in compliance with the design regulations applicable for construction in the metropolitan area of Incheon, as specified by the Ministry of Land, Transportation, and Infrastructure (MOLIT) in the Republic of Korea.

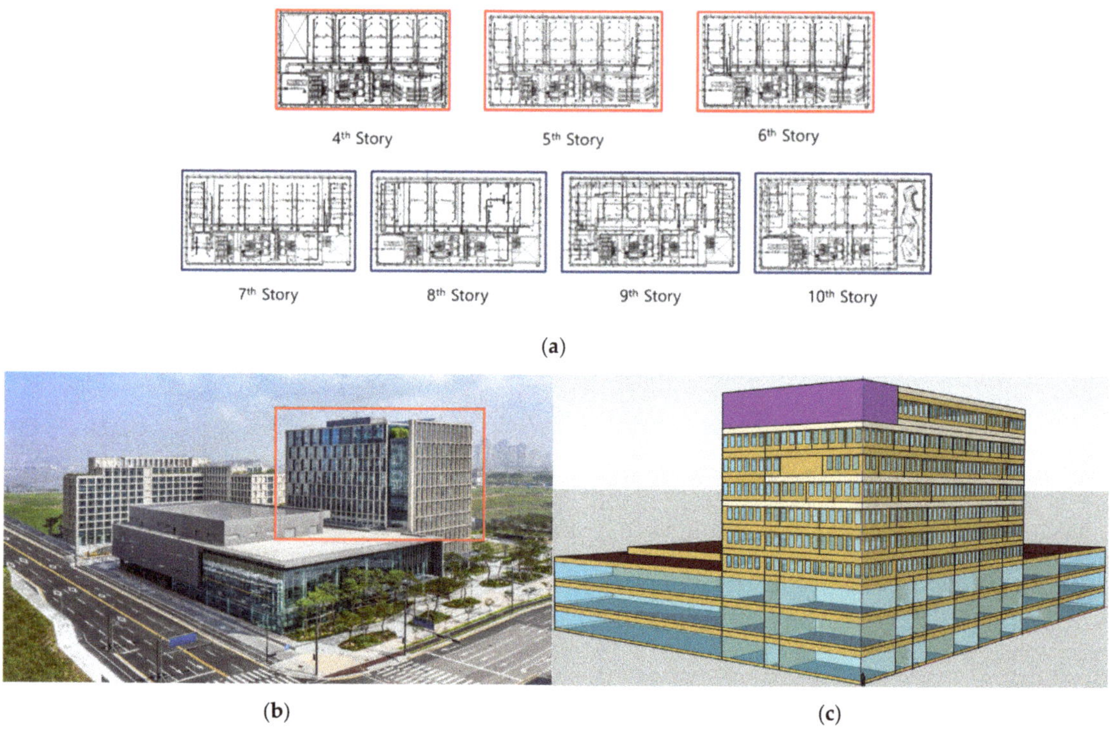


**Figure 1.** Environment of the test site: (**a**) floor-plans of the test sites; (**b**) illustration of the test site; (**c**) test site reconstructed as a digital twin.

The actual material properties of the education hall considered in the simulation are listed in Table 5. Because of the wide variety of material properties at the site, the material properties most commonly used in each space type, considered thermal zones as occupant spaces, were considered representative properties in each space in the simulation.

The properties of each material were obtained from references provided by the material standards established by MOLIT in the Republic of Korea. The specifications of the doors were not considered, except for the thickness and types of material, because the material properties were absent. The thickness represented in Table 5 is not the value from the regulation but was directly measured from the site using laser-guided measurement tools.

## 2.3. Internal Heat Gain

The input parameters in Table 6 were obtained directly from the actual parameters applied for the initial design phase of the subject education hall. Computer laboratories and coffee stations were considered part of classrooms and cafeterias in space types.

For replacement by the ASHRAE Baseline 90.1-2016 criteria, the coffee stations and computer laboratories considered to be cafeterias and classrooms were reconfigured separately as the space types of lounges and laboratories. Currently, no applicable or relevant ventilation standards exist for the operation of buildings in the Republic of Korea regarding the concentration level of $CO_2$. Thus, recommendations from municipal health organizations were introduced. As the library was originally designed for an administrative office, the design density used for the initial phase analysis was the same as the parameters for the administrative office.

Figure 2 shows the occupation history of the entire subject education hall from July 2020 to December 2021. As illustrated, the occupancies were concentrated in the 4th to 10th stories. Regarding the events in the 4th to 10th stories, the frequencies were considered equal for each story because the central HVAC operators were requested to run the facility at the occupants' request as long as the inhabitants resided in the stories. Furthermore, considering the purpose of the 7th to 9th stories, they were connected with open staircases. The 7th to 10th stories were utilized for administrative work. In contrast, the 1st to 3rd stories were primarily for the lobby with small meeting rooms in an open space with broad access to outdoor air and were not often occupied. The number of occupancy events effective for outdoor-air-handling-unit (OHU) schedules was postulated to be the day excluding official holidays in the Republic of Korea.

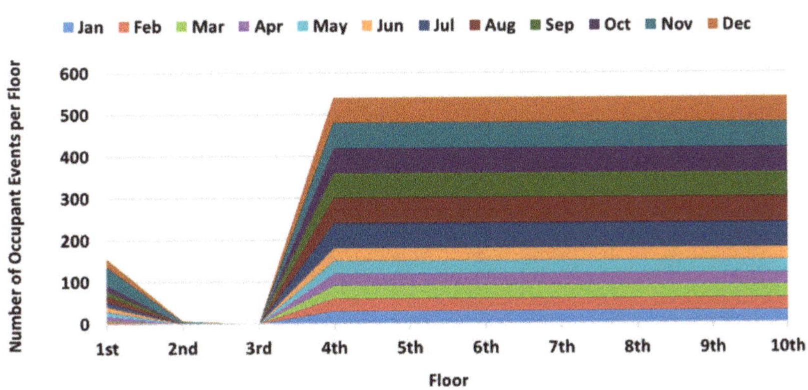


**Figure 2.** Occupant schedules in binary values from 2020 to 2021.

Because of the absence of occupational histories with hourly granularity, references provided by the Department of Energy (DOE) were considered. Therefore, the simulation considered ruleset schedules applicable to colleges from references provided by the DOE. Additionally, the occupant comfort model of ASHRAE 55 [51] with clothing schedules was considered in the simulation to be comparable with LEED criteria.

### 2.4. Central HVAC System

The flow of the heat source installed in the system is shown in Figure 3. Water resources for the education hall have been supplied through two separate supply pipelines. The cold water for domestic use has been supplied by the local water resources entity supplied by Korea Water Resources Corporation (K-water), and the hot water at 115 °C has been supplied by Korea District Heating Corporation (KDHC). Furthermore, because the primary purpose of the present study was not to discover the energy-saving capability through a simple postulation by replacing the heat sources for ease of simulation, an HVAC system with absorption chillers was considered to maximize the applicability of the present model in compliance with local regulations.

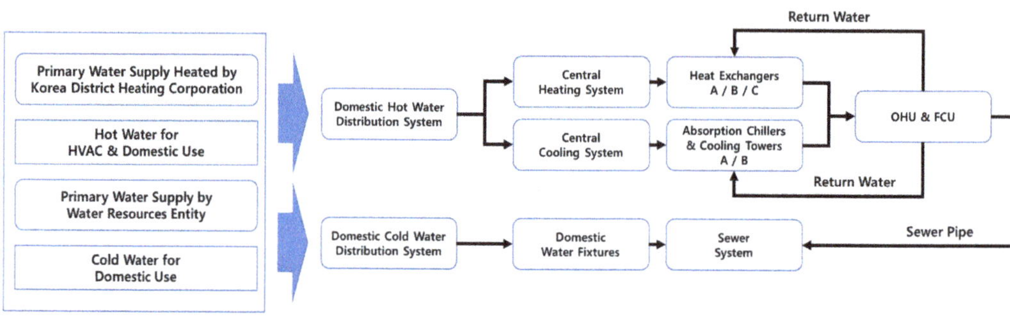


**Figure 3.** Flow of the central heat sources in the central system.

In the current operation, the flow meters installed at the entrances of the supply and discharge pipes are the only measurements available in the current system. However, the operators did not record the history, and access to it was limited to the simulation. Thus, the intraday amount of water resources required was postulated to be auto-sized by EnergyPlus based on the heating and cooling demands of the subject education hall.

In addition, for the digital twin to be eligible for LEED certification, the HVAC system must be auto-sized. The central conditioning system separately processes the hot water supplied primarily by KDHC, whether the water resources are for air conditioning or domestic water use. The HVAC system, operated by medium-hot and chilled water in the subject education hall, consists of separate water circulation loops for each fan coil unit (FCU), OHU, and domestic water resource.

Considering the local operational instructions for the central HVAC system, the hot water loop (HWL) supplying hot water to the OHUs and FCUs was scheduled to be off from June to August. The intraday operation window was set to shut down the system at 21:00 and turn it on at 06:00, considering the intraday end of the occupancy schedules and preheating procedures that generally turn on the HVAC system an hour ahead.

### 2.5. Utility Environment

The utility tariff applied in the simulation consisted of a tariff on water resources regulated by the KDHC. For the utilization prices of water resources for the conditioning processes, the end-user may select the options to participate in the time-variant tariff or the uniform tariff system depending on the type of HVAC system, whether it is a centralized or decentralized system, as summarized in Table 7. Table 8 presents seasonal time-of-use (TOU) tariff standards for the electricity consumption.

**Table 7.** Utility environment for the district heat sources in the simulation [59].

| Criteria | Applicable Elements in the Simulation |
| --- | --- |
| Utility Company | Korea District Heating Corporation |
| Resource | Conditioning |
| Seasonality | Uniform |
| Demand Charge | 361.98 KRW/(Mcal/h) |
| Utilization Price | KRW/Mcal |

**Table 8.** Tariff criteria for the electric load applicable to the subject education hall [60].

| Utility Company | Season | Hours | Utilization Price | Demand Charge |
| --- | --- | --- | --- | --- |
| Korea Electric Power Corporation | Spring Fall Summer | Peak Load (A–B) A: 10:00–12:00 B: 13:00–17:00 | 80.1 KRW/kWh (Spring/Fall) 155.8 KRW/kWh (Summer) | 6980 KRW/kW |
| | | Mid-peak Load (A–C) A: 09:00–10:00 B: 12:00–13:00 C: 17:00–23:00 | 89.9 KRW/kWh (Spring/Fall) 59.6 KRW/kWh (Summer) | |
| | | Light Load (A–B) A: 10:00–12:00 B: 13:00–17:00 | 45.2 KRW/kWh | |
| | Winter | Peak Load (A–C) A: 10:00–12:00 B: 17:00–20:00 C: 22:00–23:00 | 127.1 KRW/kWh | |
| | | Mid-peak Load (A–C) A: 09:00–10:00 B: 12:00–17:00 C: 20:00–22:00 | 88.4 KRW/kWh | |
| | | Light Load (A) A: 23:00–09:00 | 49.2 KRW/kWh | |

## 3. Results and Discussion

This section analyzes the simulation results for the CAV HVAC system at the test site equipped with an absorption chiller and heat exchangers. Given the primary function of the test site on the university campus, the work efficiency schedule was set to zero. Section 3.1 evaluates the economic viability of the HVAC system, including its energy efficiency and impact on thermal comfort. Section 3.2 assesses thermal comfort breaches, detailing hourly and annual fluctuations in Mean Air Temperature (MAT) and RH. Section 3.3 describes the IAQ.

It should be acknowledged that simulation-based results may deviate from actual building performance due to uncertainties in input assumptions, occupant behavior, and operational schedules. In the absence of post-occupancy validation data, the outcomes presented in this study should be interpreted as indicative estimates rather than precise predictions. Nonetheless, the modeling approach followed established standards and design documentation, ensuring a reasonable degree of representativeness. Future studies incorporating monitored data would help narrow the performance gap and further enhance the reliability of simulation-based assessments.

### 3.1. Economic Feasibility

Figure 4 shows the energy consumption from the thermal and electric loads differentiated by black and white segments for thermal and electrical energy, respectively. The stacked bars indicate the energy usage across various case scenarios. Case 1, the reference scenario, recorded the highest energy consumption of 7,323,502 kWh, whereas Case 3 had the lowest energy consumption of 3,978,263 kWh owing to internal heat parameter adjustments. The data across all cases indicate the need for revisions in building codes in the Republic of Korea owing to the observed reductions in total energy consumption.

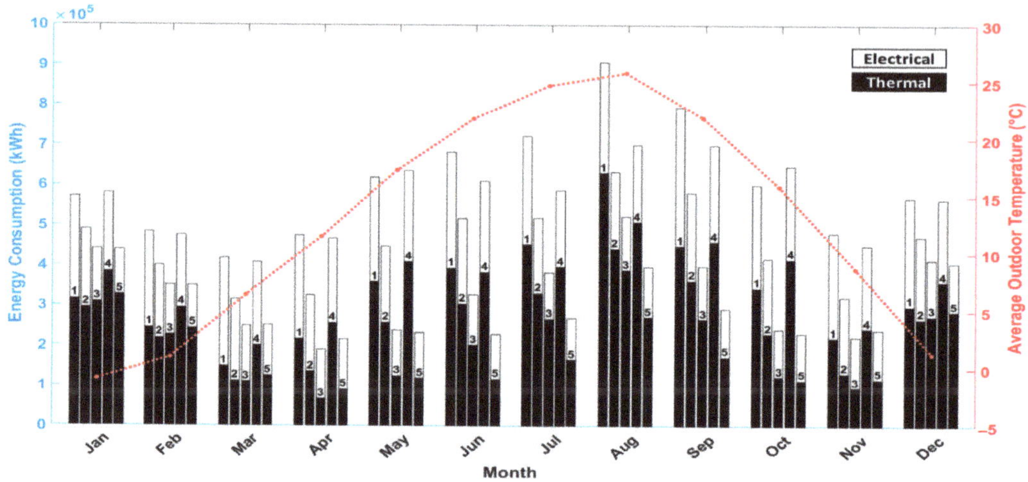


**Figure 4.** Monthly energy consumption by the simulated HVAC operations.

Excluding Case 5, in which the comprehensive LEED v4.1 design conditions were implemented, the ratio of thermal to electrical consumption was maintained at approximately 60% to 40%. In contrast, a significant shift was observed in Case 5, where 46% of the energy consumption was attributed to thermal use and 54% to electrical use, resulting in a 36.21% reduction in annual energy consumption. The most substantial energy reductions were recorded in Case 5 during the summer peak season, which is typically the most challenging period for system operators, with a 59.19% decrease compared with the reference case. From the perspective of electricity use, reductions of 22% and 53.09% were achieved in Cases 5 and 3, respectively, even in the absence of demand response measures.

During the peak summer periods, the lowest electrical energy consumption was recorded in Case 5, with a reduction of 58.16%. In terms of thermal energy, the lowest consumption was also demonstrated by Case 5, with 436,228 kWh consumed, in contrast to 903,519.45 kWh in Case 4 and 654,794.17 kWh in Case 3, both of which were characterized by adjusted design parameters. The importance of holistic design modifications in achieving optimal energy efficiency was underscored.

Table 9 shows the annual energy requirements for the heat rejection processes in the cooling tower across different scenarios and the correlation between consumption and monthly outdoor air temperature (OAT). Heat rejection involves dissipating excess thermal energy from the cooling system, which serves as an indicator of HVAC system efficiency. The effectiveness of the cooling tower in expelling thermal energy was determined by the energy received from the chiller and the energy used by its fans. This relationship mirrors the fluctuations in the annual energy consumption.

**Table 9.** Correlation between monthly outdoor temperature and energy consumption.

| Energy from Cooling Tower | Case 1 | Case 2 | Case 3 | Case 4 | Case 5 |
|---|---|---|---|---|---|
| Cooling Tower Fan Electricity Energy (kWh) | 91,563 | 60,918 | 39,266 | 92,219 | 25,520 |
| Chiller Condenser Heat Transfer Energy (kWh) | 4,645,374 | 3,194,140 | 2,195,203 | 4,441,787 | 1,527,306 |
| **Correlation between OAT** | **Case 1** | **Case 2** | **Case 3** | **Case 4** | **Case 5** |
| Electricity Consumption | 0.517 | 0.572 | 0.168 | 0.607 | −0.389 |
| Thermal Energy Consumption | 0.756 | 0.166 | −0.410 | 0.367 | −0.061 |

The monthly thermal-energy demand is shown in Figure 4. Notably, Case 5 achieved a significant reduction in thermal energy dissipation, approximately 67.12% less annually than in Case 1. Additionally, the energy used by the fans at the cooling tower in Case 5 was 72.13% lower than that in Case 1, indicating marked improvements in energy efficiency.

Case 5, in which all LEED v4.1 criteria were implemented, was demonstrated to be the most robust against fluctuations in the OAT, whereas Case 1 was found to be highly sensitive. In Case 4, the thermal energy consumption was less dependent on the OAT owing to the improvement in ventilation through the active utilization of outdoor air. However, this approach resulted in increased electricity consumption by primary HVAC components.

Cases 2 and 3, which incorporated envelope retrofits and internal heat gain design densities based on local code requirements, respectively, indicated that the effectiveness of the design adaptations is strongly influenced by these regulatory codes. Modifications of the envelope alone were insufficient to maximize the energy performance benefits without the inclusion of additional retrofitting measures. Overall, the correlation data from Cases 2 to 4 suggested that improved robustness to OAT variations was achieved. Nonetheless, the findings emphasize that a combination of passive and active strategies is essential to fully optimize energy performance outcomes.

Table 10 presents the heat transfer data observed during the summer period from May to August and the winter periods from January to April and September to December, thereby illustrating the influence of geological and climatic conditions on the building energy performance. Effective heat management strategies were identified, including expulsion of heat during summer and retention of heat during winter. The internal heat gains resulting from the density of occupants and electronic devices were found to have minimal impact on the envelope heat transfer, as evidenced by the limited seasonal variation observed in Cases 3 and 4.

**Table 10.** Monthly average heat transfers through external surfaces in MWh.

| Transfer Type | Season | Surface Type | Case 5 | Case 4 | Case 3 | Case 2 | Case 1 |
|---|---|---|---|---|---|---|---|
| Heat Gain | Summer | External Window | 21.33 | 43.87 | 44.62 | 21.63 | 44.34 |
| | | External Wall | 8.00 | 1.33 | 1.52 | 8.03 | 1.40 |
| | Winter | External Window | 19.04 | 44.29 | 46.32 | 18.91 | 46.27 |
| | | External Wall | 8.07 | 0.29 | 0.42 | 8.01 | 0.36 |
| Heat Loss | Summer | External Window | 8.54 | 6.74 | 5.73 | 8.66 | 6.17 |
| | | External Wall | 8.45 | 7.72 | 6.45 | 8.49 | 7.05 |
| | Winter | External Window | 29.71 | 22.89 | 18.25 | 30.34 | 19.51 |
| | | External Wall | 13.71 | 26.24 | 21.15 | 13.31 | 22.79 |

In Case 5, which included comprehensive renovation measures, heat transfer was optimized most effectively. During summer, the exterior window in Case 5 was found to gain 51.89% less heat than that in Case 1, whereas its exterior surface demonstrated a heat absorption capacity that was five times greater. These results indicated that windows, which are key contributors to the greenhouse effect, play a more significant role than walls in regulating heat gain. During the winter, the materials used for the exterior wall in Case 5 retained 39.84% more heat than those in Case 1. In conclusion, the seasonal difference between heat gain and heat loss was optimally balanced in Case 5, corresponding with the primary operational objectives designated for each season.

### *3.2. Thermal Comfort in the Primary Breathing Zones*

Figure 5 presents the simulation results of the PMV during operational hours in lecture halls with moderate occupant densities and regular schedules, with thermal comfort evaluated according to ASHRAE 55 standards. The X and Y axis in each heat map represent hours from 00 to 24 and months from January to February. The lines illustrated in the section dedicated for average relative humidity represent average zonal air temperature, whereas the lines in sky-blue represent average zonal relative humidity.

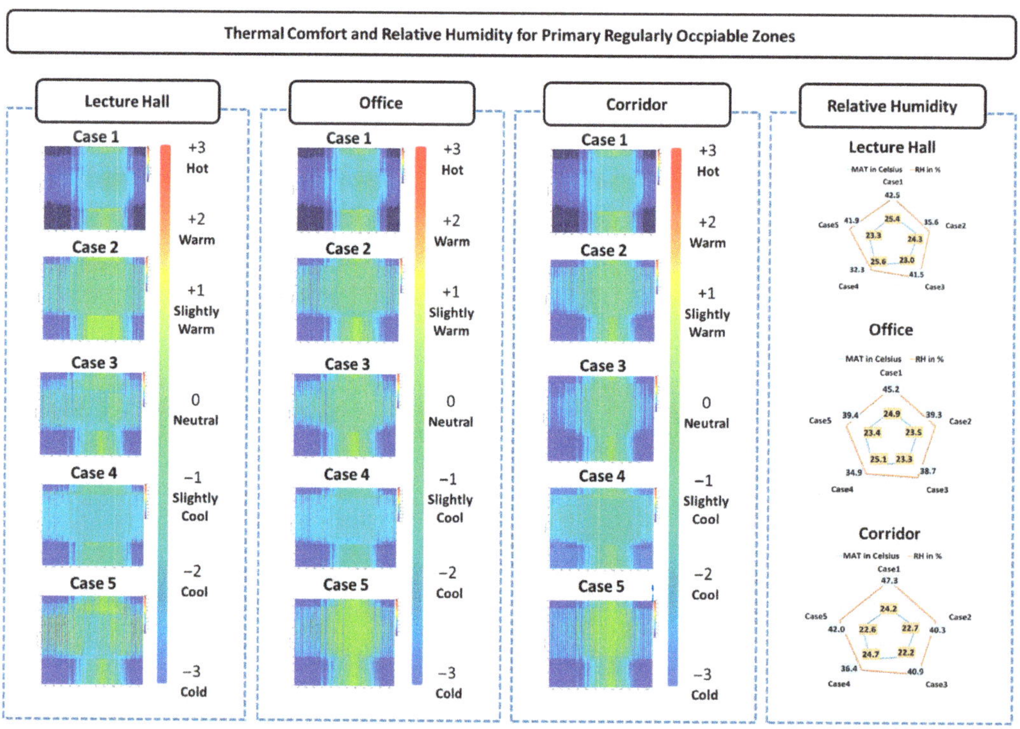


**Figure 5.** PMV and RH for the zones with the regular occupant schedules.

Lecture halls, offices, and corridors were selected based on design considerations. The lecture hall was characterized by the highest design occupant density, fixed occupancy schedules, and significant internal electric loads. Offices and corridors were categorized as regularly occupied zones with fixed schedules and moderate occupant densities. Across all cases, satisfactory comfort levels were achieved, with a closer alignment to the thermal neutrality index observed in cases where optimized ventilation strategies were applied, particularly in Case 5. Case 3 was notably influenced by the OAT owing to higher ventilation demand, resulting in significant indoor temperature fluctuations—cooler conditions during winter and warmer conditions during summer—owing to increased conditioning loads. This effect was also evident in summer operations, during which the central HVAC system remained active overnight to enhance operational efficiency.

According to ASHRAE Standard 55 for thermal comfort, a thermal index value closer to the neutral index indicates a better comfort condition for occupants. Given the regularity of occupancy schedules, lecture rooms were designed to accommodate the highest occupant densities. As illustrated in Figure 5, it is evident that thermal comfort conditions were satisfied in all simulation cases.

A comparison between Case 1 and Case 5 illustrates that the frequency of occurrence for the neutral thermal index increases, as shown in the heat map. In Case 3, where a higher ventilation rate was applied without complementary adjustments to other influencing factors, the hourly indoor air temperature was more strongly affected by outdoor air infiltration. This led to lower indoor temperatures during winter and increased cooling demands during summer. As a result, the central HVAC system was simulated to operate during the night cycle in summer to maintain efficiency.

Further analysis in Figure 5 confirms that thermal zones complied with ASHRAE 90.1 comfort standards, maintaining MATs between 20 and 26 °C and RH between 30% and 50%. Ventilation adjustments were found to significantly affect RH levels; increased ventilation in Cases 2, 3, and 5 resulted in RH reductions, whereas in Case 4, RH increased solely owing to ventilation modifications.

Temperature variations based on the scale distribution between 17 and 40 °C were most noticeable in key areas such as lecture halls, offices, and corridors with regular occupancy schedules, where temperature decreased by 8.3%, 6.0% and 6.4% with enhanced passive and active measures in Case 5, whereas the annual MAT in Case 4 considered that only the modification of ventilation increased by 0.8%, 0.8%, and 2.3% in the lecture halls, offices, and corridors, respectively. These findings highlight that LEED criteria implementation does not compromise thermal comfort but effectively reduces energy consumption. Significant thermal load adjustments were evident in lecture rooms for Case 3, where modifications in occupant density from 0.5 to 1.5 ppl/m$^2$ resulted in a 9.4% temperature rise and a 2.4% drop in RH, underscoring the critical interaction between design regulations and thermal management.

### 3.3. *Indoor Air Quality*

Figure 6 depicts the indoor air quality from the perspective of occupant health. According to the standards for indoor $CO_2$ concentrations set by ASHRAE and the WHO in Europe, Japan, and the Republic of Korea, concentrations should be maintained below 1000, 920, 1000, and 1000 ppm, respectively.

Consequently, the simulation, which assumed the renovation of the building from design to operational schemes, did not indicate potential violations of the design criteria of the building regarding $CO_2$ levels. In the simulation, the occupant schedule for the library was set to be the same as that of the classroom because the original design purpose of the spaces allocated for the library was classrooms.

Because the concentration level of outdoor $CO_2$ varied roughly between 426.3 and 475.7 ppm, a smaller difference between them could be perceived as a better indoor environment for the occupants' health. The average concentration difference in Case 5 was simulated to be the lowest in every space type designed to be breathing zones, whereas the differences from Cases 1 to 4 were greater in every space type, regardless of the operation schedule of the building. Notably, IAQ management in Cases 5 and 3 was determined to be the lowest, whereas IAQ management in Case 4, which was expected to be the lowest with changes in the ventilation requirement, tended to function less effectively with greater differences in concentrations simulated in the corridor, laboratory, and lecture hall.

| Case 5 | Classroom | Lounge | Computer Lab | Coneference Room | Corridor | Lobby | Lecture Room | Library | Medium Office | Small Office |
|---|---|---|---|---|---|---|---|---|---|---|
| January | 502.28 | 476.35 | 513.67 | 563.26 | 568.27 | 517.11 | 512.50 | 506.56 | 469.79 | 497.95 |
| February | 505.90 | 478.48 | 517.54 | 569.92 | 576.56 | 521.59 | 516.07 | 510.16 | 471.44 | 501.06 |
| March | 510.03 | 481.48 | 522.89 | 577.01 | 588.10 | 527.21 | 517.31 | 515.67 | 473.65 | 504.56 |
| April | 503.11 | 475.45 | 515.94 | 569.19 | 582.76 | 521.25 | 507.63 | 509.98 | 467.02 | 498.26 |
| May | 497.29 | 472.09 | 508.98 | 557.90 | 575.80 | 515.26 | 499.62 | 504.58 | 463.12 | 491.66 |
| June | 457.37 | 452.01 | 453.35 | 477.98 | 525.09 | 458.08 | 437.40 | 437.40 | 452.67 | 469.77 |
| July | 451.28 | 445.80 | 447.88 | 472.64 | 517.43 | 451.82 | 431.41 | 431.41 | 446.31 | 463.59 |
| August | 447.47 | 441.89 | 444.02 | 468.10 | 513.42 | 448.08 | 427.80 | 427.80 | 442.85 | 459.61 |
| September | 492.27 | 464.66 | 504.27 | 557.27 | 580.64 | 513.93 | 494.28 | 501.79 | 454.34 | 484.41 |
| October | 502.95 | 477.96 | 513.45 | 563.13 | 579.77 | 520.46 | 505.97 | 509.05 | 468.65 | 496.57 |
| November | 516.54 | 488.38 | 529.33 | 582.94 | 595.19 | 534.09 | 521.30 | 522.12 | 480.16 | 510.88 |
| December | 533.20 | 504.50 | 546.03 | 600.67 | 608.83 | 550.50 | 542.34 | 538.55 | 497.06 | 528.24 |
| **Case 4** | **Classroom** | **Lounge** | **Computer Lab** | **Coneference Room** | **Corridor** | **Lobby** | **Lecture Room** | **Library** | **Medium Office** | **Small Office** |
| January | 512.53 | 477.64 | 544.52 | 565.71 | 570.40 | 527.28 | 494.50 | 529.77 | 499.10 | 528.33 |
| February | 516.86 | 479.93 | 549.95 | 572.18 | 578.03 | 532.62 | 497.10 | 535.34 | 502.65 | 532.88 |
| March | 520.22 | 483.28 | 554.40 | 574.76 | 588.06 | 538.44 | 496.12 | 539.76 | 506.22 | 534.19 |
| April | 513.99 | 477.92 | 548.23 | 566.69 | 582.16 | 533.79 | 486.24 | 533.22 | 500.19 | 526.09 |
| May | 507.30 | 475.11 | 539.13 | 555.22 | 574.37 | 529.65 | 478.02 | 526.78 | 495.79 | 517.16 |
| June | 461.53 | 454.44 | 460.75 | 482.43 | 523.74 | 461.64 | 437.40 | 437.40 | 474.91 | 492.87 |
| July | 455.22 | 448.28 | 456.05 | 478.26 | 519.96 | 455.83 | 431.41 | 431.41 | 468.65 | 490.58 |
| August | 451.15 | 443.95 | 451.40 | 472.68 | 513.89 | 451.49 | 427.80 | 427.80 | 464.05 | 484.37 |
| September | 504.96 | 469.74 | 537.17 | 557.14 | 580.03 | 530.73 | 471.18 | 524.76 | 491.80 | 512.74 |
| October | 512.18 | 480.53 | 539.20 | 556.83 | 578.57 | 533.70 | 482.81 | 528.58 | 500.51 | 518.26 |
| November | 527.74 | 490.42 | 561.09 | 581.11 | 595.51 | 546.02 | 500.24 | 544.75 | 512.93 | 539.18 |
| December | 543.84 | 506.04 | 578.41 | 600.69 | 610.57 | 561.97 | 522.18 | 563.73 | 529.68 | 559.93 |
| **Case 3** | **Classroom** | **Lounge** | **Computer Lab** | **Coneference Room** | **Corridor** | **Lobby** | **Lecture Room** | **Library** | **Medium Office** | **Small Office** |
| January | 503.11 | 477.00 | 511.41 | 560.12 | 571.31 | 517.33 | 509.39 | 506.16 | 470.24 | 496.52 |
| February | 506.53 | 479.08 | 515.45 | 566.80 | 579.15 | 521.55 | 512.74 | 509.68 | 471.95 | 499.55 |
| March | 510.13 | 480.97 | 519.80 | 572.57 | 589.35 | 526.07 | 514.10 | 514.76 | 473.61 | 502.38 |
| April | 503.73 | 474.17 | 514.16 | 567.67 | 583.80 | 520.41 | 509.98 | 509.71 | 466.87 | 496.51 |
| May | 495.93 | 470.25 | 505.60 | 554.23 | 574.66 | 516.51 | 503.00 | 506.18 | 463.55 | 490.70 |
| June | 457.14 | 451.46 | 452.94 | 476.57 | 523.41 | 457.90 | 437.40 | 437.40 | 452.89 | 468.71 |
| July | 450.90 | 445.65 | 447.54 | 471.64 | 515.74 | 451.74 | 431.41 | 431.41 | 446.60 | 463.09 |
| August | 446.87 | 441.53 | 443.34 | 466.60 | 510.17 | 447.70 | 427.80 | 427.80 | 442.69 | 458.48 |
| September | 490.85 | 463.39 | 500.36 | 551.12 | 578.83 | 515.37 | 491.86 | 502.29 | 455.35 | 482.53 |
| October | 502.86 | 476.15 | 510.31 | 559.37 | 579.66 | 520.12 | 508.38 | 509.01 | 468.83 | 493.98 |
| November | 517.32 | 487.48 | 526.15 | 579.90 | 597.11 | 532.98 | 521.65 | 520.88 | 480.07 | 508.32 |
| December | 533.76 | 504.78 | 543.01 | 596.25 | 611.64 | 550.08 | 538.43 | 537.81 | 497.44 | 526.32 |
| **Case 2** | **Classroom** | **Lounge** | **Computer Lab** | **Coneference Room** | **Corridor** | **Lobby** | **Lecture Room** | **Library** | **Medium Office** | **Small Office** |
| January | 509.60 | 477.38 | 541.60 | 562.98 | 560.13 | 516.65 | 497.62 | 534.59 | 498.75 | 531.46 |
| February | 513.36 | 479.51 | 546.75 | 569.10 | 567.26 | 520.73 | 500.50 | 540.03 | 502.09 | 536.20 |
| March | 518.09 | 483.02 | 553.74 | 575.79 | 576.32 | 526.00 | 500.61 | 545.85 | 505.63 | 540.08 |
| April | 511.51 | 476.09 | 548.49 | 569.07 | 570.33 | 520.52 | 492.01 | 539.44 | 499.31 | 533.04 |
| May | 502.61 | 471.63 | 535.15 | 553.10 | 561.91 | 515.38 | 483.59 | 533.07 | 493.99 | 521.25 |
| June | 459.41 | 451.96 | 459.26 | 477.73 | 514.49 | 457.89 | 437.40 | 437.40 | 473.18 | 490.89 |
| July | 453.14 | 445.75 | 454.01 | 473.45 | 509.05 | 451.66 | 431.41 | 431.41 | 466.53 | 486.49 |
| August | 449.19 | 441.70 | 449.80 | 468.83 | 504.11 | 447.72 | 427.80 | 427.80 | 462.22 | 481.45 |
| September | 498.01 | 464.32 | 533.11 | 549.50 | 564.19 | 514.31 | 472.98 | 531.67 | 489.16 | 513.09 |
| October | 509.47 | 477.45 | 540.26 | 559.50 | 566.64 | 519.96 | 489.46 | 536.50 | 499.38 | 526.13 |
| November | 524.97 | 489.62 | 561.18 | 582.49 | 583.19 | 532.37 | 505.99 | 552.23 | 512.59 | 546.85 |
| December | 541.13 | 505.92 | 576.32 | 599.00 | 598.51 | 549.38 | 525.44 | 568.72 | 528.95 | 563.34 |
| **Case 1** | **Classroom** | **Lounge** | **Computer Lab** | **Coneference Room** | **Corridor** | **Lobby** | **Lecture Room** | **Library** | **Medium Office** | **Small Office** |
| January | 500.24 | 496.73 | 518.63 | 545.91 | 556.71 | 515.58 | 500.65 | 503.74 | 521.34 | 496.24 |
| February | 503.37 | 500.07 | 522.72 | 551.18 | 563.56 | 519.91 | 503.53 | 507.53 | 525.42 | 499.33 |
| March | 505.14 | 502.02 | 524.82 | 552.46 | 571.68 | 523.53 | 503.57 | 511.15 | 526.37 | 501.88 |
| April | 497.59 | 495.04 | 517.74 | 544.63 | 564.78 | 517.24 | 495.39 | 505.13 | 519.75 | 495.39 |
| May | 491.39 | 490.03 | 510.46 | 536.80 | 557.09 | 512.00 | 489.19 | 500.89 | 513.64 | 490.94 |
| June | 455.21 | 457.49 | 453.21 | 467.17 | 507.91 | 456.52 | 437.40 | 437.40 | 482.74 | 469.31 |
| July | 449.07 | 451.02 | 447.77 | 461.46 | 501.09 | 450.34 | 431.41 | 431.41 | 475.94 | 462.25 |
| August | 445.29 | 447.04 | 443.83 | 457.32 | 496.36 | 446.40 | 427.80 | 427.80 | 471.23 | 458.18 |
| September | 483.80 | 483.39 | 503.17 | 531.73 | 557.08 | 510.72 | 479.55 | 497.79 | 504.87 | 485.10 |
| October | 497.33 | 496.06 | 514.20 | 540.60 | 561.83 | 516.64 | 494.01 | 504.38 | 515.73 | 495.96 |
| November | 510.89 | 507.90 | 529.59 | 556.82 | 577.77 | 530.59 | 507.99 | 517.54 | 530.81 | 508.17 |
| December | 529.79 | 526.33 | 549.65 | 579.07 | 594.58 | 548.03 | 529.22 | 534.79 | 552.09 | 526.03 |

**Figure 6.** Monthly average $CO_2$ concentration level in ppm simulated in the indoor environment.

As shown in Figure 6, the concentration differences during break periods with less pressurization of the building, including Case 5 with all possible modifications and Case 3 with the modified internal heat parameters, exhibited the lowest concentration level from September to December. Notably, IAQ management in Cases 5 and 3 was determined to be the lowest, whereas IAQ management in Case 4, which was expected to be the lowest with changes in the ventilation requirement, tended to function less effectively with greater differences in concentrations simulated in the corridor, laboratory, and lecture hall.

## 4. Conclusions

This study examined the impact of passive design modifications, specifically internal heat gains, building envelope, and ventilation strategies, on thermal comfort, indoor air quality, and HVAC performance in a multipurpose education building based on a real floor plan.

By introducing the LEED baseline, 36.21% of the potential energy-saving opportunities for thermal load in building operations were estimated in comparison with the thermal energy requirements simulated based on the current passive design criteria in the commission. Despite the passive design criteria being replaced by the LEED baseline, it did not exhibit any significant difference in the indoor air quality, thermal comfort of the occupants, unmet hours, or ventilation requirements, but rather a decrease in the thermal energy required to condition the building in the proper ranges.

Because the building envelope and factors related to the internal heat gains are the design criteria heavily relevant to the local construction regulations, the passive elements in building design in consideration of energy efficiency were the most critical design philosophy before the introduction of active energy-saving measures for building operations, whereas the ventilation strategy could be considered as one of the active and flexible measures for building management after commissioning, which was simulated not to be as influential as the results from changes in the passive criteria.

Hence, including standardized regulations in the local regulation of building designs applicable to buildings intended to participate in smart cities or green button initiatives could be reasonable and critical to maximizing energy-saving opportunities without sacrificing the voluntary will of participation. However, the most critical outcome of the present study is that all possible measures must be taken together without simply modifying several design parameters to maximize the benefits of LEED v4.1 and actualize the authentic purpose of green buildings.

Although this study does not include a detailed assessment of economic trade-offs such as retrofit costs or LEED implementation expenses, it is important to note that the observed improvements in thermal comfort and indoor air quality were achieved without the need for significant structural modifications or investment in active systems. Apart from the building envelope, no additional implementation measures were introduced. This indicates that passive design strategies, when appropriately applied and simulated, can yield substantial environmental benefits at minimal cost, particularly in existing buildings where financial resources may be constrained. These findings highlight the practical feasibility of applying non-invasive strategies to improve building performance, and they may serve as a valuable reference for future low-cost retrofit initiatives.

One notable limitation of this study is the absence of validation against actual energy or water consumption data. Although the simulation results offer valuable insights into the performance of passive design strategies, empirical validation would have strengthened the credibility and applicability of the findings. Unfortunately, the building operators did not systematically record flow meter or sub-metering data during the monitoring period, making direct comparison with real-world performance infeasible. This limitation has been acknowledged in the discussion, and future research should incorporate measured consumption data to enhance the robustness of the analysis.

Although this study is based on a single building, the underlying approach that combines passive design strategies with a simulation-based assessment of thermal comfort and indoor air quality offers a replicable framework that can be applied to a wide range of building types with similar functional and environmental requirements. The methodology is particularly adaptable to education facilities that seek to enhance indoor environmental performance through non-mechanical, climate-responsive interventions. By adjusting design inputs and climate data, this framework could be extended to other typologies and geographic regions, allowing for comparative analysis across diverse climatic and architectural contexts. Furthermore, the proposed methodology may serve as a foundation for future integration with digital twin technologies, wherein real-time sensor data can be continuously fed into the simulation environment. This integration would not only support ongoing performance monitoring but also enable hourly or sub-hourly optimization of indoor environmental parameters, thereby facilitating adaptive building operation and more efficient energy management throughout the building lifecycle.

**Author Contributions:** Conceptualization, J.L.; methodology, J.L.; software, J.L.; validation, J.L.; formal analysis, J.L.; investigation, J.L.; resources, J.L.; data curation, J.L.; writing—original draft preparation, J.L.; writing—review and editing, J.S.; visualization, J.L.; supervision, J.S.; project administration, J.S.; funding acquisition, J.S. All authors have read and agreed to the published version of the manuscript.

**Funding:** This research was supported by the KERI Primary Research Program of MSIT/NST (No. 25A01074).

**Data Availability Statement:** The original contributions presented in this study are included in the article. Further inquiries can be directed to the corresponding author.

**Acknowledgments:** During the preparation of this manuscript/study, the authors used OpenStudio 3.6.0 for the purpose of energy simulation. The authors have reviewed and edited the output and take full responsibility for the content of this publication.

**Conflicts of Interest:** The authors declare no conflicts of interest.

## Abbreviations

The following abbreviations are used in this manuscript:

| | |
|---|---|
| RTS | Radiant time series |
| DOAS | Dedicated outdoor air system |
| HVAC | Heating, ventilation, and air conditioning |
| LEED | Leadership in Energy and Environmental Design |
| MOLIT | Ministry of Land, Infrastructure and Transportation |
| ASHRAE | American Society of Hearing, Refrigerating and Air Conditioning Engineers |
| WHO | World Health Organization |
| IAQ | Indoor air quality |
| RH | Relative humidity |
| OAT | Outdoor air temperature |
| CAV | Constant air volume |
| MAT | Mean air temperature |
| PMV | Predicted mean vote |
| OHU | Outdoor air handling unit |
| FCU | Fan coil unit |
| AHU | Air handling unit |
| KDHC | Korea District Heating Corporation |
| KEPCO | Korea Electric Power Corporation |
| TC | Thermal comfort |
| IEQ | Indoor environmental quality |
| EE | Energy efficiency |
| K-water | Korea Water Resources Corporation |
| BD+C | Building Design and Construction |
| EBOM | Existing Building Operation and Maintenance |
| CEN | European Committee for Standardization |
| CO2 | Carbon dioxide |
| PF | Phenolic foam |
| GHG | Greenhouse gas |
| WNA | World Nuclear Association |

## Appendix A

Table A1 below represents the classification of ASHRAE climate zones with environmental conditions based on latitudes. In the case of the present simulation environment, the climate zone can be classified as climate zone (CZ) 4. The letter codes noted by A, B, and C represent the atmospheric conditions that are humid, dry, and marine, respectively. Heating degree day (HDD) and cooling degree day (CDD) represent the days required to heat and cool the buildings to maintain thermal comfort for the inhabitants. The other types of buildings willing to simulate and create LEED report may utilize the information provided by the climate conditions in Table A1.

**Table A1.** Classification of climate zones after ASHRAE codes for the simulation.

| Climate Condition | Zonal Index | Thermal Condition |
|---|---|---|
| Cool | 5A/B/C | CDD_10 < 3500 and 2000 < HDD_18 < 4000 |
| Mixed | 4A/B/C | CDD_10 < 3500 and 2000 < HDD_18 < 3000 |
| Warm | 3A/B/C | CDD_10 < 3500 and HDD_18 < 2000 |
| Hot | 2A/B | 3500 < CDD_10 < 5000 |

Figure A1a,b illustrate the buffer zone specified for the climate condition considered not to vary significantly and the location of the subject building. Because the subject building is located within the buffer, the cooling and heating degree days can be calculated by averaging the sum of the differences between the average temperature on the target day and the standard temperature specified by HDD and CDD.

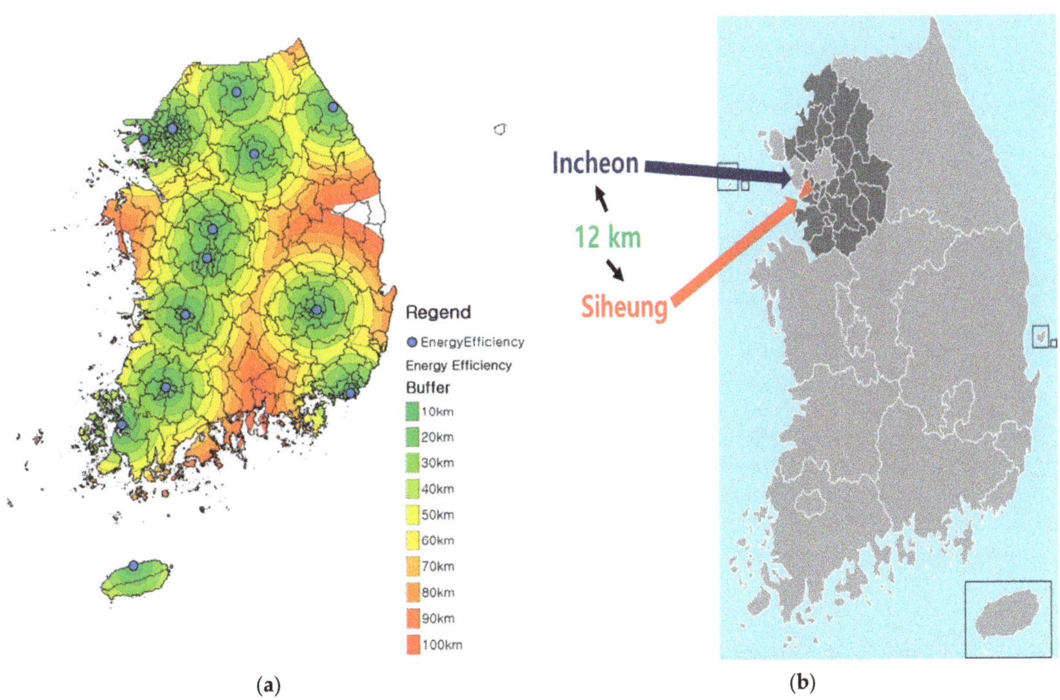


**Figure A1.** Illustration of the simulation environment on the atmospheric condition: (**a**) buffer zones for ASHRAE climate zones [61]; (**b**) location of the reference climate zone in the Republic of Korea.

Regarding the atmospheric conditions in Siheung, where the candidate building is located, the number of CDD and HDD was defined by the municipal climate center as 50.1 and 2646.4, respectively. Thus, the subject location was concluded to belong to ASHRAE Climate Zone 4C, representing the marine environment surrounding the building. For the zone referenced in Figure A1a, the climate conditions in the subject area were postulated not to differ significantly from the conditions in Incheon, where the reference data are available for EnergyPlus.

## References

1. Carbon Dioxide Emissions from Electricity. World Nuclear Association. Available online: https://world-nuclear.org/information-library/energy-and-the-environment/carbon-dioxide-emissions-from-electricity (accessed on 4 May 2025).
2. Angelidou, M.; Politis, C.; Panori, A.; Bakratsas, T.; Fellnhofer, K. Emerging Smart City, transport and energy trends in urban settings: Results of a pan-European foresight exercise with 120 experts. *Technol. Forecast. Soc. Change* **2022**, *183*, 121915. [CrossRef]
3. Mosannenzadeh, F.; Di Nucci, M.R.; Vettorato, D. Identifying and prioritizing barriers to implementation of Smart Energy City Projects in Europe: An empirical approach. *Energy Policy* **2017**, *105*, 191–201. [CrossRef]
4. Mousavi, S.; Gheibi, M.; Wacławek, S.; Behzadian, K. A novel smart framework for optimal design of green roofs in buildings conforming with energy conservation and Thermal Comfort. *Energy Build.* **2023**, *291*, 113111. [CrossRef]
5. Almeida, R.; Georgieva, P.; Martins, N. Heating setpoint recommendation strategy for Thermal Comfort and energy consumption optimization. *Energy Build.* **2023**, *296*, 113406. [CrossRef]
6. Wang, C.; Zhu, Y.; Guo, X. Thermally responsive coating on building heating and cooling energy efficiency and indoor comfort improvement. *Appl. Energy* **2019**, *253*, 113506. [CrossRef]
7. Picard, T.; Hong, T.; Luo, N.; Lee, S.H.; Sun, K. Robustness of energy performance of zero-net-energy (ZNE) homes. *Energy Build.* **2020**, *224*, 110251. [CrossRef]
8. Wang, W.; Zmeureanu, R.; Rivard, H. Applying multi-objective genetic algorithms in green building design optimization. *Build. Environ.* **2005**, *40*, 1512–1525. [CrossRef]
9. Rabani, M.; Madessa, H.B.; Ljungström, M.; Aamodt, L.; Løvvold, S.; Nord, N. Life Cycle Analysis of GHG emissions from the building retrofitting: The case of a Norwegian office building. *Build. Environ.* **2021**, *204*, 108159. [CrossRef]
10. Shi, L.; Qi, X.; Yang, Z.; Tao, L.; Li, Y.; Qiu, J.; Jiang, X. Comparative study of greenhouse gas emission calculations and the environmental impact in the life cycle assessment of buildings in China, Finland, and the United States. *J. Build. Eng.* **2023**, *70*, 106396. [CrossRef]
11. Shingler, J. *Long Term Study of Pennsylvania's Low Income Usage Reduction Program: Results of Analyses and Discussion*; Penn State University: University Park, PA, USA, 2009.
12. Apostolopoulos, V.; Mamounakis, I.; Seitaridis, A.; Tagkoulis, N.; Kourkoumpas, D.S.; Iliadis, P.; Angelakoglou, K.; Nikolopoulos, N. An Integrated Life Cycle Assessment and life cycle costing approach towards sustainable building renovation via a dynamic online tool. *Appl. Energy* **2023**, *334*, 120710. [CrossRef]
13. Echarri-Iribarren, V.; Echarri-Iribarren, F.; Rizo-Maestre, C. Ceramic panels versus aluminium in buildings: Energy consumption and environmental impact assessment with a new methodology. *Appl. Energy* **2019**, *233–234*, 959–974. [CrossRef]
14. Eicker, U.; Monien, D.; Duminil, É.; Nouvel, R. Energy performance assessment in urban planning competitions. *Appl. Energy* **2015**, *155*, 323–333. [CrossRef]
15. Alidadi, M.; Sharifi, A. The extent of inclusion of smart city indicators in existing Urban Sustainability Assessment Tools. In *Urban Climate Adaptation and Mitigation*; Elsevier: Amsterdam, The Netherlands, 2023; pp. 175–198. [CrossRef]
16. O'Dwyer, E.; Pan, I.; Acha, S.; Shah, N. Smart Energy Systems for Sustainable Smart Cities: Current Developments, trends and Future Directions. *Appl. Energy* **2019**, *237*, 581–597. [CrossRef]
17. Rastogi, A.; Choi, J.-K.; Hong, T.; Lee, M. Impact of different LEED versions for Green Building Certification and Energy Efficiency Rating System: A multifamily Midrise case study. *Appl. Energy* **2017**, *205*, 732–740. [CrossRef]
18. Carrera, B.; Peyrard, S.; Kim, K. Meta-regression framework for energy consumption prediction in a smart city: A case study of Songdo in South Korea. *Sustain. Cities Soc.* **2021**, *72*, 103025. [CrossRef]
19. Ahvenniemi, H.; Huovila, A.; Pinto-Seppä, I.; Airaksinen, M. What are the differences between sustainable and Smart Cities? *Cities* **2017**, *60*, 234–245. [CrossRef]
20. Abu-Rayash, A.; Dincer, I. Development of Integrated Sustainability Performance Indicators for better management of Smart Cities. *Sustain. Cities Soc.* **2021**, *67*, 102704. [CrossRef]
21. Obata, S.H.; Agostinho, F.; Almeida, C.M.V.B.; Giannetti, B.F. LEED certification as booster for sustainable buildings: Insights for a Brazilian context. *Resour. Conserv. Recycl.* **2019**, *145*, 170–178. [CrossRef]

22. Balachandra, P.; Ravindranath, D.; Ravindranath, N.H. Energy efficiency in India: Assessing the policy regimes and their impacts. *Energy Policy* **2010**, *38*, 6428–6438. [CrossRef]
23. Vaisi, S.; Varmazyari, P.; Esfandiari, M.; Sharbaf, S.A. Developing a multi-level Energy Benchmarking and certification system for office buildings in a cold climate region. *Appl. Energy* **2023**, *336*, 120824. [CrossRef]
24. Ferreira, A.; Pinheiro, M.D.; de Brito, J.; Mateus, R. Retail Buildings' sustainability assessment tools: A critical analysis of LEED, BREEAM and dgnb. *SSRN Electron. J.* **2022**. [CrossRef]
25. Luo, N.; Pritoni, M.; Hong, T. An overview of data tools for representing and managing building information and Performance Data. *Renew. Sustain. Energy Rev.* **2021**, *147*, 111224. [CrossRef]
26. Neukomm, M.; Nubbe, V.; Fares, R. *Grid-Interactive Efficient Buildings Technical Report Series: Overview of Research Challenges and Gaps*; National Renewable Energy Laboratory: Golden, CO, USA, 2019. [CrossRef]
27. Hult, E.; Granderson, J.; Mathew, P. *Energy Management in Small Commercial Buildings: A Look at How HVAC Contractors Can Deliver Energy Efficiency to This Segment*; Lawrence Berkeley National Laboratory: San Francisco, CA, USA, 2014.
28. Homod, R.Z.; Gaeid, K.S.; Dawood, S.M.; Hatami, A.; Sahari, K.S. Evaluation of energy-saving potential for optimal time response of HVAC control system in Smart Buildings. *Appl. Energy* **2020**, *271*, 115255. [CrossRef]
29. Taheri, S.; Hosseini, P.; Razban, A. Model predictive control of heating, ventilation, and air conditioning (HVAC) systems: A state-of-the-art review. *J. Build. Eng.* **2022**, *60*, 105067. [CrossRef]
30. Müller, F.L.; Jansen, B. Large-scale demonstration of precise demand response provided by residential heat pumps. *Appl. Energy* **2019**, *239*, 836–845. [CrossRef]
31. Menezes, A.C.; Cripps, A.; Buswell, R.A.; Wright, J.; Bouchlaghem, D. Estimating the energy consumption and power demand of small power equipment in office buildings. *Energy Build.* **2014**, *75*, 199–209. [CrossRef]
32. Massidda, L.; Marrocu, M. Total and thermal load forecasting in residential communities through probabilistic methods and causal machine learning. *Appl. Energy* **2023**, *351*, 121783. [CrossRef]
33. Jung, W.; Jazizadeh, F. Human-in-the-loop HVAC operations: A Quantitative Review on occupancy, comfort, and energy-efficiency dimensions. *Appl. Energy* **2019**, *239*, 1471–1508. [CrossRef]
34. Razmara, M.; Maasoumy, M.; Shahbakhti, M.; Robinett, R.D. Optimal exergy control of building HVAC system. *Appl. Energy* **2015**, *156*, 555–565. [CrossRef]
35. Kong, M.; Dong, B.; Zhang, R.; O'Neill, Z. HVAC Energy Savings, thermal comfort and air quality for occupant-centric control through a side-by-side experimental study. *Appl. Energy* **2022**, *306*, 117987. [CrossRef]
36. Huang, K.-T.; Hwang, R.-L. Future trends of residential building cooling energy and passive adaptation measures to counteract climate change: The case of Taiwan. *Appl. Energy* **2016**, *184*, 1230–1240. [CrossRef]
37. Chen, X.; Yang, H. Integrated energy performance optimization of a passively designed high-rise residential building in different climatic zones of China. *Appl. Energy* **2018**, *215*, 145–158. [CrossRef]
38. Kwak, Y.; Shin, H.; Kang, M.; Mun, S.H.; Jo, S.K.; Kim, S.H.; Huh, J.H. Energy modeling of pig houses: A South Korean feasibility study. *Energy Strategy Rev.* **2021**, *36*, 100672. [CrossRef]
39. Abdel-Aal, M.F.; Maarouf, I.; El-Sayary, S. Wakala buildings of Mamluk era in Cairo, Egypt and how far they meet the rating criteria of LEED V4. *Alex. Eng. J.* **2018**, *57*, 3793–3803. [CrossRef]
40. Abuseif, M.; Dupre, K.; Michael, R.N. The effect of green roof configurations including trees in a subtropical climate: A co-simulation parametric study. *J. Clean. Prod.* **2021**, *317*, 128458. [CrossRef]
41. Song, R.; Liu, D.; Pan, Y.; Cheng, Y.; Meng, C. Container FARMS: Energy modeling considering crop growth and energy-saving potential in different climates. *J. Clean. Prod.* **2023**, *420*, 138353. [CrossRef]
42. Edun, A.; Hachem-Vermette, C. Post-consumer waste recycling for high performance building envelopes in cold climates: Assessing Energy and Environmental impacts. *J. Clean. Prod.* **2022**, *372*, 133686. [CrossRef]
43. Chen, J.; Gong, Q.; Lu, L. Evaluation of passive envelope systems with radiative sky cooling and thermally insulated glazing materials for cooling. *J. Clean. Prod.* **2023**, *398*, 136607. [CrossRef]
44. Navia-Osorio, E.G.; Porras-Amores, C.; Mazarrón, F.R.; Cañas, I. Impact of climate change on sustainable production of sherry wine in nearly zero energy buildings. *J. Clean. Prod.* **2023**, *382*, 135260. [CrossRef]
45. Queiroz, N.; Westphal, F.S.; Ruttkay Pereira, F.O. A performance-based design validation study on EnergyPlus for daylighting analysis. *Build. Environ.* **2020**, *183*, 107088. [CrossRef]
46. Zhang, R.; Lam, K.P.; Yao, S.; Zhang, Y. Coupled EnergyPlus and computational fluid dynamics simulation for natural ventilation. *Build. Environ.* **2013**, *68*, 100–113. [CrossRef]
47. Xie, X.; Luo, Z.; Grimmond, S.; Sun, T. Impact of building density on natural ventilation potential and cooling energy saving across Chinese climate zones. *Build. Environ.* **2023**, *244*, 110621. [CrossRef]
48. Camblong, H.; Zapirain, I.; Curea, O.; Ugartemendia, J.; Boussaada, Z.; Zamora, R. Rule-Based Energy Management System to Enhance PV Self-Consumption in a Building: A Real Case. *Energies* **2024**, *17*, 6099. [CrossRef]

49. Pérez-Carramiñana, C.; Sabatell-Canales, S.; González-Avilés, Á.B.; Galiano-Garrigós, A. Influence of Spanish Energy-Saving Standard on Thermal Comfort and Energy Efficiency Owing to the War in Ukraine: Case Study of an Office Building in a Dry Mediterranean Climate. *Buildings* **2023**, *13*, 2102. [CrossRef]
50. United States Green Building Council. *LEED v4.1 Building Design and Construction: Getting Started Guide for Beta Participants*; USGBC: Washington, DC, USA, 2020.
51. *ASHRAE Standard-55;* Thermal Environment Conditions for Human Occupancy. American Society of Heating, Refrigerating and Air-Conditioning Engineers: Peachtree Corners, GA, USA, 2017.
52. *ASHRAE 62.1*; Ventilation for Acceptable Indoor Air Quality. American Society of Hearing, Refrigerating and Airconditioning Engineers: Peachtree Corners, GA, USA, 2016.
53. Worden, K.; Hazer, M.; Pyke, C.; Trowbridge, M. Using LEED Green rating systems to promote Population Health. *Build. Environ.* **2020**, *172*, 106550. [CrossRef]
54. "LEED v4.1," LEED v4.1 | U.S. Green Building Council. Available online: https://www.usgbc.org/leed/v41 (accessed on 11 August 2025).
55. Prototype building models [WWW Document], n.d. Prototype Building Models | Building Energy Codes Program. Available online: https://www.energycodes.gov/prototype-building-models#Commercial (accessed on 11 August 2025).
56. Calculations [WWW Document], n.d. Minimum Indoor Air Quality Performance Prerequisite Ref Guide EBOM | U.S. Green Building Council. Available online: https://www.usgbc.org/node/2767360 (accessed on 11 August 2025).
57. United States Green Building Council. *LEED v4.1 Building Design and Construction*; USGBC: Washington, DC, USA, 2023.
58. *Design Specification of the Education Hall of Seoul National University at Siheung*; Seoul National University: Siheung, Republic of Korea, 2018.
59. Korean District Heating Corp. Available online: https://www.kdhc.co.kr/eng/main/main.do (accessed on 4 May 2025).
60. Kepco. Available online: https://home.kepco.co.kr/kepco/EN/F/htmlView/ENFBHP00101.do?menuCd=EN060201 (accessed on 4 May 2025).
61. Kim, H.-M.; Park, W.-J.; Kim, Y.-C.; Kim, H.-J. Comparative climate reference point analysis of domestic energy efficiency system and ASHRAE standard by GIS. *KIEAE J.* **2019**, *19*, 47–52. [CrossRef]

MDPI AG
Grosspeteranlage 5
4052 Basel
Switzerland
Tel.: +41 61 683 77 34

*Buildings* Editorial Office
E-mail: buildings@mdpi.com
www.mdpi.com/journal/buildings

www.ingramcontent.com/pod-product-compliance
Lightning Source LLC
Chambersburg PA
CBHW061840210926
51850CB00006B/932

* 9 7 8 3 7 2 5 8 5 4 2 5 7 *